BIOTECHNOLOGY IN AGRICULTURE, INDUSTRY AND MEDICINE

BIOTECHNOLOGY

HEALTH, FOOD, ENERGY AND ENVIRONMENT APPLICATIONS

BIOTECHNOLOGY IN AGRICULTURE, INDUSTRY AND MEDICINE

Additional books in this series can be found on Nova's website
under the Series tab.

Additional E-books in this series can be found on Nova's website
under the E-book tab.

BIOTECHNOLOGY IN AGRICULTURE, INDUSTRY AND MEDICINE

BIOTECHNOLOGY

HEALTH, FOOD, ENERGY AND ENVIRONMENT APPLICATIONS

LUIS G. TORRES BUSTILLOS
AND
INES GARCIA PEÑA
EDITORS

nova publishers
New York

For permission to use material from this book please contact us:
Telephone 631-231-7269; Fax 631-231-8175
Web Site: http://www.novapublishers.com

Library of Congress Cataloging-in-Publication Data

ISBN 978-1-62081-071-2

Library of Congress Control Number: 2012931746

Published by Nova Science Publishers, Inc. † New York

CONTENTS

FOREWORD

We are living in a period where the time required to pass from discoveries and inventions to their applications are becoming increasingly shorter. While it took more than 60 years after the first laboratory trials in 1933 to market the first hybrid corn, only 5 years were necessary since the first reports of transgenic insulin to its commercial production. In other fields, such as the environment, bioremediation techniques which were unknown 40 years, have rapidly integrated to the physicochemical methods to solve persistent problems in the biosphere of previously unknown molecules. In a world of declining fossil energy, biotechnology explores a multiple products and processes hidden in the rich genome of microbes, plants and animals. One of the reasons for the renewed spirit of biotechnology is supported by its capacity to swiftly incorporate the multidisciplinary advances in basic and applied sciences.

This volume presents the wide range of biotechnology research by the UPIBI community, teachers and students in the past 24 years. Novel methodologies, basic results and new processes are reported in several of the most dynamic fields in biotechnology such as health, energy, food and environment. This effort will certainly be an inspiration for students wishing to join research teams, for academic colleagues in search of new knowledge and for professionals exploring of novel ideas or innovative solutions.

Sergio Revah Moiseev (PhD)
Universidad Autónoma Metropolitana-Cuajimalpa
Mexico

ACKNOWLEDGMENTS

The editors thank the invaluable help of the following colleagues, who helped in the review process of the chapters contained in this book:

Juan Aranda Barradas
Patricia Alvarez Berber
Marco Aurelio Brito
Isaac Chairez Oria
Ines Garcia Peña
Sergio Huerta Ochoa
Marcia Morales Ibarría
Alejando Muñoz Diosdado
Itzia Padilla Martínez
Rosario Peralta-Perez
Guadalupe Ramirez Sotelo
Edgar Salgado Manjarrez
Luis G. Torres Bustillos
Wilverth Villatoro Monzon

In: Biotechnology
Editors: L. G. Torres Bustillos and I. Garcia Peña
ISBN 978-1-62081-071-2

Chapter 1

UPIBI AND THE BIOTECHNOLOGY

Juan Aranda[1], Inés Peña[2] and Luis G. Torres[2*]
[1]Bioengineering Department, Mexico
[2]Bioprocesses Departament, UPIBI-IPN, Mexico

THE CREATION OF THE UPIBI-POLYTECHNIC INSTITUTE OF MEXICO

One of the interdisciplinary fields of knowledge most strongly bound and developed in the last two or three decades worldwide is Biotechnology, which comprises the fusion of techniques, methods and knowledge from biological and medical sciences, with the engineering principles for the design of systems, processes and products related to food production, health care and environmental protection.

For the purpose of being at the forefront of the national biotechnology education and development, the National Polytechnic Institute (IPN) of Mexico founded the Professional Interdisciplinary Unit of Biotechnology (UPIBI) as an academic unit of higher education and graduate studies to help solving problems in the food, health and environment areas through the application of biotechnology and bioengineering principles. The UPIBI project started up at the National Polytechnic Institute in 1987 when the Institute's General Advisory Council approved the institutional decree formalizing the creation of UPIBI. Since then, UPIBI-IPN is an engineering school and a research institution, aiming at training professionals for bioprocess and bioproducts design as well as for applied research, including the development of basic and detail engineering for the integration of technology packages transferable to industry.

* Email: LTorresB@ipn.mx

The Academic Programs in UPIBI-IPN

UPIBI offers since it was created 25 years ago, the following Bachelor's degrees:

- Environmental Engineering,
- Biomedical Engineering,
- Biotechnological Engineering
- Food Engineering
- Pharmaceutical Engineering

In addition, UPIBI offers since 2004 when the Graduate and Research section was founded, two graduate programs:

- Master of Science in Bioprocess, and
- Ph. D. in Bioprocess.

The training of engineers in UPIBI is characterized by a strong foundation of biology, physics, chemistry, mathematics and engineering principles. Our graduates acquire and develop, during their stay in UPIBI, teamwork skills and decision making abilities that explain the relative ease with which they are incorporated into many different levels in a variety of sectors in the work market related to food and health industries and to environment protection agencies and enterprises as well.

The effective professional skills of our students have been supported by the development of a number of research programs and projects dealing with different areas of science and engineering, including industrial microbiology and biochemical engineering; molecular biology and genetics; environmental technology; modeling, simulation, control and optimization of bioprocesses; and nanotechnology.

An Overview of the Main Research Programs in UPIBI-IPN

The research programs currently in progress at UPIBI comprise:

- Bioprocesses for biofuels production.This program deals with the use of microorganisms, cells or parts thereof for the generation biofuels (primarily hydrogen, methane, ethanol and diesel).
- Bioprocesses for the treatment of effluents and solid wastes. The main objective in this program is the production of technological solutions to eliminate and revaluate effluents and wastes, by means of microbial cells in anaerobic and aerated systems, in order to reduce pollution, increase effluents and wastes biodegradability, and to induce the reutilization of revalorized solid wastes.
- Bioprocesses for the production of pharmaceutical products and foodstuffs. The central aim in this research program is to obtain compounds of pharmaceutical interest or food using microorganisms, cells or parts thereof.

Also, some other initiatives regarding Nanosciences and Micro-Nano- technologies are currently being under development to add to the research main streams in bioprocesses actually under taken in UPIBI.

SOME RESEARCH RESULTS AFTER 25 YEARS OF WORK IN UPIBI-IPN

In the year 2012, UPIBI celebrates its 25th anniversary. The robust advances achieved by UPIBI's staff and alumni regarding the education and development of Biotechnology in Mexico are clearly remarkable, as a review on historic and statistical data from the institution could reveal. However, in this book, we present a summarized outlook that reflects the efforts and products resulting from the research programs in UPIBI.

The chapters that follow this introduction are contributions from UPIBI's researchers in the Biotechnology field, basically in the areas of Food, Health, Energy and Environment. There are also two papers that deal both with the Medical Engineering and Chemistry fields, since they are basic scientific works from UPIBI research staff that indirectly impact on potentially important bioprocess development.

The main goal in every chapter is to show the state of the art of the approached subject, but also, to sum up the bachelors and postgraduate thesis works, journal publications, books or book chapters, project reports and written extended works presented in congresses and meetings, that every research project since 2004 has reported.

UPIBI is proud of the achievements presented in this book, because even if the research activities in organized programs have started only recently, research results are now being gradually recognized at the national and international scales. Our efforts are continuously focusing on developing biotechnology and bioengineering for bioprocess principles and technology, but only the time will show the true potential of UPIBI in the international frame.

The first goal of UPIBI will always be to contribute to the solution of the national problems concerning food, health, energy and environmental issues, and we can clearly and proudly say that we are on that road.

In: Biotechnology
Editors: L. G. Torres Bustillos and I. Garcia Peña
ISBN 978-1-62081-071-2

Chapter 2

Chloroplast Transformation in Plants for the Production of Vaccines

Jesús Agustín Badillo-Corona* * *and Nuria González-Rábade
Deapartamento de Bioprocesos,
Unidad Profesional Interdisciplinaria de Biotecnología
Instituto Politécnico Nacional
Mexico City, Mexico

Abstract

Plants offer several advantages for the production of subunit vaccines, including ease of scalability, low production cost, reduced risk of contamination with animal or bacterial pathogens, relative ease of genetic manipulation, and complex protein modification machinery. In addition, when the production of these vaccines occurs through the genetic manipulation of the chloroplast, high levels of expression can be achieved and gene spread controlled, as transgenes are not transmitted through pollen. Our main focus, at the Laboratory of Molecular Biotechnology in UPIBI, is to examine the expression of several antigens with potential as vaccine components for various diseases, including HIV and diabetes, in plants like lettuce, tomato and alfalfa, through the genetic transformation of chloroplasts via particle bombardment. We have generated several vectors that are specific for the transformation of lettuce and alfalfa, encoding both diabetes (gad) and HIV (p24 and nef) antigens, and are currently working on the optimization of tissue culture regeneration methods for these crops. Our most important aims are to achieve high levels of expression, obtaining stable proteins, and to optimize the choice of plant species for production of, ideally, oral edible vaccines. Production at a large scale of these chloroplast-derived vaccines in the edible tissue of plants would combine production and delivery systems.

* Email: jbadilloc@ipn.mx

Introduction

Recombinant protein production systems used today range from prokaryotic systems such as *Escherichia coli* and *Bacillus subtilis*, to eukaryotic systems such as yeast, *Aspergillus niger*, mammalian and insect cell cultures, transgenic animals, and plants. Out of these systems, transgenic plants are potentially one of the most economical systems for large-scale production of recombinant proteins for industrial and pharmaceutical uses. Their advantages include: the low cost of growing plants on a mass scale and the ease of scale-up, with agricultural systems already in place; the availability of natural protein storage organs such as seeds and tubers, further reducing the costs of production and potentially augmenting the half-life of the protein; the established practices for efficient harvesting, transporting, storing, and processing, utilizing the available food processing technologies; the elimination of the purification requirement, specially if the protein is derived from an edible plant; and the possibility of compartmentalization in different organelles, such as chloroplasts, preventing rapid degradation in the cytosol or interference with cell metabolism (Kusnadi *et al.*, 1997).

Currently, production of commercially available heterologous proteins is dominated by the bacterial expression system but has the drawback of not being able to produce highly complex proteins that fold correctly and undergo forms of post-translational modification, like N-glycosylation (Carter *et al.*, 2002).

Furthermore, there is always a risk of presence of endotoxins and other pyrogens, which complicate their use for vaccine production (Petsch and Anspach, 2000) and increase the cost of purification, as the end product has to be certified for the absence of such unwanted components. Yeasts, as eukaryotes, can assemble complex recombinant proteins much more efficiently than bacteria, but the glycosylation of proteins is very different from that found in mammals (Coughlan and Brodsky, 2005). An obvious option to overcome these problems would be to use mammalian cells. However, factors such as the cost of infrastructure and consumables and the risk of pathogen contamination or the presence of oncogenic agents make this platform of production extremely costly for proteins required both on a large and a small scale.

Finally, insect cells, which share many features with mammalian cells and have advantages like a less expensive culture medium and high productivity, have also limited scalability.

The Use of Plants as Bioreactors

Plants offer a clear solution to all these problems, making them a serious candidate for future large-scale biopharmaceutical production, especially since they do not harbor human pathogens (colonizing bacteria can be removed by appropriate sanitary measures), and need only more land or greenhouse space to augment the scale of production (Yusibov and Rabindran, 2008). Plant-expressed proteins produced so far are varied and range from diagnostic proteins (Hood *et al.*, 1997), industrial and pharmaceutical enzymes (Hood *et al.*, 2003; Woodard *et al.*, 2003), food additives (Lamphear et al., 2005), to therapeutic proteins (Ruhlman *et al.*, 2007), antibodies (Larrick *et al.*, 1998), and vaccine antigens (Streatfield and Howard, 2003).

Genes encoding foreign proteins can be introduced into plants in various ways. The most common one is nuclear transformation, either through *Agrobacterium*-mediated transformation, where the transgene sequences are first introduced into an *Agrobacterium* strain that then infects the host plant tissue and transfers the transgenes contained in a plasmid, or via particle bombardment, using tungsten or gold particles that carry the transgenes in a plasmid. Plant viral expression vectors have also been used to generate transgenic plants (Yusibov *et al.*, 2002; Karasev *et al.*, 2005; Perez-Filgueira *et al.*, 2004). In this case, the antigen is expressed from a modified plant viral genome.

After inoculation of the virus, the antigen is synthesized in the plant cell cytoplasm, although it can be targeted to different compartments. Another method of transformation is chloroplast transformation, where the expressed antigen is encoded by a sequence integrated into the chloroplast genome either by particle bombardment or by polyethylene glycol (PEG) treatment. This method is currently used in our laboratory and is described below.

THE POTENTIAL USE OF ORAL VACCINES

Vaccination is an effective and cost-efficient method for fighting human and animal diseases, preventing their spread and, in some cases, completely eradicating some infectious diseases.

The concept of vaccination involves stimulation of the immune system with exogenous antigen to induce immune responses in order to protect from a particular infection (Puls and Emery, 2006), as shown in Figure 1. Ideally, a vaccine should be inexpensive, safe, stable, easy to administer and able to confer protective immunity with few side effects. When a vaccine is administered, it should provoke an immune response capable of neutralizing the pathogen it targets, when exposed to it. There are more than 25 vaccines licensed for use in humans with many more in the development pipeline (Yusibov and Rabindran, 2008). Using partially processed plant material containing a vaccine, applied either orally or topically, could circumvent many of these challenges.

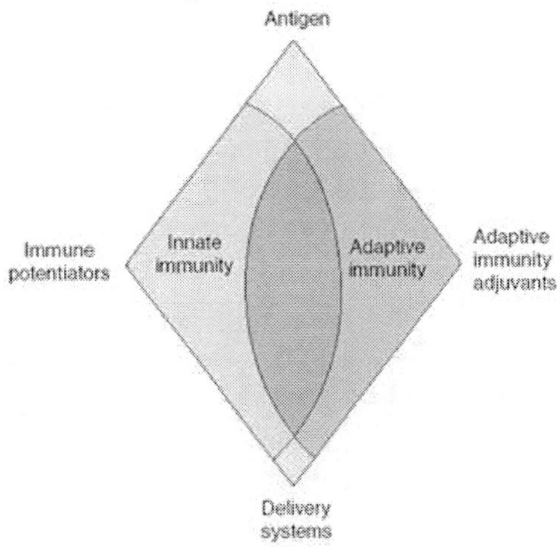

Figure 1. Essential components of a vaccine. Immune potentiators acting on the innate compartment of the immune system and adjuvants specific for the adaptive immune response are co-administered with antigen through appropriate delivery systems in order to elicit a strong immune response to a vaccine (Chiarella *et al.*, 2007).

There might not be a need for a cold chain or for complex infrastructure and highly specialized personnel. Certain plant tissues, like seeds, are incredibly stable and could function as a means of long-term storage. Large amounts of antigens can be produced at relatively low cost in an agriculturally based system, where the potential for scale-up is most evident (Obregon *et al.*, 2006). The choice of the plant species used to produce the selected antigen should allow for oral delivery in the form of an edible vaccine (Streatfield, 2005). Manufacture of these vaccines in the edible tissue of plants (e.g. leaves or fruits) would combine production and delivery systems. Immunization might be possible simply through food consumption (Hansen, 1997); however, the need for controlled doses may make freeze drying and encapsulation preferable ways to achieve this. The high expression of antigens in chloroplasts, for example, would reduce the amount of plant material required for vaccination and may permit encapsulation of freeze-dried material or pill formation (Molina *et al.*, 2004). Furthermore, plant-expressed antigens have been shown to be able to induce mucosal and serum immune responses when administered parenterally or orally to experimental animals (Daniell *et al.*, 2001b).

CHLOROPLAST TRANSFORMATION

The chloroplast is the most prominent member of the plastid family. Plastids have arisen from an internalized cyanobacterial ancestor, where a primary photosynthetic endosymbiosis between a cyanobacterium and an eukaryotic host occurred (Moreira *et al.*, 2000). Although they maintain some of that ancestral genome, the majority of the endosymbiotic genes have been transferred to the nucleus, encoding proteins that are imported post-translationally (Harris *et al.*, 1994). Chloroplasts have their own genome, containing between 120 and 135 genes, concerned mostly with encoding components of the four thylakoid photosynthetic complexes (Lopez-Juez and Pyke, 2005), plus Rubisco, the most abundant protein on Earth, which is responsible for the incorporation of atmospheric CO_2 into plant structures. They have very high rates of transcription and translation, facilitating the accumulation of large amounts of soluble proteins (Heifetz and Tuttle, 2001). In a mature leaf cell, in which there may be more than 100 plastids, each containing 10-100 genomes, it is possible to find 10,000 or more identical copies of each plastid gene (Bendich, 2004). The chloroplast's primary role is photosynthesis, utilizing light energy and atmospheric CO_2 to synthesize carbohydrates. Also, chloroplast signals communicate the state of the organelle to the nucleus and modulate light-controlled gene expression accordingly (Koussevitzky *et al.*, 2007).

The first plastid genome transformation of a higher plant, tobacco, was accomplished in 1990 by the introduction of the gene that codes for the aminoglycoside 3"-adenylyltransferase (*aadA*) which conferred resistance to the antibiotic spectinomycin (Svab *et al.*, 1990). Plastid transformation involves insertion of a transgene into one or a few ptDNA copies, followed by gradual dilution of plastids carrying non-transformed copies on a selective medium (Maliga, 2003). The latter is accomplished by sub-culturing the plant 2-3 times *in vitro* in the presence of the antibiotic used for selection of transformants. Chloroplast transformation usually uses two flanking sequences that, through homologous recombination, insert foreign DNA into the spacer region between functional genes of the chloroplast genome, thus targeting the foreign genes to a precise location (Daniell *et al.*, 2001a). Because the plastid genome is highly polyploid, transformation of chloroplasts allows for the presence of thousands of copies of foreign genes per plant cell, and may generate extraordinarily high levels of foreign proteins (Daniell *et al.*, 2002). Expression levels in chloroplasts have been shown to be 10 to 100 times higher than those obtained through nuclear transgene expression in plants (Bock, 2001). Chloroplast in which foreign genes have been introduced have proven to be able to process complex eukaryotic proteins, including interferons, vaccine antigens and human somatotropin with correct folding of subunits, aided by chaperones, and formation of disulphide bridges (Daniell *et al.*, 2005).There is an increasing world demand for low cost vaccines that can be readily administered to the population, especially in economically less developed regions. A promising concept is the production of vaccines in plants that could be grown locally with high productivity. As discussed before, plants offer an advantageous protein production system for subunit vaccines to combat a wide range of diseases.

Table 1 presents a list of antigens expressed in plant's chloroplasts with the potential to become vaccine components. It also includes some examples of the most advanced follow-up studies, where the antigens produced have shown to elicit an immune response in mice and work efficiently when exposed to pathogenic attack, such as the TetC antigen and the B cholera subunit.

It is important to note that some antigens derived from transgenic plants obtained through nuclear transformation have already undergone Phase I clinical studies and have shown to be immunogenic, well-tolerated and without the need for a buffer or vehicle other than the plant cell, which allows for natural encapsulation. This is the case of potato-derived *E. coli* LT-B and Norovirus capsid protein (NVCP) (Mason et al., 1996), potato and lettuce-derived hepatitis B subunit vaccine (Thanavala et al., 2005; Kapusta et al., 2001), and tobacco-derived fragment C of tetanus toxin (Tregoning et al., 2005).

As can also be seen in Table 1, many plant species have been used to produce recombinant proteins via the chloroplast genome, including tobacco (Li *et al.*, 2006; Tregoning *et al.*, 2003; Leelavathi and Reddy, 2003; Chebolu and Daniell, 2007; Millan *et al.*, 2003; Koya *et al.*, 2005; Molina *et al.*, 2004), tomato (Zhou *et al.*, 2008), carrot (Kumar *et al.*, 2004) and lettuce (Ruhlman *et al.*, 2007). However, the use of this technology is still routinely performed only in tobacco, as the hurdle imposed by the difficulty of regenerating the whole plant from tissue culture *in vitro* is a considerable problem.

To exploit the chloroplast compartment for the production of pharmaceuticals that could be orally delivered or readily purified, it is important to extend chloroplast transformation to edible crops.

Antigens that will be used in active vaccination could be produced or targeted to plant tissue that could be administered without further processing. Lettuce (*Lactuca sativa*) is a commercially important crop belonging to the *Asteraceae* family. Humans consume the leaves of this crop raw and the time from sowing seed to edible biomass is only weeks compared to months for crops such as tomato or potato.

The successful transformation of lettuce plastids offers the potential to develop lettuce both as a production and a delivery system for human therapeutic proteins such as subunit vaccines. Lettuce has been shown to express transgene-encoded green fluorescent protein (GFP) to ~36 % TSP (Kanamoto *et al.*, 2006), while tomato has been successfully transformed to express the HIV-1 fusion protein p24-Nef at ~4 % TSP (Zhou *et al.*, 2008), suggesting these systems, among others, are a viable alternative for producing edible vaccines via chloroplast transformation.

Table 1. Vaccine antigens and biopharmaceutical proteins expressed via the chloroplast genome in plants

Antigen	Expression system	Expression level (TSP)	Reference
Cholera toxin B	Tobacco Lettuce	4 – 12 % 5 – 9 %	Davoodi-Semiromi *et al.*, 2009
Tetanus toxin C	Tobacco	7 – 27 %	Tregoning *et al.*, 2005
Anthrax antigen (Pag)	Tobacco	4 – 14 %	Koya *et al.*, 2005
Lyme disease (OspA)	Tobacco	1 – 10 %	Glenz *et al.*, 2006
Plague F1-V	Tobacco	15 %	Arlen *et al.*, 2008
E. coli enterotoxin B	Tobacco	2 %	Rosales-Mendoza *et al.*, 2009
Canine parvovirus (CTB)			Molina *et al.*, 2005
Swine fever virus (CSFV)	Tobacco	1 – 2 %	Shao *et al.*, 2008
Human papillomavirus (L1)	Tobacco	20 – 26 %	Fernandez-San Millan *et al.*, 2008
Rotavirus (VP6)	Tobacco	0.6 – 3 %	Birch-Machin *et al.*, 2004
Hepatitis C virus	Tobacco	2	Daniell *et al.*, 2005
SARS protein	Tobacco	0.2	Li *et al.*, 2006
HIV p24 and p24-Nef	Tobacco Tomato	4 – 40 % 0.5 – 2.5 %	McCabe *et al.*, 2008; Zhou *et al.*, 2008
Amoebiasis (LecA)	Tobacco	7 %	Chebolu and Daniell, 2007
Malaria (CTB)	Tobacco Lettuce	8 - 12 % 5 - 9 %	Davoodi-Semiromi *et al.*, 2009
Diabetes type 1 (CTB and GAD65)	Tobacco Lettuce Chlamydomona s	16 % 2 % 0.3 %	Ruhlman *et al.*, 2007; Wang *et al.*, 2008
Interferon □2b	Tobacco	2 – 21 %	Arlen *et al.*, 2007
Insulin-like growth factor	Tobacco	32 %	Daniell *et al.*, 2009a; 2009b
Human alpha1-antitrypsin	Tobacco	2 %	Nadai *et al.*, 2008
Antimicrobial peptide	Tobacco	30 %	Oey *et al.*, 2009

TSP: Total soluble protein. Table edited from Daniell *et al.*, 2009.

DIABETES

The estimated number of people suffering from diabetes according to the WHO and the International Diabetes Federation (IDF) has been reported to be 246 million people worldwide, its incidence is increasing rapidly, and it is estimated that by the year 2030, this number will double. Of those 246 million, 5%-10% are estimated to be diabetes mellitus type 1 cases (http://www.who.com, http://www.idf.org). Although diabetes mellitus occurs throughout the world, it is more common in developed countries, demonstrating the need for a safe and effective vaccine.

Diabetes mellitus is a human chronic disease that occurs when the pancreatic Islets of Langerhans, specifically the β-cells contained in them, do not produce enough insulin or when the body cannot effectively use the insulin it produces (Raz *et al.*, 2005). The disease is characterized by the presence of humoral and cellular immune responses to pancreatic islet proteins. These proteins include the islet cell antigen (ICA) (Botazzo *et al.*, 1974), insulin and pro-insulin (Kuglin *et al.*, 1988; Palmer *et al.*, 1983), glutamic acid decarboxylase (GAD) (Baekkeskov *et al.*, 1990) and the tyrosine phosphatase-like enzyme (IA-2) (Christie *et al.*, 1992). The immune response against these antigens results in the destruction of all insulin-producing β-cells. Furthermore the immune response to these antigens predates the clinical onset of diabetes (Narendran *et al.*, 2005), giving further support for an autoimmune etiology of the disease. Although the initial events that trigger the destruction of β-cells are not completely understood, it has been demonstrated that the 65-kDa isomer of GAD (GAD65) plays a critical role in the destruction of pancreatic islets (Yoon *et al.*, 1999). Previous studies showed that 80% of pre-diabetics and most recent-onset diabetics have autoantibodies directed against GAD65 (Baekkeskov *et al.*, 1990; Hagopian *et al.*, 1993). In a previous work by Badillo-Corona (2007) it was demonstrated that it is possible to introduce the gene that codes for a chimeric form of GAD65 (composed of the region coding for the first 87 amino acids from GAD67 from *Rattus norvegicus* and the region coding for the rest of the amino acids for human GAD65) but encountered a problem as the protein that accumulated was a truncated form (45 kDa) of the enzyme.

AIDS

AIDS is another major fatal disease, with more than 33 million people living with the virus, particularly in developing countries where more than 90% of infections occur. As a retrovirus, HIV-1 integrates its genes into the target cell DNA, thereby quickly establishing a lifelong infection if not stopped at the time of initial exposure (Gallo, 2005). In the absence of treatment, the infection results in slow destruction (5-10 years) of the immune system cells; more precisely, the loss of helper T lymphocytes and destruction of lymphatic tissues leads to a weakened immune system, with the addition of becoming more susceptible to so-called opportunistic infections (Pope and Haase, 2003).

Vaccine candidates should be antigens that do not undergo extensive mutations, as this would provide a means for viral persistence. Subunit vaccines offer the advantage of targeting specific epitopes that lie within conserved areas of the virus (Johnson and Kalams, 1998). A couple of such potential antigens are the 24-kDa core protein p24 and the 31-kDa regulatory protein Nef. In other viral infections, like hepatitis B and influenza, immune responses directed at the viral core protein have shown to be protective (Iwarson *et al.*, 1985; Murray *et al.*, 1987; Russel and Liew, 1980). It has been observed that anti-p24 antibodies have been correlated with the CD4+ cell count (Andrieu *et al.*, 1988) and with a delayed progression of AIDS (Salk, 1987). The realization that p24 is expressed on the surface of infected cells (Laurent *et al.*, 1989) implies that p24 may be potentially important in the development of vaccines. Nef, on the other hand, is considered to be an indispensable regulatory protein for HIV, making it also a clear target for a multi-component vaccine. Even though they are buried within the virus particle, their level of conservation and the importance of their roles in the replication of the virus make them promising candidates as components of an HIV vaccine. These antigens have been successfully produced in chloroplasts of tobacco and tomato, achieving levels of 0.5-40% of the total soluble protein of the plant. Also, they have been shown to adopt the correct 3-D structure, as can be seen by western blot and ELISA experimental techniques (Zhou *et al.*, 2008). Furthermore, subcutaneous immunization with purified chloroplast-derived p24 has been shown to elicit a strong antigen-specific serum IgG response, comparable to that produced by *E. coli*-derived p24 (Gonzalez-Rábade *et al.*, 2011). Oral administration of a partially purified preparation of chloroplast-derived p24-Nef fusion protein, used as a booster after subcutaneous injection with either p24 or Nef, also elicited strong antigen-specific serum IgG responses (Gonzalez-Rábade *et al.*, 2011). These promising results encouraged us to try to express these antigens in an edible plant, such as lettuce.

METHODS

The Particle Bombardment Method

Genetic transformation of chloroplasts is commonly done by particle bombardment, which involves coating inert metal powder such as gold or tungsten with foreign DNA and shooting it into the leaves (Bock, 2001). Although direct gene transfer using the biolistic method is currently the most widely used technology for plastid transformation, stable introduction of foreign DNA into the chloroplast genome has also been demonstrated using a chemical treatment of protoplasts with polyethylene glycol (PEG) in the presence of vector DNA (Golds *et al.*, 1993; O'Neill *et al.*, 1993). Stable transformation in plastids has also been achieved in *Chlamydomonas*, an algae commonly used for chloroplast expression of foreign genes (Kindle *et al.*, 1991; Goldschmidt-Clermont, 1991).

The transformation frequency of either of these two alternative treatments, however, is significantly lower than the biolistic protocol. A depiction of different stages in the transformation of tobacco plastids by particle bombardment of leaves followed by selection of resistant lines can be seen in Figure 2.

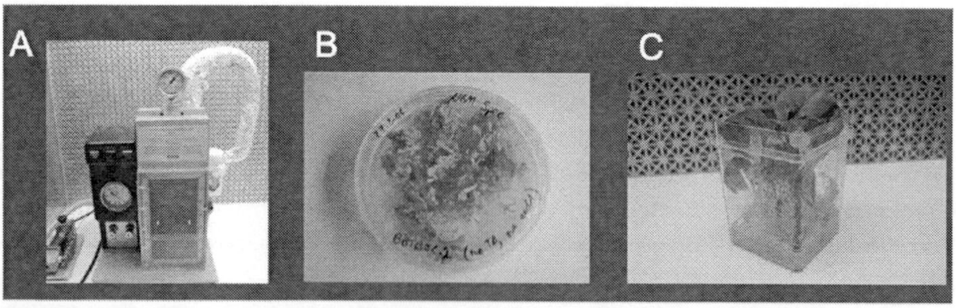

Figure 2. Generation of transplastomic tobacco plants. A. BioRad particle bombardment gun used for shooting tungsten or gold particles coated with the desired DNA into wild-type leaves. B. After incubation for four weeks, selection of transformed shoots on spectinomycin-containing plant regeneration medium, supplemented with phytohormones that induce shoot formation (naphthalene acetic acid (NAA) and benzylaminopurine (BAP) in the case of tobacco). Untransformed shoots would be completely bleached due to effective inhibition of plastid protein synthesis by spectinomycin. C. Transplastomic plant grown from a spectinomycin-resistant shoot in a Magenta box with spectinomycin-containing agar medium.

One of the most widely used plant models for expression of foreign proteins, and one of the models used in our laboratory, is tobacco, a non-food crop in which leaf tissue can be harvested for protein production prior to flowering and seed set (Fischer *et al.*, 2004). Despite its alkaloid content, tobacco is relatively easy to manipulate, grows very well under greenhouse conditions and produces a considerable amount of biomass. Furthermore, alkaloids can be easily, but costly, separated from proteins during purification and processing (Watson *et al.*, 2004).

Also, tobacco is more attractive in terms of transgene containment than other systems based on food crops, such as maize and rice, where proteins are extracted from the seed. Chloroplast transformation, in particular, provides further containment due to maternal inheritance of the organelle (Ruf *et al.*, 2007).

This is a particularly important factor to consider when focusing on HIV antigen genes, for example (McCabe *et al.*, 2008).

The levels of protein accumulation obtained through chloroplast transformation are high enough to be commercially viable, as is the case of tetanus toxin C, which accumulates to ~25% TSP in leaves (Tregoning *et al.*, 2003).

Vectors Used for Chloroplast Transformation

The vector that we use for chloroplast transformation encoding for HIV antigens is based on the pZF series (Zhou *et al.*, 2008), where the gene of interest is inserted between the *trnfM* and *trnG* genes of the chloroplast genome (Figure 3).

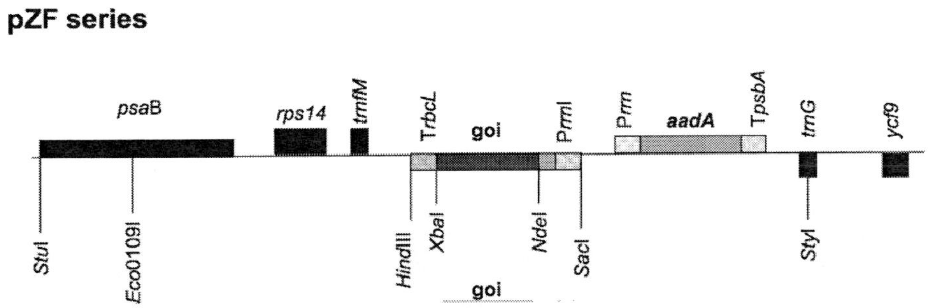

Figure 3. pZF series of constructs for the expression of HIV-1 antigens *p24* and *Nef* in chloroplasts: Codon-optimized inserts targeting the *trnfM/trnG* region of tobacco chloroplasts (Zhou *et al.*, 2008). All vectors carry an expression cassette for the selectable marker *aadA* and an expression cassette for the HIV gene of interest (goi), driven by the *rrn* promoter with the T7g10 5′ UTR.

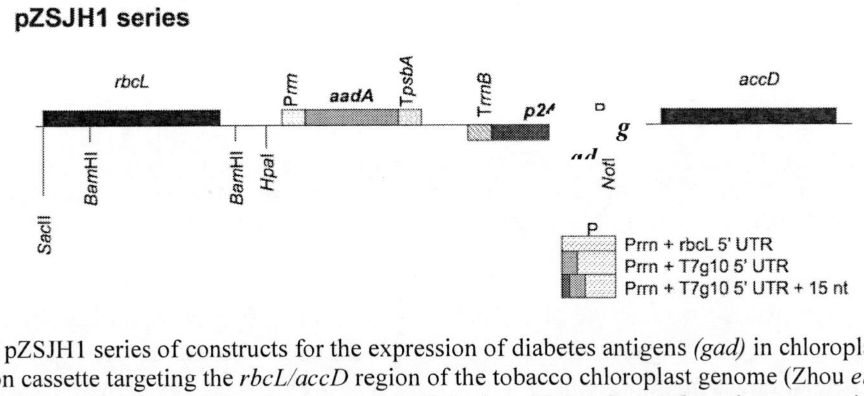

Figure 4. pZSJH1 series of constructs for the expression of diabetes antigens *(gad)* in chloroplasts: expression cassette targeting the *rbcL/accD* region of the tobacco chloroplast genome (Zhou *et al.*, 2008). All vectors carry an expression cassette for the selectable marker *aadA* and an expression cassette for the gene of interest (*gad*), driven by a P*rrn* promoter (shown in P legend).

Also, the *aadA* selectable marker gene (conferring resistance to spectinomycin) is flanked by *loxP* series to allow for future excision.

The construct that we use for chloroplast transformation encoding for diabetes antigens is based on the pZSJH1 series (Zhou *et al.*, 2008), where the gene of interest is inserted between the *rbcL* and *accD* genes of the chloroplast genome (Figure 4). Also, this vector contains an *aadA* selectable marker gene.

Experimental Background

Previously, during his PhD studies, Badillo-Corona (2007) used vectors from the series pZSJH1 to introduce the gene *gad*, coding for a glutamic acid decarboxylase (GAD - a human protein with potential to be used as a vaccine against diabetes), to produce transplastomic plants. The stable integration and transfer of the gene to the progeny was achieved. The presence of GAD in transplastomic plants was then assessed by western blotting as can be seen in Figure 5. The monoclonal antibody detected a protein of apparent molecular mass of 45 kDa in the protein extracts from lines ZSJH1gad 10, 11 and ZSJH1gad mut 21, 22 and 23. A faint band, corresponding to a protein of molecular mass of 34 kDa, was visible in the protein extracts from lines ZSJH1gad 10 and ZSJH1gad mut 22 and 23.

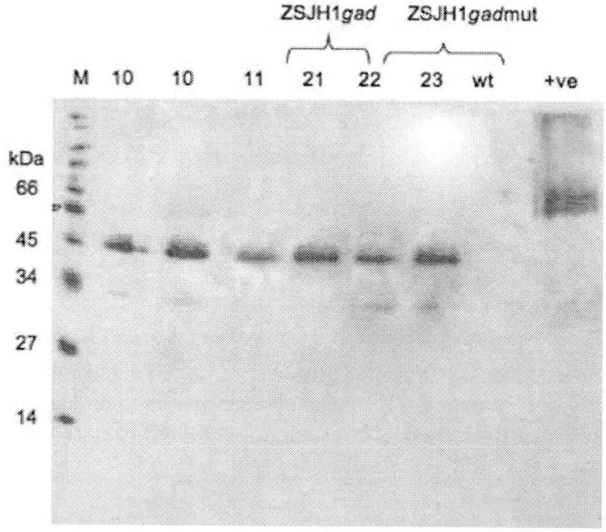

Figure 5. Western blot of the analysis of transplastomic lines with anti-GAD monoclonal antibodies. Total soluble protein from wild-type tobacco (wt) and transplastomic lines ZSJH1gad 10, 11, ZSJH1gad mut 21, 22 and 23 were separated by SDS-12% PAGE and electroblotted onto a nitrocellulose membrane. Anti-GAD monoclonal antibodies were used to detect the presence of GAD and GADmut by ECL. M: marker; +ve: 10 ug of whole brain-cell lysate from human brain.

The antibody bound to a protein of apparent molecular mass of 34 kDa in the wild-type extract. This explains the presence of the band corresponding to a protein of 34 kDa in lines ZSJH1gad 10 and ZSJH1gad mut 22 and 23, which could be the result of non-specific binding to endogenous plant proteins.

The antibody detected the presence of GAD65 and GAD67 in the whole brain-cell lysate used as a positive control. Although the 45-kDa protein detected in the transplastomic lines was not present in the wild-type lane, it was not of the expected size for GAD65 (65 kDa).

The high specificity of the monoclonal antibody used suggested that the protein detected could be a truncated version or degradation product. Before further investigating this matter we decided to analyze the sequence of the *gad* gene used for transformation. We found that within the sequence coding for the gene there was a sequence identical to the ribosome binging site (RBS), contained in the 5´UTR, which intended to serve as a translation control. This additional RBS is followed by an ATG codon at an appropriate distance to serve as a start codon for translation. We speculated then that this alternative RBS was being used to initiate translation and somehow was being favored over the RBS site that was supposed to generate the 65-kDa protein, probably due to a more stable 3-D configuration of the mRNA transcript.

With respect to the HIV antigen-encoding transplastomic plants, González-Rábade (2010) sought to investigate the effect of leaf age on the accumulation of the antigen(s) in tobacco leaves as part of her PhD research project. Since there have been reports of the accumulation of other antigens, such as rotavirus VP6, in which protein accumulation seemed to be dependent on leaf age (Birch-Machin *et al.*, 2004), this seemed like a logical first step toward the characterization of these plants. A transplastomic plant was grown until 12 leaves had appeared and then the amount of soluble p24 present in all 12 leaves was determined by p24-specific ELISA using the extracts previously analyzed by immunoblotting (Figure 6).

The protein of interest was detected in all leaves, even though the largest amount of p24 was detected in leaves 6-8, with the HIV-1 antigen accumulating at levels of up to 450 mg/kg fresh weight (~4.5 % TSP).

The accumulation pattern seen, with slightly less protein in leaves 1-3 and 10-12 can be explained by normal cell turn over, reflecting an increase in protein accumulation, not in protein synthesis, in mature leaves.

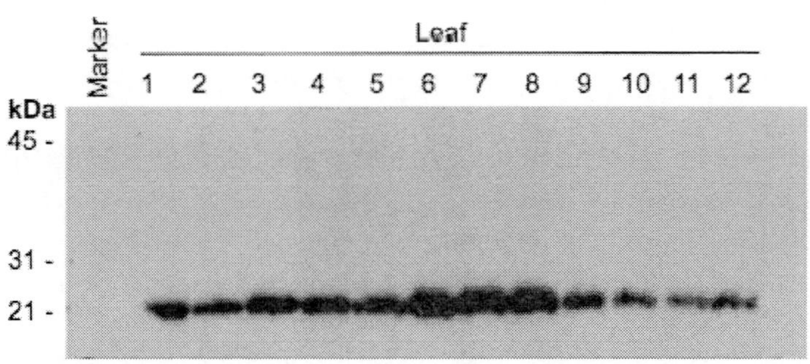

Figure 6. Western blot showing leaves 1-12 of a 12-week-old transplastomic plant with anti-p24 polyclonal antibodies. Total soluble protein from leave extracts was separated by SDS-12% PAGE and electroblotted onto a nitrocellulose membrane. Anti-p24 polyclonal antibodies were used to detect the presence of p24 by ECL.

González-Rábade (2010) also investigated the accumulation of the fusion protein p24-Nef and thus carried out an experiment in the same way but with a plant comprising 15 leaves. In this case, a band for a protein of ~50 kDa could be seen in all the leaves, accounting for the fusion protein, as well as a faint 24 kDa band, probably due to degradation or simply the p24 monomer (Figure 7). Surprisingly, in this case the youngest leaf, leaf 15, accumulated the highest quantity of the fusion protein, showing that the protein was highly stable and less susceptible to natural age-related protein turn over in the older leaves.

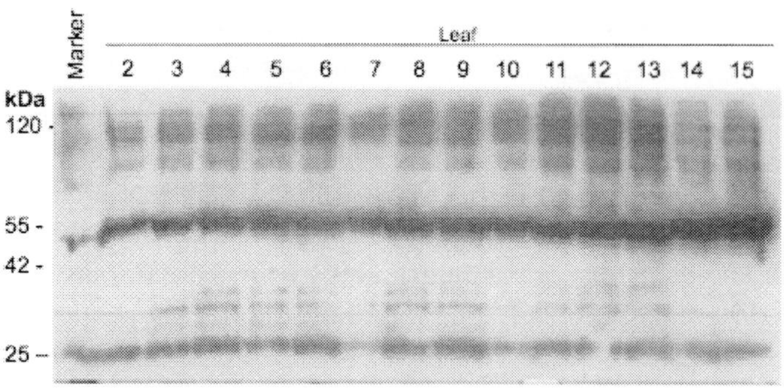

Figure 7. Western blot showing leaves 1-15 of a 12-week-old transplastomic plant with anti-p24 polyclonal antibodies. Total soluble protein from leave extracts was separated by SDS-12% PAGE and electroblotted onto a nitrocellulose membrane. Anti-p24 polyclonal antibodies were used to detect the presence of p24-Nef by ECL.

Current Research in Our Laboratory

In order to generate GAD65 proteins of the right size, a couple of new vectors were designed by our research group at the Laboratory of Molecular Biotechnology at the Professional Interdisciplinary Unit of Biotechnology of the National Polytechnic Institute (UPIBI-IPN).

These vectors, also from the series pZSJH1, contained expression cassettes for GAD65 carrying the Prrn promoter (Zhou *et al.*, 2008) and the TrrnB terminator. However, in these vectors a silent mutation was introduced to eliminate the RBS (Ribosome Binding Site) site that, we believed, was being used as an alternative site to start translation. This mutation would avoid problems of alternative initiation and would, hopefully, generate an un-truncated version of GAD65.

Since we believed the problem in the size of the protein arose because of the presence of this RBS site, we modified this region by site directed mutagenesis to change a couple of bases within the RBS. The original sequence was changed from TTGGAGG to TTAGAAG. The mutation was confirmed by DNA sequencing, where the sequence of the original vector was compared to the sequence of the vector where a mutation in the RBS took place (Figure 8).

The change in this bases are silent mutations and thus do not change the amino acid sequence of the protein. We have used the newly generated vector to transform tobacco plants and are in the process of analyzing these lines to evaluate if a full size GAD65 protein can be produced in plants.

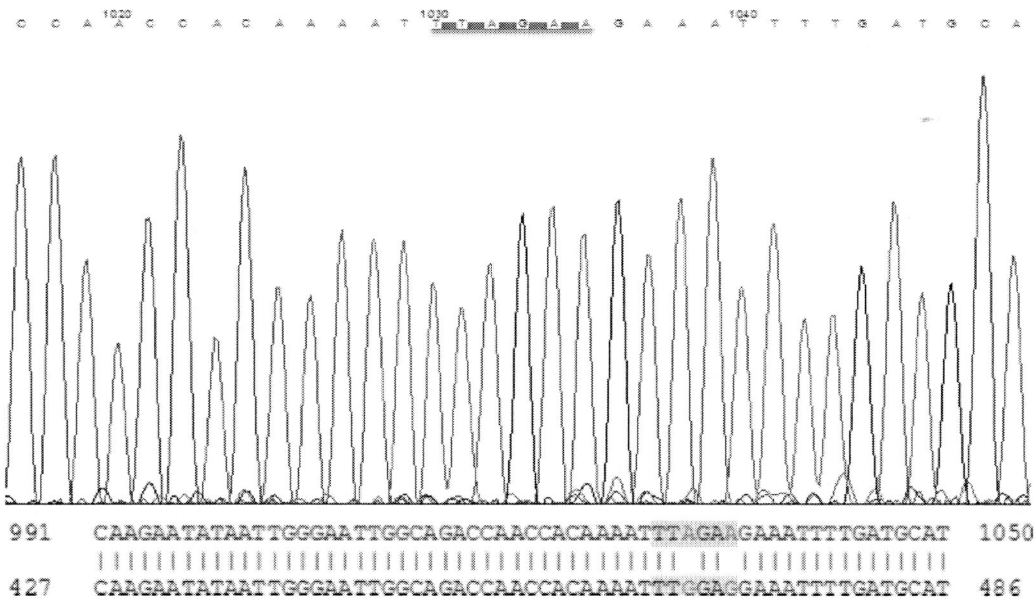

Figure 8. DNA sequence comparison showing the coding region for the protein of interest, indicating the mutation that took place within the alternative RBS site. The upper sequence is present in the vector generated by silent mutagenesis while the lower one is the sequence in the original vector. The bases that were originally G and were changed to A are shown in red.

As mentioned before, the expression of *p24* and *nef* in tobacco and tomato chloroplasts has been demonstrated (Zhou *et al.*, 2008), as well as the immunogenicity of the antigens in mice (Gonzalez-Rabade *et al.*, 2011). The main drawback for antigen production in tobacco, however, is its high alkaloid content, which is the main reason why we are keen on expanding chloroplast transformation to other crops besides tobacco. In this respect, we are currently working on chloroplast transformation of lettuce and alfalfa in our laboratory. Regarding the HIV-1 antigens p24 and Nef, we have generated a variety of constructs, specific for lettuce and alfalfa chloroplast transformation. The interesting feature of these new constructs is that knowledge obtained in other crops to increase accumulation of foreign proteins in chloroplasts has been incorporated into these vectors. The construct for lettuce chloroplast transformation was based on pLCV2, targeting genes to the *trnA*/*trnI* intergenic region of the lettuce plastid genome (Lelivelt *et al.*, 2005). All our constructs carry an expression cassette for an antibiotic-resistance gene (*aadA*) for selection of transformed lines. We are currently working on transformation of lettuce with this new construct and optimization of conditions for plant tissue regeneration to obtain transplastomic plants with high efficiency. The tissue culture conditions for the regeneration of alfalfa are also being studied. If recombinant protein expression levels in lettuce and/or alfalfa are high, these systems could be very promising for the development of an edible vaccine.

Future Perspectives and Conclusions

One of the many advantages of the chloroplast transformation technology is the possibility of producing large quantities of heterologous proteins at a low cost (Maliga, 2003).

The technology has been used to successfully produce human and viral proteins with therapeutic properties at higher levels than those achieved by nuclear transformation (Birch-Machin *et al.*, 2004; Fernandez-San Millan *et al.*, 2003; Staub *et al.*, 2000; Tregoning *et al.*, 2005). As discussed above, the interest of our laboratory is to use this valuable technology to improve the levels of accumulation of various important antigens in a crop that could facilitate their use as platforms of production. Even though current chloroplast transformation vectors allow for the accumulation of high quantities of protein, the stability of the proteins is still a significant problem since some proteins seem to degrade as the plant ages. This degradation might be a process tied to the senescence of the plant, to the amino acid sequence of the protein or be species-specific. No study has so far analyzed the stability of heterologous proteins in plants other than tobacco.

An alternative to the use of crops as production systems would be the use of photosynthetic eukaryotic microalgae, such as *Chlamydomonas reinhardtii*. This alga contains chloroplasts that are amenable for genetic transformation and, like plant chloroplasts, are capable of accumulating great quantities of foreign proteins. The advantage of *C. reinhardtii* over plants is that its life cycle is much shorter and its cultivation and downstream processing are similar to current mature and well-characterized prokaryotic expression systems. The drawback is that expression levels in the chloroplast are not as high as in plants. This gives our research group a window of opportunity to explore promoter and translation elements that would allow for the accumulation of large quantities of foreign proteins, making algae systems commercially more attractive.

In transplastomic plants, HIV-1 p24 and p24-Nef have been shown to accumulate to very high levels and to produce immunogenic responses (Zhou *et al.*, 2008; Gonzalez-Rabade *et al.*, 2011). In the future, it would be ideal to have transplastomic edible plants, such as lettuce or tomato, in order to decrease the purification costs of these types of antigens and potentially produce oral vaccines. The concept of oral vaccines has changed over time and, even though it has not been discarded completely, the idea of vaccinating people by feeding them an antigen-containing fruit such as an apple or banana is now considered unfeasible because of concerns about dose control. Therefore, the term "edible vaccines" refers more to an antigen that can be produced in an edible part of a plant, which would automatically decrease purification costs, as there would be no need to eliminate human pathogens (as in the case of animal cell production systems) or bacterial toxins (as in the case of bacterial expression systems). Also, the edible part of the plant could be lyophilized and the concentration of the antigen determined to produce encapsulated pills that could then be transported to remote places without the need of a cold chain. Hence, the possibility of producing high levels of foreign antigens in edible plants becomes a very attractive option from a commercial point of view. In this respect, our laboratory is actively trying to implement chloroplast transformation in lettuce and alfalfa, both edible crops. Even though these two species have been transformed, their regeneration *in vitro* is not well established and seems to be highly species-specific. These tissue culture constrains are currently being addressed by our research group.

Finally, the reported finding that a fusion protein could be more stable than a single protein and that there are protein stability determinants in chloroplasts (González-Rábade, 2010; Apel *et al.*, 2010) could prove very useful as a guide, first for the identification of the sequences that can confer this stability and then for their use to further increase the protein stability of our antigens of interest. Whether the stability of some proteins changes across different plant species remains an unknown aspect that also deserves attention. The expression of *p24* and *nef* genes (and potentially *gad*) in lettuce could help answer this question, as well as functioning as a stepping stone towards the production of stable edible vaccines.

REFERENCES

Andrieu, J. M., Eme, D., Venet, A., Audroin, C., Tourani, J. M., Stern, M., Israelbiet, D., Beldjord, K., Driss, F., and Even, P. (1988). Serum HIV antigen and anti-p24 antibodies in 200 HIV seropositive patients correlation with CD4 and CD8 lymphocyte subsets. *Clin. Exp. Immun.*, 73: 1–5.

Apel, W., Schulze, W.X. and Bock, R. (2010). Identification of protein stability determinants in chloroplasts. *Plant J.*, 63: 636-650.

Arlen, P.A., Falconer, R., Cherukumilli, S., Cole, A., Cole, A.M., Oishi, K.K., and Daniell, H. (2007) Field production and functional evaluation of chloroplast-derived interferon-alpha2b. *Plant Biotechnol. J.,* 5: 511-525.

Arlen, P.A., Singleton, M., Adamovicz, J.J., Ding, Y., Davoodi-Semiromi, A., and Daniell, H. (2008) Effective plague vaccination via oral delivery of plant cells expressing FI-V antigens in chloroplasts. *Infect. Immun.*, 76: 3640-3650.

Badillo-Corona, J.A. (2007) Expression of HIV-1 p24 in tobacco chloroplasts. PhD Thesis. University of Cambridge, UK.

Baekkeskov, S., Aanstoot, H. J., Christgau, S., Reetz, A., Solimena, M., Cascalho, M., Folli, F., Richterolesen, H., and Camilli, P. D. (1990). Identification of the 64K autoantigen in insulin-dependent diabetes as the GABA-synthesizing enzyme glutamic acid decarboxylase. *Nature*, 347: 151–156.

Bendich, A.J. (2004). Circular chloroplast chromosomes: the grand illusion. *Plant Cell*, 16: 1661–1666.

Birch-Machin, I., Newel, C. A., Hibberd, J. M., and Gray, J. C. (2004). Accumulation of rotavirus VP6 protein in chloroplasts of transplastomic tobacco is limited by protein stability. *Plant Biotechnol. J.,* 2: 261-270.

Bock, R. (2001). Transgenic plastids in basic research and plant biotechnology. *J. Mol. Biol.,* 312: 425–438.

Botazzo, G., Florin-Christensen, A., and Doniach, D. (1974). Islet autoantibodies in diabetes mellitus with autoimmune polyendocrin deficiencies. *Lancet*, 2: 1279–1287.

Carter, J. E. and Langridge, W. H. R. (2002). Plant-based vaccines for protection against infectious and autoimmune diseases. *Crit. Rev. Plant Sci.*, 21: 93–109.

Chebolu, S. and Daniell, H. (2007). Stable expression of Gal/GalNAc lectin of *Entamoeba histolytica* in transgenic chloroplasts and immunogenicity in mice towards vaccine development for amoebiasis. *Plant Biotechnol. J.,* 5: 230–239.

Chiarella, P., Massi, E., Robertis, M.D., Signori, E. and Fazio, V.M. (2007). Adjuvants in vaccines and for immunization: current trends. *Expert. Opin. Biol. Th.,* 7: 1551-1562.

Christie, M. R., Tun, R. Y. M., Lo, S. S. S., Cassidy, D., Brown, T. J., Hollands, J., Shattock, M., Bottazzo, G. F., and Leslie, R. D. G. (1992). Antibodies to GAD and tryptic fragments of islet 64K antigen as distinct markers for development of IDDM - studies with identical-twins. *Diabetes*, 41: 782–787.

Coughlan, C.M. and Brodsky, J.L. (2005). Use of yeast as a model system to investigate protein conformational diseases. *Mol. Biotechnol.*, 30: 171–180.

Daniell, H., Chebolu, S., Kumar, S., Singleton, M. and Falconer, R. (2005). Chloroplast-derived vaccine antigens and other therapeutic proteins. *Vaccine*, 23: 1779–1783.

Daniell, H., Khan, M.S. and Allison, L. (2002). Milestones in chloroplast genetic engineering: an environmentally friendly era in biotechnology. *Trends Plant Sci.*, 7: 84–91.

Daniell, H., Muthukumar, B., and Lee, S. B. (2001a). Marker free transgenic plants: engineering the chloroplast genome without the use of antibiotic selection. *Curr. Genet.*, 39: 109–116.

Daniell, H., Ruiz, G., Denes, B., Sandberg, L., and Langridge, W. (2009a) Optimization of codon composition and regulatory elements for expression of human insulin like growth factor-1 in transgenic chloroplasts and evaluation of structural identity and function. *BMC Biotechnol.*, 9: 33-38.

Daniell, H., Singh, N.D., Mason, H., and Streatfield, S.J. (2009b). Plant-made vaccine antigens and biopharmaceuticals. *Trends Plant Sci. Rev.*, 14: 669-679.

Daniell, H., Streatfield, S.J. and Wycoff, K. (2001b). Medical molecular farming: production of antibodies, biopharmaceuticals and edible vaccines in plants. *Trends Plant Sci.*, 6: 219–226.

Davoodi-Semiromi, A., Samson, N., and Daniell, H. (2009) The green vaccine: a global strategy to combat infectious and autoimmune diseases. *Hum. Vaccine,* 5: 488-493.

Fernandez-San Millan, A., Mingo-Castel, A., Miller, M., and Daniell, H. (2003). A chloroplast transgenic approach to hyper express and purify human serum albumin, a protein highly susceptible to proteolytic degradation. *Plant Biotechnol. J.,* 1: 71-79.

Fernandez-San Millan, A., Ortigosa, S.M., Hervas-Stubbs, S., Corral-Martinez, P., Segui-Simarro, J.M., Gaetan, J., Coursaget, P., and Veramendi, J. (2008) Human papillomavirus L1 protein expressed in tobacco chloroplasts self-assembles into virus-like particles that are highly immunogenic. *Plant Biotechnol. J.,* 6: 427-441.

Fischer, R., Stoger, E., Schillberg, S., Christou, P. and Twyman, R.M. (2004). Plant- based production of biopharmaceuticals. *Curr. Opin. Plant Biol.*, 7: 152–158.

Glenz, K., Bouchon, B., Stehle, T., Wallich, R., Simon M.M., and Warzecha, H. (2006) Production of a recombinant bacterial lipoprotein in higher plant chloroplasts. *Nat. Biotechnol.*, 24: 76-77.

Golds, T.J., Maliga, P. and Koop, H. (1993). Stable plastid transformation in PEG-treated protoplasts of *Nicotiana tabacum*. *Biotechnol.*, 11: 95–97.

Goldschmidt-Clermont, M. (1991). Transgenic expression of aminoglycoside adenine transferase in the chloroplast: a selectable marker for site-directed transformation of chlamydomonas. *Nucleic Acids Res.*, 19: 4083–4089.

González-Rábade, N. (2010) Immunological analysis of chloroplast-derived HIV-1 antigens. PhD Thesis. University of Cambridge, UK.

Gonzalez-Rabade, N., McGowan, E. G., Zhou, F., McCabe, M. S., Bock, R., Dix, P. J., Gray, J. C. and Ma, J. K.-C. (2011). Immunogenicity of chloroplast-derived HIV-1 p24 and a p24-Nef fusion protein following subcutaneous and oral administration in mice. *Plant Biotechnol. J.,* 9: 1-35.

Hagopian, W. A., Chelsen, B. M., Karlsen, A. E., Larsen, F., Moody, A., Grubin, C. E., Rowe, R., Petersen, J., McEvoy, R., and Lernmark, A. (1993). Autoantibodies in IDDM primarily recognize the 65,000-Mr rather than the 67,000-Mr isoform of glutamic-acid decarboxylase. *Diabetes*, 42: 631–636.

Hansen, E. (1997). Production of recombinant antigens in plants for animal and human immunization. *Braz. J. Genet.*, 20: 1–15.

Harris, E.H., Boynton, J.E. and Gillham, N.W. (1994). Chloroplast ribosomes and protein-synthesis. *Microbiol. Rev.*, 58: 700–754.

Heifetz, P.B. and Tuttle, A.M. (2001). Protein expression in plastids. *Plant Biotechnol. J.,* 4: 157–161.

Hood, E.E., Witcher, D.R., Maddock, S., Meyer, T., Baszczynski, C., Bailey, M., Flynn, P., Register, J., Marshall, L., Bond, D., Kulisek, E., Kusnadi, A., Evan- gelista, R., Nikolov, Z., Wooge, C., Mehigh, R.J., Hernan, R., Kappel, W.K., Ritland, D., Ritland, D., Li, C.P. and Howard, J.A. (1997). Commercial production of avidin from transgenic maize: characterization of transformant, production, processing, extraction and purification. *Mol. Breeding*, 3: 291–306.

Hood, E.E., Bailey, M.R., Beifuss, K., Magallanes-Lundback, M., Horn, M.E., Callaway, E., Drees, C., Delaney, D.E., Clough, R. and Howard, J.A. (2003). Criteria for high-level expression of a fungal laccase gene in transgenic maize. *Plant Biotechnol. J.,* 1: 129–140.

Iwarson, S., Tabor, E., Thomas, H.C., Snoy, P. and Gerety, R.J. (1985). Protection against hepatitis B virus infection by immunization with hepatitis B core antigen. *Gastroenterology*, 88: 763–767.

Johnson, R.P. and Kalams, S. (1998). The science of HIV vaccine development. HIV InSite, 1-27, http://hivinsite.ucsf.edu/InSite?page=kb–02–01–06. 33, 38.

Kanamoto, H., Yamashita, A., Asao, H., Okumura, S., Takase, H., Hattori, M., Yokota, A. and Tomizawa, K. (2006). Efficient and stable transformation of *Lactuca sativa* L. cv. Cisco (lettuce) plastids. *Transgenic Res*., 15: 205–217.

Kapusta, J., Modelska, A., Pniewski, T., Figlerowicz, M., Jankowski, K., Lisowa, O., Plucienniczak, A., Koprowski, H. and Legocki, A.B. (2001). Oral immunization of human with transgenic lettuce expressing hepatitis B surface antigen. *Adv. Exp. Med. Biol.,* 495, 299–303.

Karasev, A.V., Foulke, S., Wellens, C., Rich, A., Shon, K.J., Zwierzynski, I., Hone, D., Koprowski, H. and Reitz, M. (2005). Plant based HIV-1 vaccine candidate: Tat protein produced in spinach. *Vaccine*, 23, 1875–1880.

Kindle, K.L., Richards, K.L. and Stern, D.B. (1991). Engineering the chloroplast genome: techniques and capabilities for chloroplast transformation in *Chlamydomonas reinhardtii*. *Proc. Natl. Acad. Sci*. USA, 88: 1721–1725.

Koussevitzky, S., Nott, A., Mockler, T.C., Hong, F., Sachetto-Martins, G., Surpin, M., Lim, J., Mittler, R. and Chory, J. (2007). Signals form chloroplasts converge to regulate nuclear gene expression. *Science*, 316: 715–719.

Koya, V., Moayeri, M., Leppla, S.H. and Daniell, H. (2005). Plant-based vaccine: mice immunized with chloroplast-derived anthrax protective antigen survive anthrax lethal toxin challenge. *Infect. Immun*., 73: 8266–8274.

Kuglin, B., Gries, F. A., and Kolb, H. (1988). Evidence of IgG autoantibodies against human proinsulin in patients with IDDM before insulin-treatment. *Diabetes*, 37: 130–132.

Kumar, S., Dhingra, A., and Daniell, H. (2004). Plastid-expressed betaine aldehyde dehydrogenase gene in carrot cultured cells, roots, and leaves confers enhanced salt tolerance. *Plant Physiol*., 136: 2843–2854.

Kusnadi, A. R., Nikolov, Z. L., and Howard, J. A. (1997). Production of recombinant proteins in transgenic plants: Practical considerations. Biotechnol. *Bioeng*., 56: 473–484.

Lamphear, B.J., Barker, D.K., Brooks, C.A., Delaney, D.E., Lane, J.R., Beifuss, K., Love, R., Thompson, K., Mayor, J., Clough, R., Harkey, R., Poage, M., Drees, C., Horn, M.E., Streatfield, S.J., Nikolov, Z., Woodard, S.L., Hood, E.E., Jilka, J.M. and Howard, J.A. (2005). Expression of the sweet protein brazzein in maize for production of a new commercial sweetener. *Plant Biotechnol. J.,* 3: 103–114.

Larrick, J.W., Yu, L., Chen, J., Jaiswal, S. and Wycoff, K. (1998). Production of antibodies in transgenic plants. *Res. Immunol*., 149: 603–608.

Laurent, A. G., Krust, B., Rey, M. A., Montagnier, L., and Hovanessian, A. G. (1989). Cell-surface expression of several species of human immunodeficiency virus type-1 major core protein. *J. Virol*., 63: 4074–4078.

Leelavathi, S. and Reddy, V.S. (2003). Chloroplast expression of His-tagged GUS-fusions: a general strategy to overproduce and purify foreign proteins using transplastomic plants as bioreactors. *Mol. Breeding*, 11: 49–58.

Lelivelt, C.L.C., McCabe, M.S., Newell, C.A., Bastiaan de Snoo, C., Van Dun, K.M.P., Birch-Machin, I., Gray, J.C., Mills, K.H.G. and Nugent, J.M. (2005). Stable plastid transformation in lettuce (*Lactuca sativa* L.). *Plant Mol. Biol.,* 58: 763-774.

Li, H., Ramalingam, S. and Chye, M. (2006). Accumulation of recombinant SARS-CoV spike protein in plant cytosol and chloroplasts indicate potential for development of plant-derived oral vaccines. *Exp. Biol. Med.*, 231: 1346–1352.

Lopez-Juez, E. and Pyke, K.A. (2005). Plastids unleashed: their development and their integration in plant development. Int. J. Dev. Biol., 49: 557–577.

Maliga, P. (2003). Progress towards commercialization of plastid transformation technology. *Trends Biotechnol.*, 21: 20–28.

Mason, H.S., Ball, J.M., Shi, J.J., Jiang, Z., Estes, M.K. and Arntzen, C.J. (1996). Expression of Norwalk virus capsid protein in transgenic tobacco and potato and its oral immunogenicity in mice. *Proc. Natl. Acad. Sci. USA*, 93, 5335–5340.

McCabe, M.S., Klaas, M., Gonzalez-Rabade, N., Poage, M., Badillo-Corona, J.A., Zhou, F., Karcher, D., Bock, R., Gray, J.C. and Dix, P.J. (2008). Plastid transformation of high-biomass tobacco variety Maryland Mammoth for production of human immunodeficiency virus type 1 (HIV-1) p24 antigen. *Plant Biotechnol. J.,* 6: 914–929.

Millan, A.F.S., Mingo-Castel, A., Miller, M. and Daniell, H. (2003). A chloroplast transgenic approach to hyperexpress and purify Human Serum Albumin, a protein highly susceptible to proteolytic degradation. *Plant Biotechnol. J.,* 1: 71–79.

Molina, A., Hervas-Stubbs, S., Daniell, H., Mingo-Castel, A.M. and Veramendi, J. (2004). High-yield expression of viral peptide animal vaccine in transgenic tobacco chloroplasts. *Plant Biotechnol. J.,* 2: 141–153.

Molina, A., Veramendi, J., and Hervas-Stubbs, S. (2005) Induction of neutralizing antibodies by a tobacco chloroplast-derived vaccine based on a B cell epitope from canine parvovirus. *Virology,* 342: 266-275.

Moreira, D., Le Guyader, H. and Philippe, H. (2000). The origin of red algae and the evolution of chloroplasts. *Nature*, 405: 69–72.

Murray, K., Bruce, S.A., Wingfield, P., van Eerd, P., de Reus, A. and Schellekens, H. (1987). Protective immunisation against hepatitis B with an internal antigen of the virus. *J. Med. Virol.*, 23: 101–107.

Nadai, M., Baily, J., Vitel, M., Job, C., Tissot, G., Botterman, J., and Dubald, M. (2008) High-level expression of active human alpha1-antitrypsin in transgenic tobacco chloroplasts. *Transgenic Res.*, 18: 173-183.

Narendran, P., Estella, E., and Fourlanos, S. (2005). Immunology of type 1 diabetes. *QJM-Int. J. Med.*, 98: 547–556.

Obregon, P., Chargelegue, D., Drake, P., Prada, A., Nuttall, J., Frigerio, L., and Ma, J. K. C. (2006). HIV-1 p24-immunoglobulin fusion molecule: a new strategy for plant-based protein production. *Plant Biotechnol. J.,* 4: 195–207.

Oey, M., Lohse, M., Scharff, L.B., Kreikemeyer, B., and Bock, R. (2009) Plastid production of protein antibiotics against pneumonia via a new strategy for high-level expression of antimicrobial proteins. *Proc. Natl. Acad. Sci. USA,* 106: 6579-6584.

O'Neill, C., Horvath, G. V., Horvath, E., Dix, P. J., and Medgyesy, P. (1993). Chloroplast transformation in plants - polyethylene-glycol (PEG) treatment of protoplasts is an alternative to biolistic delivery systems. *Plant J.,* 3: 729–738.

Palmer, J. P., Asplin, C. M., Clemons, P., Lyen, K., Tatpati, O., Raghu, P. K., and Paquette, T. L. (1983). Insulin-antibodies in insulin-dependent diabetics before insulin-treatment. *Science*, 222: 1337–1339.

Perez-Filgueira, D., Brayfield, B., S.Phiri, Borca, M., Wood, C. and Morris, T. (2004). Preserved antigenicity of HIV-1 p24 produced and purified in high yields from plants inoculated with a tobacco mosaic virus (TMV)-derived vector. *J. Virol. Methods*, 121: 201– 208.

Petsch, D. and Anspach, F.B. (2000). Endotoxin removal from protein solutions. *J. Biotechnol.*, 76: 97–119.

Pope, M. and Haase, A. T. (2003). Transmission, acute HIV-1 infection and the quest for strategies to prevent infection. Nature Med., 9: 847–852.

Puls, R.L. and Emery, S. (2006). Therapeutic vaccination against HIV: current progress and future possibilities. *Clin. Sci.*, 110: 59–71.

Raz, I., Eldor, R., and Naparstek, Y. (2005). Immune modulation for prevention of type 1 diabetes mellitus. *Trends Biotechnol.*, 23: 128–134.

Rosales-Mendoza, S., Alpuche-Solis, A.G., Soria-Guerra, R.E., Moreno-Fierros, L., Martinez-Gonzalez, L., Herrera-Diaz, A., and Korban, S.S. (2009) Expression of an Escherichia coli antigenic fusion protein comprising the heat labile toxin B subunit and the heat stable toxin, and its assembly as a functional oligomer in transplastomic tobacco plants. *Plant J.,* 57: 45-54.

Ruf, S., Karcher, D. and Bock, R. (2007). Determining the transgene containment level provided by chloroplast transformation. *Proc. Natl. Acad. Sci. USA,* 104: 6998–7002.

Ruhlman, T., Ahangari, R., Devine, A., Samsam, M. and Daniell, H. (2007). Expression of cholera toxin B-proinsulin fusion protein in lettuce and tobacco chloroplasts - oral administration protects against development of insulitis in non-obese diabetic mice. *Plant Biotechnol. J.,* 5: 495–510.

Salk, J. (1987). Prospects for the control of AIDS by immunizing seropositive individuals. *Nature*, 327: 473–476.

Shao, H.B., He, D.M., Qian, K.X, Shen, G.F., and Su, Z.L. (2008) The expression of classical swine fever virus structural protein E2 gene in tobacco chloroplasts for applying chloroplasts as bioreactors. *C. R. Biol.*, 331: 179-184.

Staub, J. M., Garcia, B., Graves, J., Hajdukiewicz, P. T. J., Hunter, P., Nehra, N., Paradkar, V., Schlittler, M., Carroll, J. A., Spatola, L., Ward, D., Ye, G. N., and Russell, D. A. (2000). High-yield production of human therapeutic protein in tobacco chloroplasts. *Nature Biotechnol.*, 18: 333-338.

Streatfield, S.J. (2005). Plant-based vaccines for animal health. *Rev. Sci. Tech. Off. Int. Epiz.*, 24: 189–199.

Streatfield, S.J. and Howard, J.A. (2003). Plant-based vaccines. *Int. J. Parasitol.*, 33: 479–493.

Svab, Z., Hajdukiewicz, P. and Maliga, P. (1990). Stable transformation of plastids in higher plants. *Proc. Natl. Acad. Sci. USA*, 87: 8526–8530.

Thanavala, Y., Mahoney, M., Pal, S., Scott, A., Richter, L., Natarajan, N., Goodwin, P., Arntzen, C.J. and Mason, H.S. (2005). Immunogenicity in humans of an edible vaccine for hepatitis B. *Proc. Natl. Acad. Sci. USA*, 102, 3378–3382.

Tregoning, J.S., Nixon, P., Kuroda, H., Svab, Z., Clare, S., Bowe, F., Fairweather, N., Ytterberg, J., Wijk, K.J.V., Dougan, G. and Maliga, P. (2003). Expression of tetanus toxin fragment C in tobacco chloroplasts. *Nucleic Acids Res.*, 31: 1174–1179.

Tregoning, J. S., Clare, S., Bowe, F., Edwards, L., Fairweather, N., Qazi, O., Nixon, P. J., Maliga, P., Douglas, D., and Hussell, T. (2005). Protection against tetanus toxin using a plant-based vaccine. *Eur. J. Immunol.*, 35: 1320-1326.

Wang, X., Brandsma, M., Tremblay, R., Maxwell, D., Jevnikar, A.M., Huner, N., and Ma, S. (2008) A novel expression platform for the production of diabetes-associated autoantigen human glutamic acid decarboxylase (hGAD65). *BMC Biotechnol.*, 8: 87.

Watson, J., Koya, V., Leppla, S.H. and Daniell, H. (2004). Expression of *Bacillus anthracis* protective antigen in transgenic chloroplasts of tobacco, a non-food/feed crop. *Vaccine*, 22: 434–4384.

Woodard, S.L., Mayor, J.M., Bailey, M.R., Barker, D.K., Love, R.T., Lane, J.R., Delaney, D.E., McComas-Wagner, J.M., Mallubhotla, H.D., Hood, E.E., Dan- gott, L.J., Tichy, S.E. and Howard, J.A. (2003). Maize (Zea mays)-derived bovine trypsin: characterization of the first large-scale, commercial protein product from transgenic plants. Biotechnol. *Appl. Biochem.*, 38: 123–130.

Yoon, J. W., Yoon, C. S., Lim, H. W., Huang, Q. Q., Kang, Y., Pyun, K. H., Hirasawa, K., Sherwin, R. S., and Jun, H. S. (1999). Control of autoimmune diabetes in NOD mice by GAD expression or suppression in beta cells. *Science*, 284: 1183–1187.

Yusibov, V., Hooper, D.C., Spitsin, S.V., Fleysh, N., Kean, R.B., Mikheeva, T., Deka, D., Karasev, A., Cox, S., Randall, J. and Koprowski, H. (2002). Expression in plants and immunogenicity of plant virus-based experimental rabies vaccine. *Vaccine*, 20: 3155–3164.

Yusibov, V. and Rabindran, S. (2008). Recent progress in the development of plant-derived vaccines. Expert Rev. *Vaccines*, 7: 1173–1183.

Zhou, F., Badillo-Corona, J.A., Karcher, D., Gonzalez-Rabade, N., Piepenburg, K., Borchers, A.M.I., Maloney, A.P., Kavanagh, T.A., Gray, J.C. and Bock, R. (2008). High-level expression of human immunodeficiency virus antigens from the tobacco and tomato plastid genomes. *Plant Biotechnol. J.,* 6: 897–913.

In: Biotechnology
Editors: L. G. Torres Bustillos and I. Garcia Peña
ISBN 978-1-62081-071-2

Chapter 3

The Role of Glycosides on Biotechnology

Marco A. Brito-Arias[*]
Dept. of Natural Sciences. UPIBI-IPN, Mexico City, Mexico

Glycosides in Biotechnology

The aim of this chapter is to provide an overview about the applications and the potential of glycosides in biotechnological related areas such as health, food and environmental chemistry. The role of glycosides in biological processes has been the subject of extensive studies and as a result of this research, different disciplines such as natural product chemistry, synthesis and pharmacology have been increasingly exploited and combined, particularly for providing novel alternatives in the treatment of antiviral, antitumoral, antifungal, or immune related disease. Also in environmental chemistry the design of glycosides with aliphatic chain fragments have been used in the preparation of biosurfactants or to produce biofuels. Moreover, the isolation from natural sources or by chemical synthesis of glycosides has been increasingly developed in food chemistry to produce sweeteners with low caloric content, antioxidants with anti aging properties, or to control altered metabolism processes such as obesity and diabetes. Therefore, we have selected some examples found in literature or from our own research, describing the preparation and use of glycosides in the topics mentioned above.

Glycosides in Therapeutics

Glycosides are useful therapeutic agents being currently used as antitumor, antiviral and antibiotic agents mainly in the form of O-, N-, and C-glycosides. Many of them have shown remarkable action against AIDS, Hepatitis, Influenza, Herpes and other virus related infections. Moreover the preparation of glycoconjugates has been a current strategy for the design of synthetic vaccines, and usually involves the preparation by chemical or enzymatic methods of the oligosaccharide moiety which provides the immune specificity. Also naturally

[*] Email: britoarias@hotmail.com

occurring glycosides are widespread compounds, many of them showing remarkable activity as citotoxic (Phifer et al., 2007), anti COX-2 (Chai et al., 2008), Monoaminoxidase inhibitors (Urbain et al., 2008), neuroprotective (Yoon et al., 2008), α-glycosidase inhibitors (Aparna et al., 2009), and antiiinflammatory (Ukiya et al., 2007) among others.

For instance, the naturally ocurring biflavonoid Ginkgetin isolated from *Ginkgo biloba* and *Cephalotaxus harringtonia* was conjugated to sialic acid, since influenza A and B viruses are known to bind themselves on the host cells via sialic acid residue of glycoconjugate receptors at the first stage of infection and tested.

The resulting glycoconjugate was found to inhibit the influenza virus sialidase in infected mice (Miki et al., 2007).

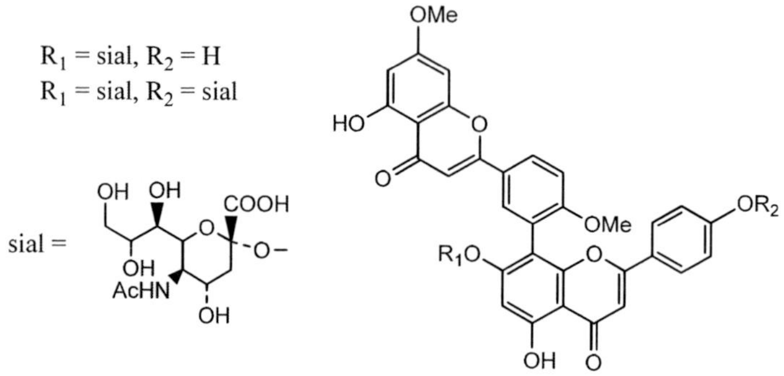

Ginkgetin

Synthetic glycosides have been design as mimics of saccharide substructures present in glycoproteins. Such is the case of the trisaccharide substructure of the ganglioside GQ1 which shows a remarkable affinity for the myelin-associated glycoprotein (MAG). The aromatic trisaccharide antagonist substructure of the ganglioside increase the relative inhibitory potency by more than 1000-fold in comparison to the reference trisaccharide. Also the aromatic mimic improved pharmacokinetic properties due to the presence of aromatic moieties, a lower molecular weight, and a reduced number of polar hydroxyl functions (Shelke et al., 2007).

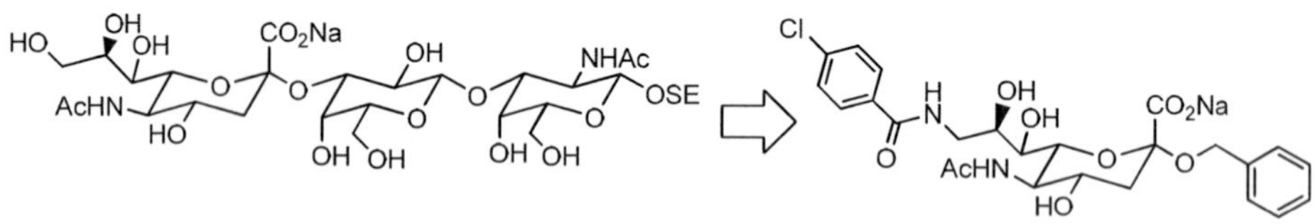

trisaccharide substructure aromatic glycoside antagonist

An efficient chemoenzymatic approach for preparing homogeneous glycoprotein synthesis was achieved by using endoglycosidase and ribonuclease B as a model system. Thus, the Endo-A could efficiently attach a preassembled oligosaccharide to a GlcNAc-containing protein in a regio- and stereospecific fashion, when the corresponding sugar oxazoline was used as the donor substrate (Li et al., 2006).

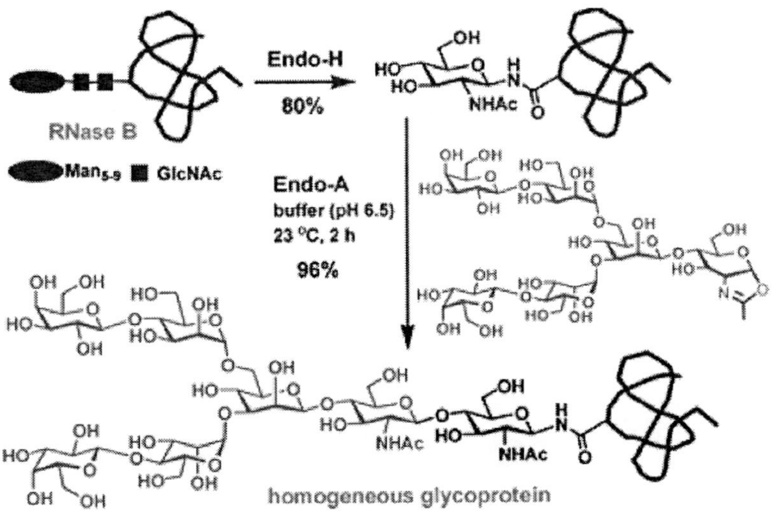

Immobilization is also a current strategy for preparing synthetic oligosaccharides and as an example of this approach, a biotin sulfone protocol is described as a new tool for preparing a library of 1→2 oligomannosides suitable for immunoanalysis, and also its application to multiple analysis profiling and surface plasmonic analysis of anti-*Candida albicans* antibody reactivity (Collot et al., 2008).

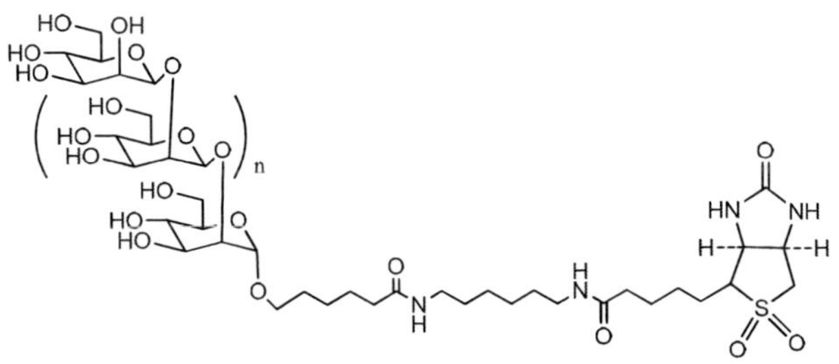

Semisynthetic derivatives of 28,29-Didehydronystatin A_1 (S44HP), a genetically engineered antifungal polyene macrolide antibiotic containing and aminosaccharide or disaccharide moiety (products of Amadori rearrangement), were reported as well as their structure–activity relationships.

These derivatives were tested against yeasts *Candida albicans, Cryptococcus humicolus*, and filamentous fungi (molds) *Aspergillus niger* and *Fusarum oxysporum* providing inhibition *in vitro* (Preobrazhenskaya et al., 2009).

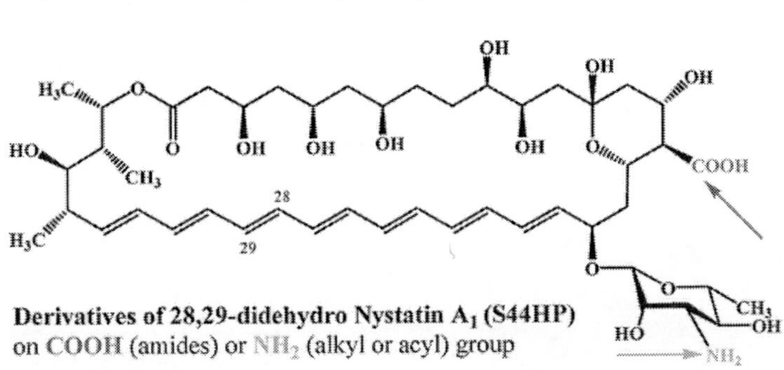

Derivatives of 28,29-didehydro Nystatin A_1 (S44HP)
on COOH (amides) or NH_2 (alkyl or acyl) group

Some structurally simple glycosides such as per-butanoyl *N*-acetyl-D-mannosamine (Bu_4ManNAc) have been evaluated in antitumoral processes, showing a reduced invasion of metastatic MDA-MB-231 breast cancer cells. The active compound Bu_4ManNAc reduced both MUC1 expression and MMP-9 activity (via down-regulation of CXCR4 transcription), whereas "inactive" per-butanoylated-D-mannose (Bu_5Man) had counterbalancing effects on these oncogenes (Campbell et al., 2009).

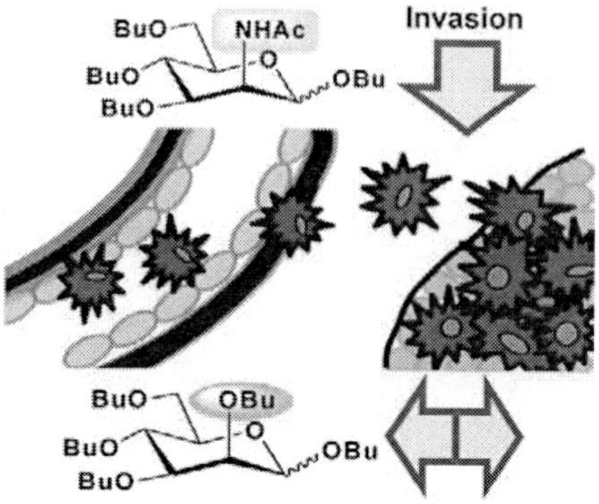

Oligonucleotides are short N-glycoside sequences that are mainly design by solid phase procedures to be ultimately used in the antisense strategy. Such approach consisting in the attachment of the antisense strand to viral, antitumoral or autoimmune related disease, and therefore inhibiting their expression.

Oligodeoxyribonucleotides containing 2′-*O*-methylribonucleotides were prepared and evaluated as TLR7 and TLR9 receptor antagonists, which may be suitable candidates for treating inflammatory and autoimmune diseases where inappropriate or uncontrolled TLR activation has been implicated such as in lupus immune disease (Wang et al., 2009).

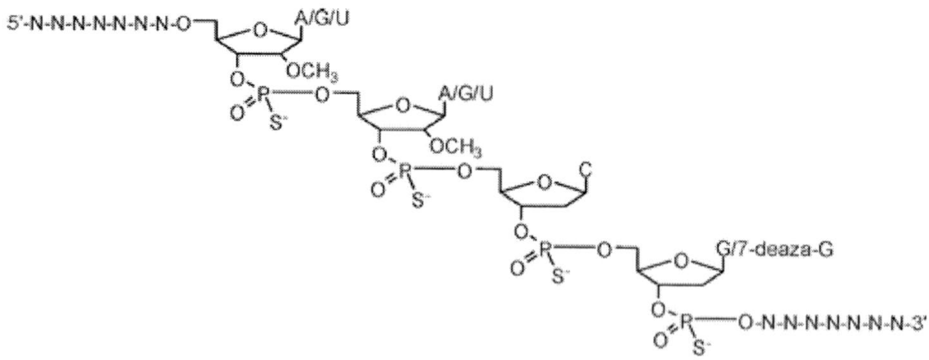

Aromatic 3,3′-diesters of thiodigalactoside were synthesized in a rapid three-step sequence and evaluated as inhibitors of cancer- and immunity-related galectins. Two of the compounds were selected for testing in cell culture and were shown to have potent antimigratory effects on human PC-3 prostate and human A549 non small-cell lung cancer cells (Delaine et al., 2008).

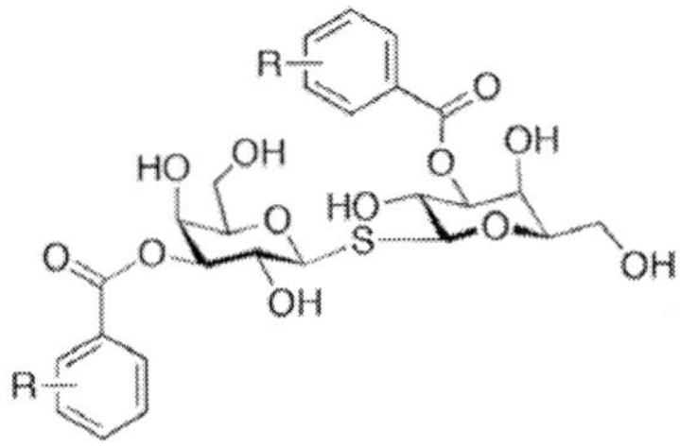

An optimized synthesis of D-glucose and D-galactose derived carriers, bearing an adipate linker and four *tert*-butoxycarbonyl protected aminopropyl groups have been prepared and proposed as vaccines against group A *Streptococcus*.

Prophylactic group A streptocuccus vaccine candidates were synthesized by conjugating multiple copies of a single M protein derived peptide antigen (either J8 or J14) onto the carbohydrate carriers.

High serum IgG antibody titers against each of the incorporated peptide epitopes were detected following subcutaneous immunization of B10.BR (H-2k) mice with the liposaccharide vaccine candidates (Simerska et al., 2008).

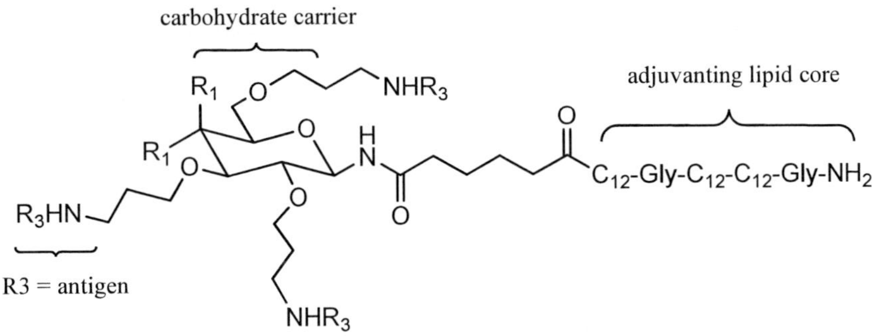

Naturally occurring glycosides containing xanthone ring as aglycone were isolated form glycosides *Gentianella amarella* ssp. *Acuta* and characterized by 2D NMR and mass spectrometry.

Xanthones veratriloside, corymbiferin 1-*O*-glucoside, swertianolin, norswertianolin, swertiabisxanthone-I were weakly active against acetylcholinesterase (AChE), except triptexanthoside C, which inhibited AChE with an IC_{50} of 13.8 ± 1.6 μM. On the other hand bellidin and bellidifolin glycosides produced inhibition of monoamine oxidases (MAO) activity at 10^{-5} M (Urbain et al., 2008).

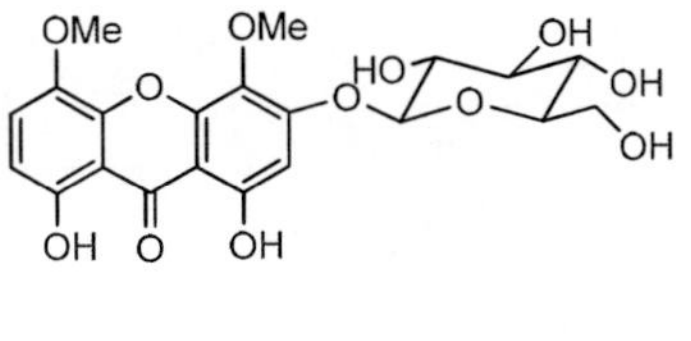

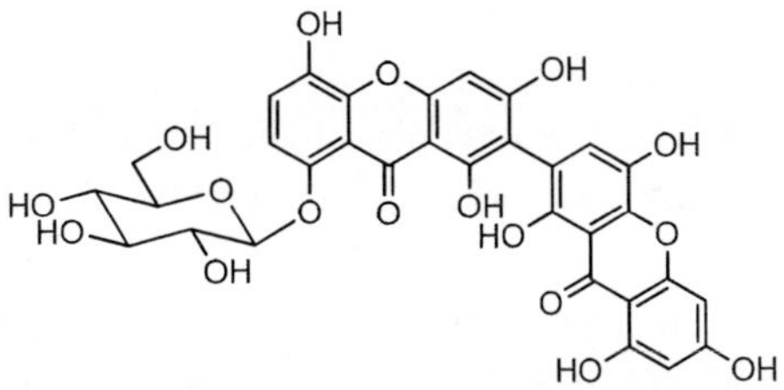

The semi synthetic homogeneous glycoprotein enzyme ribonuclease C has been developed through the efficient formation of mixed disulfides on the thiol-rich fusion protein

A followed by subsequent intern cleavage, giving the fragment B with all seven cysteines protected against oxidation. The native chemical ligation of B with synthetic glycopeptide thioesters provide glycoproteins (Piontek et al., 2009).

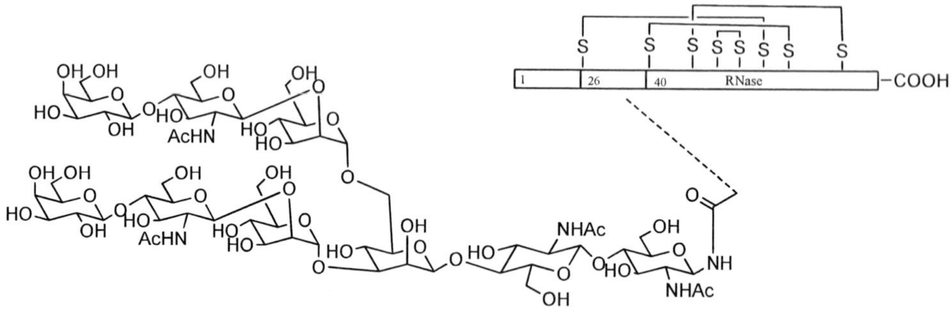

The isolation and chemical synthesis of amphipatic resin glycosides known as tricolorins have been achieved and their potential biological activities such as antimicrobial, mammalian cytotoxicity and plant toxicity were evaluated. These resin glycosides also present the ability to inhibit the growth of invasive weeds so they have been used by farmers to protect their crops (Brito-Arias et al., 2004).

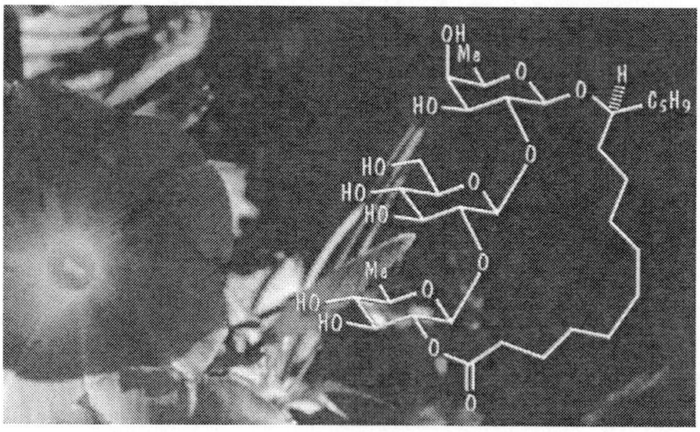

Glycosides as Prodrugs

Prodrug is defined to describe inactive compounds which after biotransformation during drug metabolism produce an active molecule responsible of the pharmacological effect.

Also the term involves the chemical modification of another biologically active molecule which will be released through enzymatic or chemical hydrolysis. Prodrugs are mainly used for improving pharmacokinetics parameters such as absorption, distribution, metabolism, or chemical stability and molecular carriers for better specificity.

One of the most effective ways to modify drugs is by attaching a sugar moiety to generate a glycoside which could eventually improve most of the aspects mentioned above. Glycoside prodrugs usually improve the partition coefficient providing a good balance between lypophilic and hydrophilic ratio which is needed for crossing membranes, for being soluble in fluids, and could be cleaved by glycosidases or in acidic media.

Therefore, carbohydrates have been attached to a significant number of active molecules and used as prodrugs in the treatment of antitumoral, antiviral, self immune disorders among others.

A number of synthetic methods for attaching active molecule to sugar moieties have been described, and in most of the methods the principle consist in the condensation of the drug with a pyranoside as glycoside donor in the presence of a promoter or an enzyme such as transferase or glycosidase.

For instance the synthesis of 2′-paclitaxel methyl 2-glucopyranosyl succinate for specific targeted delivery to cancer cells was developed, finding that the glycoside prodrug has not only improved the pharmaceutical properties of paclitaxel, such as solubility and stability, but also enhanced the specific target delivery to MCF-7 cells without the cytotoxicity against normal cells (Liu et al., 2007).

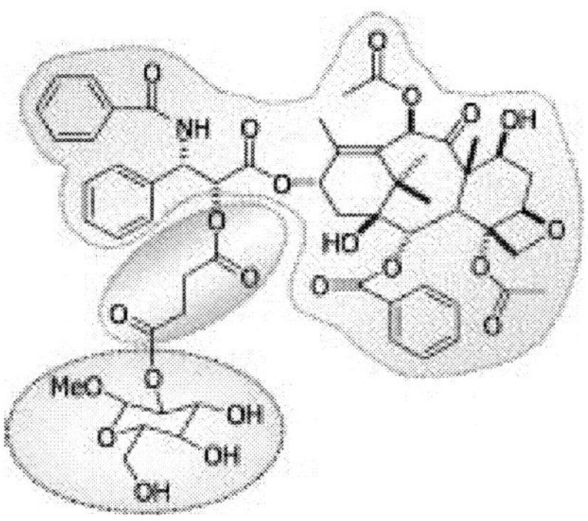

The synthesis and biological studies of different Duocarmycin based glycosidic prodrugs and their use in the antibody-directed enzyme prodrug therapy has been also described (Tietze et al., 2009).

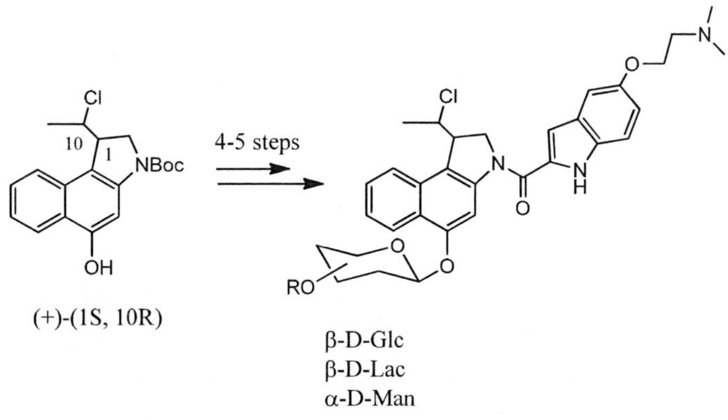

Phosphate prodrugs derived from *N*-acetylglucosamine have been shown to display enhanced chondroprotective activity in explant cultures and represent a new lead in antiosteoarthritis drug discovery (McGuigan et al., 2008).

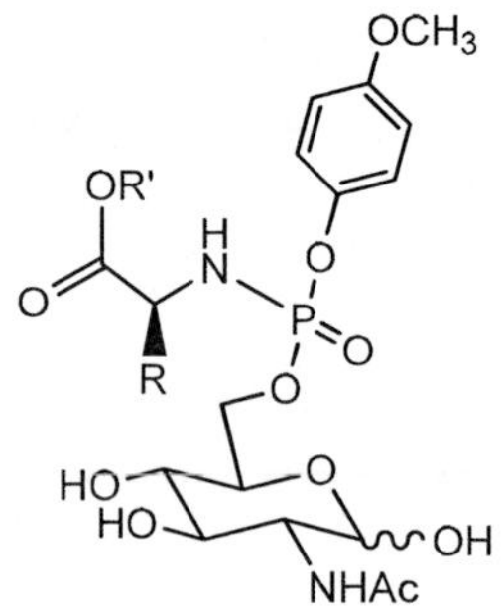

Cyanocobalamin (vitamin B_{12}) is an essential nutrient as well as a very useful carrier in drug delivery.

The synthesis of six vitamin B_{12} conjugates, and the effect of conjugation on their solubilities and stabilities in various media were carried out, observing that solubility could be enhanced as much as 19-fold (Wang et al., 2007).

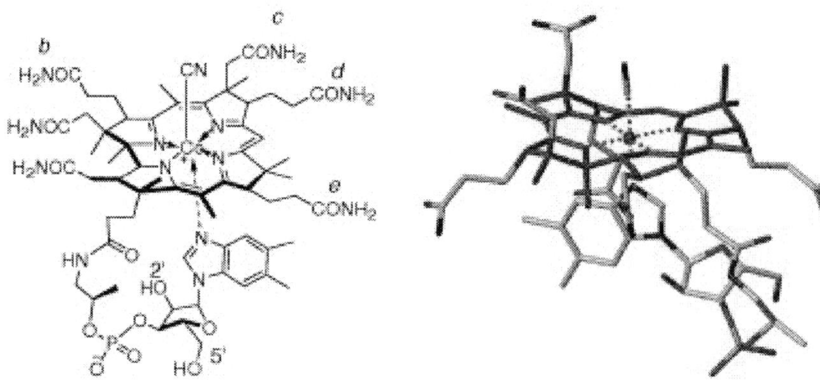

Likewise, glycosides of anti-inflammatory 5-amino salicylates have been prepared as prodrugs and considered as specific drugs in the treatment of colitis (Rodriguez-Romero et al., 2009).

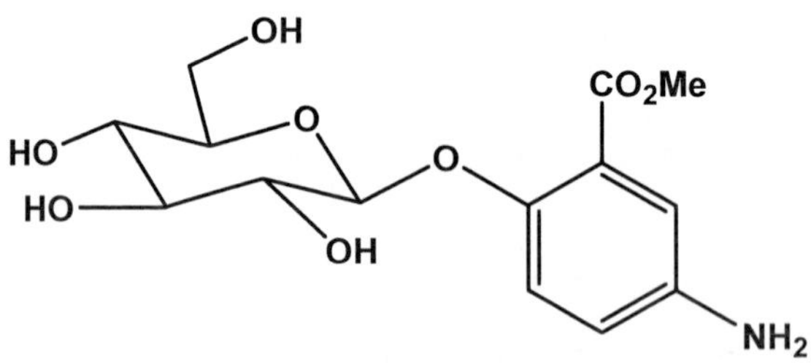

The Synthesis of dopamine and L-DOPA-α-glycosides by reaction with cyclomaltohexaose catalyzed by cyclomaltodextrin glucanyltransferase is described (Yoon et al., 2009).

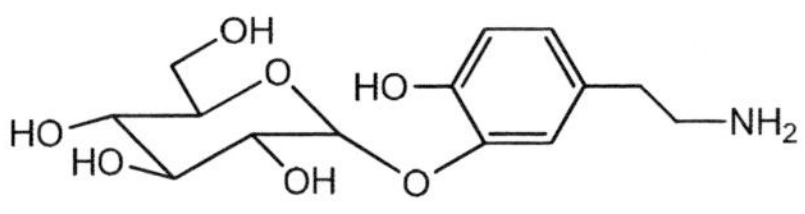

3-O-α-D-glucopyranosyl dopamine

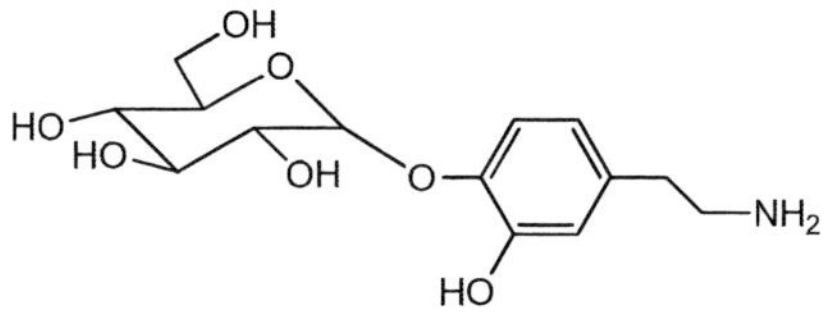

4-O-α-D-glucopyranosyl dopamine

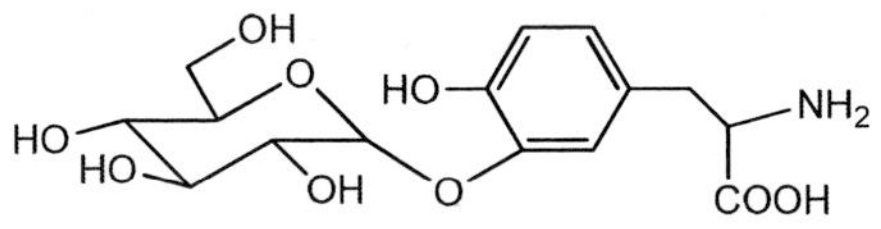

3-O-α-D-glucopyranosyl L-DOPA

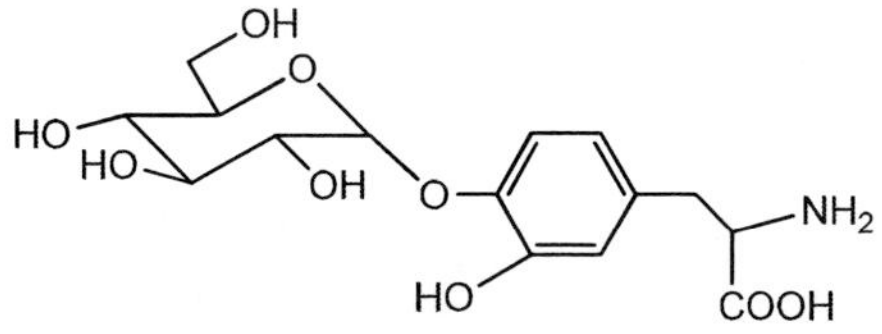

4-O-α-D-glucopyranosyl L-DOPA

A β-glucuronidase-activated prodrug approach was applied to 10-hydroxycamptothecin, an alkaloid with promising antitumor activity but poor water solubility was attached to glucuronic acid, observing that the resulting glucuronide prodrug was 80 times more soluble than 10-hydroxycamptothecin in aqueous solution at pH 4.0 and was stable in human plasma. This Prodrug may be useful for selective cancer chemotherapy by a prodrug monotherapy (PMT) or antibody-directed enzyme prodrug therapy (ADEPT) strategy (Leu et al., 2008).

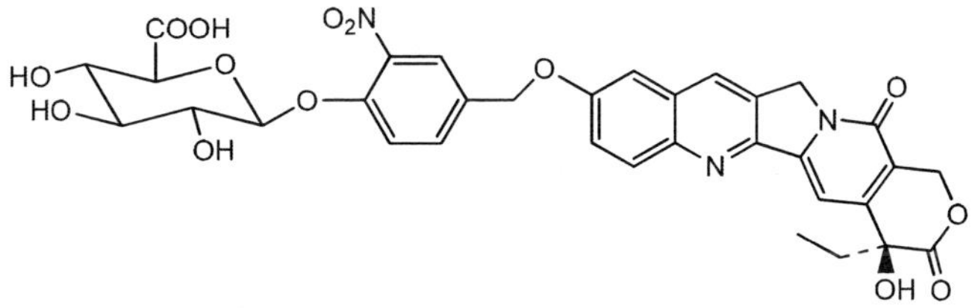

Glycosides as Fluorescence Markers

Glycosides have been widely used as substrates for enzymatic detection when the carbohydrate moiety is attached to a chromophore such as umbelliferone, fluoresceine indoxil, resorufin, and phenylazonaphtols among others.

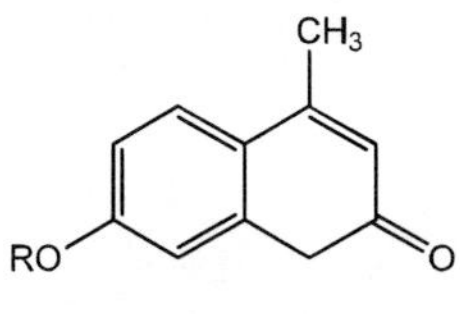

umbelliferone

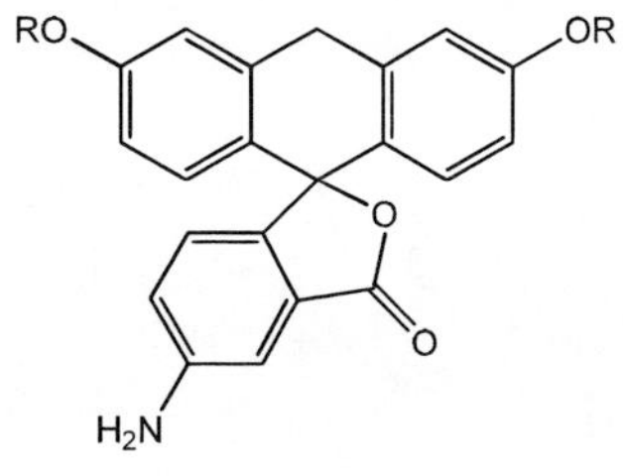

fluoresceine

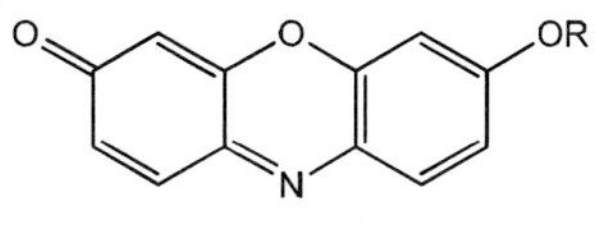

resorufin

R= glucose, galactose, glucuronic acid

The resulting O-glycoside is hydrolyzed by a glycosidase releasing the chromophore which is detected as strong fluorescence in the sites of enzymatic activity.

By following this strategy different sugar have been attached to the fluorescent compound and the protocol extended for detection of gene markers in plant mɔlecular biotechnology and microbiology.

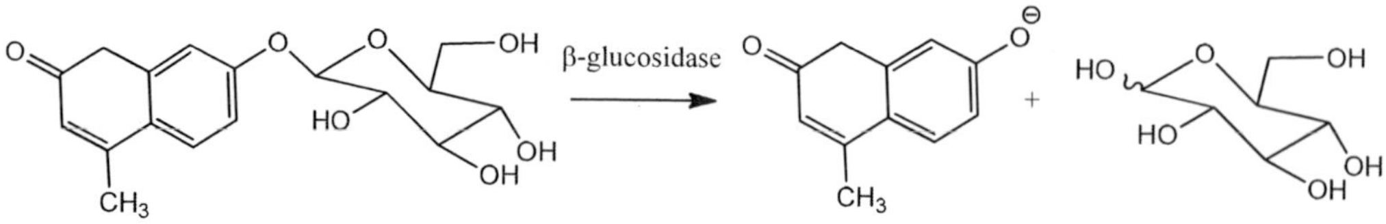

umbelliferone-β–D-O-glucopyranoside (non fluorescent) umbelliferone (highly fluorescent)

Under this concept other fluorescent analogues of glycosidic nature have been developed. Such is the case of fluorescent-labeled DNA by PCR with a thymidine nucleotide analogue bearing an acridone derivative and incorporated as a substrate for PCR using KOD Dash DNA polymerase forming a new fluorescent-labeled DNA that is useful as a DNA probe (Shoji et al., 2007).

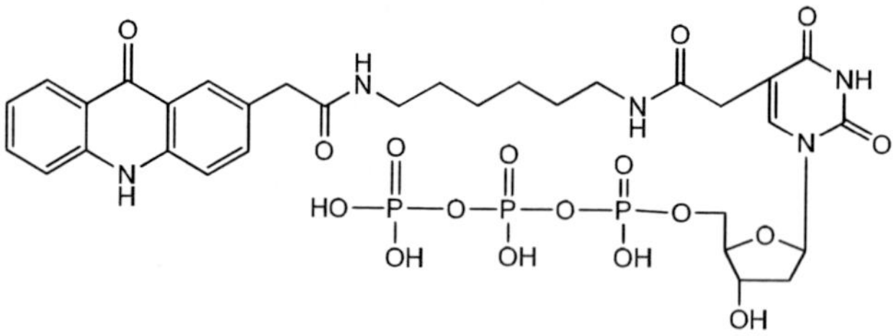

Fluorescent glycosides has been synthesized and used to detect bulged structure in DNA and RNA for screening numerous diseases. These agents show pronounced selectivity for specific bulged motifs and are able to enhance slipped DNA synthesis, a hallmark functional assay of bulge binding, assuring that these signature patterns can be harnessed as molecular probes of bulged hotspots in DNA and RNA (Jones et al., 2007).

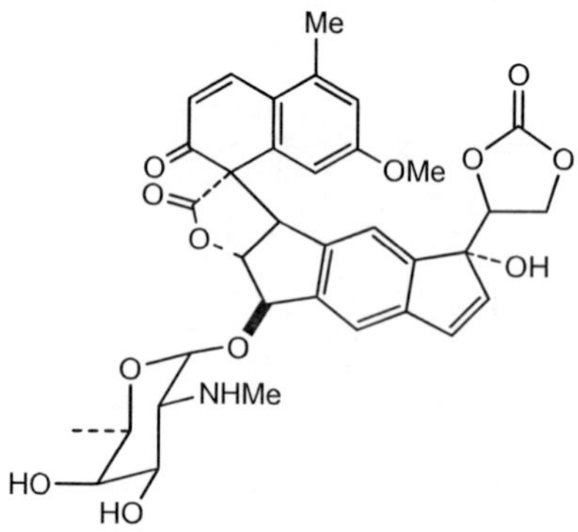

A protocol for incorporation of a photoisomerizable azobenzene moiety into synthetic stereo-enriched [R_p] and [S_p] PS-oligonucleotides has been proposed.

The azobenzene pendant is attached at pre-selected positions in internucleotidic phosphorothioate oligonucleotides of both [R_p] and [S_p] diastereomers using *N*-iodoacetyl-*p*-aminoazobenzene as reagent. It is observed that the azobenzene moiety imparts greater stability to oligomer duplexes in (*E*) — N configuration as compared to (*Z*) configuration (Patnaik et al., 2007).

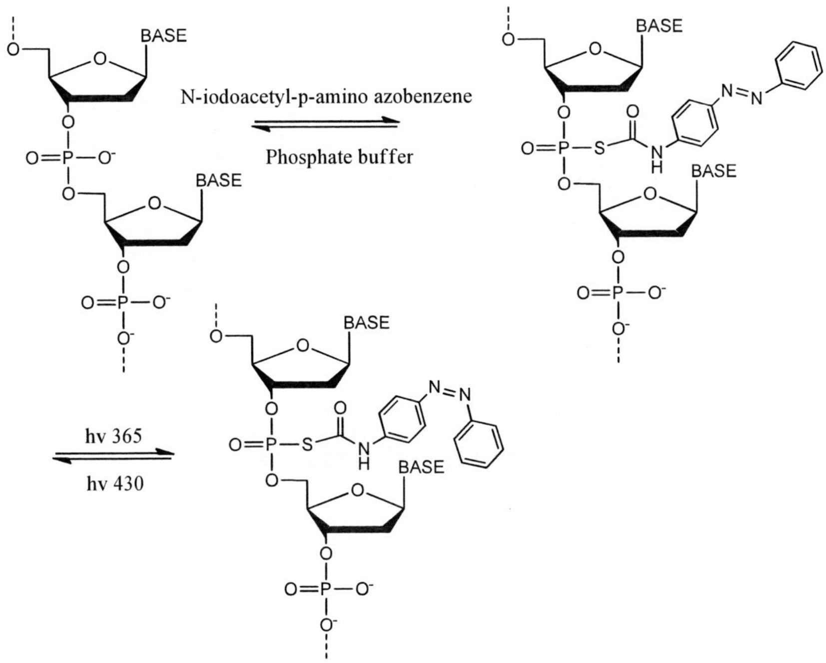

A new colorimetric anion sensor was synthesized for chiral recognition, which accommodates a combination of three different functional groups such as chromophore (azophenol dye), binding site (thiourea group), and chiral barrier (glucopyranosyl group).

The colorimetric changes of the glycosidic sensor with various α-amino carboxylates as well as chiral carboxylates such as naproxen were examined (Choi et al., 2008).

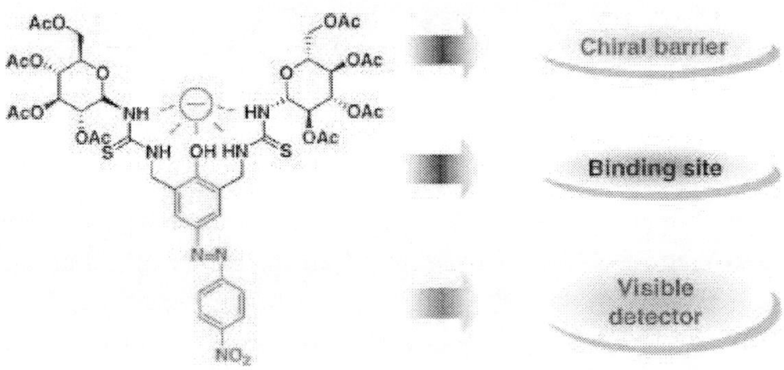

Cyclodextrins attached to fluorogenic quelating agents such as 2-pyridyl triazole were synthesized using click chemistry.

The resulting glycosidic fluorophore was assayed exhibiting shift emissions based on different solvent polarity (solvatofluorochromism), or by the presence of certain ions such as Mg^{2+} and Zn^{2+}.

This approach can be extended in the sensing of transition and alkali metal ions, amino acid and other small molecules like aromatic hydrocarbons, quinones and steroids (David et al., 2007).

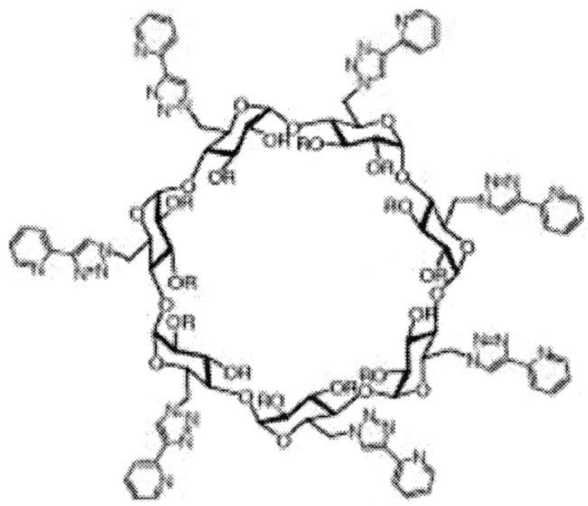

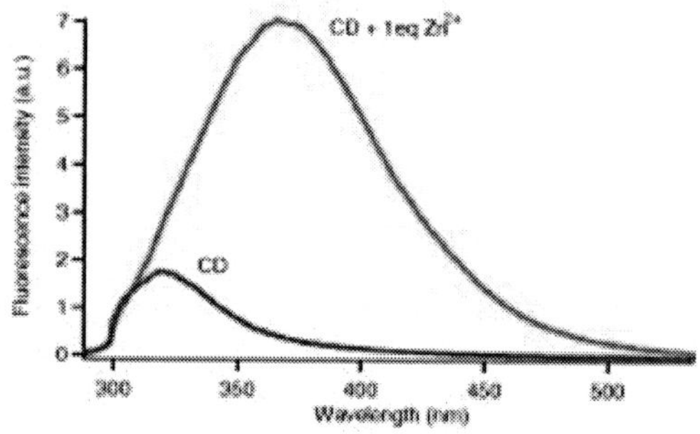

Also fluorescent glycosides have been found from natural sources as it is reported in the isolation from *Trigonotis peduncularis* of three aryldihydronaphthalene-type lignan sucrose diesters named trigonotins A−C. These sucrose fluorophores showed a strong yellow-green fluorescence emission under basic conditions. The structures of the new compounds were elucidated by means of spectroscopic methods, and the nature of their fluorescence was examined (Otsuka et al., 2008).

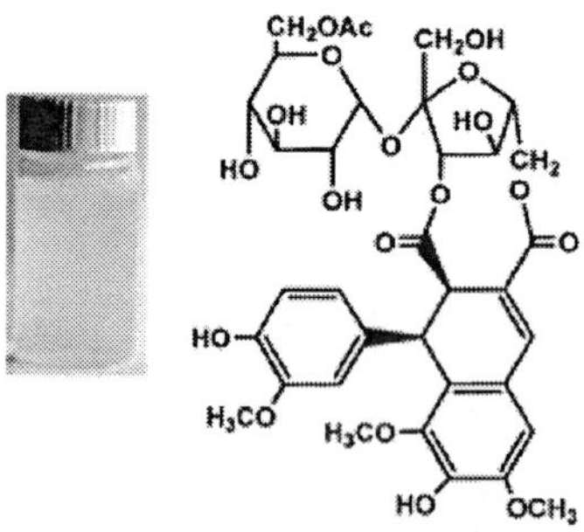

Oligodeoxynucleosides (ODNs) containing a 2'-N-(pyren-1-yl)-group on the conformationally locked nucleosides (X) and (Y) show that they can bind to complementary RNA more strongly than to the DNA. The Aze-pyr (**X**) containing ODNs with the complementary DNA and RNA duplexes showed an increase in the fluorescence intensity (measured at $\lambda_{em} \approx 376$ nm) depending upon the nearest neighbor at the 3'-end to **X** [dA (~12−20-fold) > dG (~9−20-fold) > dT (~2.5−20-fold) > dC (~6−13-fold)] while Aza-ENA-pyr (**Y**)-modified ODNs showed an enhancement of the fluorescence intensity only with the complementary DNA (1.4−3.9-fold, $\Phi_F = 0.16−0.47$) (Honcharenko et al., 2008).

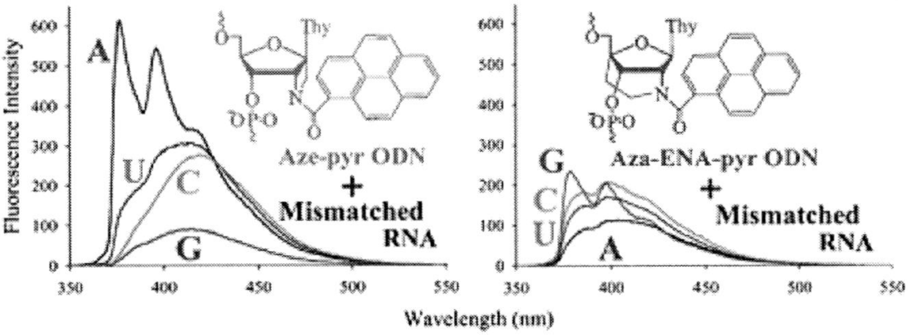

A fluorescent sensor containing two anthracenetriazolymethyl groups was prepared following the sugar-1,2,3-triazo formation strategy (click reaction), and its fluoroionophoric properties toward transition metal ions were investigated. In methanol, the sensor exhibits highly selective recognition of Cu^{2+} and Hg^{2+} ions among a series of tested metal ions. The association constant for **5**·Cu^{2+} and **5**·Hg^{2+} in methanol was calculated to be 4.0×10^5 M^{-1} and 1.1×10^5 M^{-1}, respectively. The detection limits for the sensing of Cu^{2+} and Hg^{2+} ions were 1.39×10^{-6} M and 1.39×10^{-5} M, respectively (Hsieh et al., 2009).

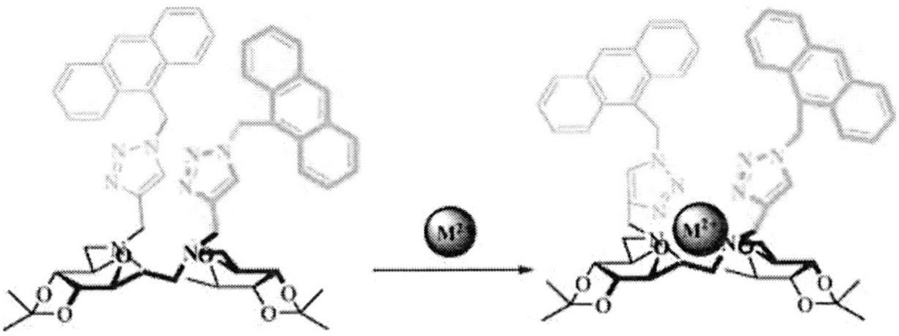

The synthesis of Sudan-β-D-glucuronides was performed by condensation of red Sudan diazo dyes with glururonide donors under Lewis acid catalyst, and the resulting substrate was used for histochemical localization of β-glucuronidase (GUS) in transgenic plants that contains the GUS reporter system. Using the Sudan IV-glucuronide substrate on transformed *Arabidopsis* plants harboring the promoter of S-adenosyl-1-methionine synthetase gene fused to the GUS coding sequence produce red precipitate in the sites of β-glucuronidase expression (Van der Eycken et al., 2000).

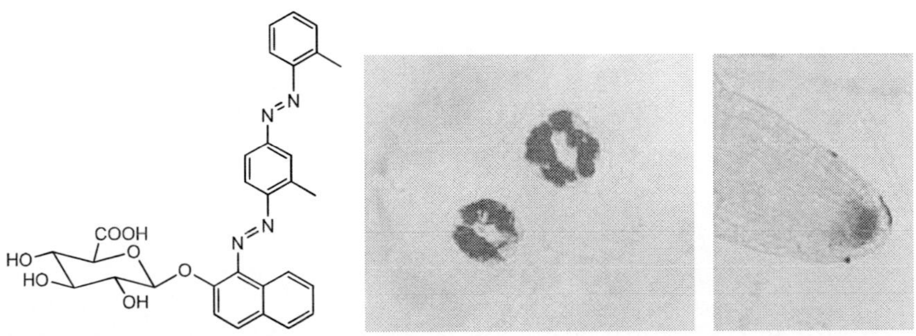

Glycosides as Biosurfactants

Surfactant of glycosidic nature is another class of compounds of high impact which are found in a wide range of applications, mainly as detergents, in bioremediation, additives, pharmaceuticals and cosmetics.

They are produced when a long chain alkyl alcohol is attached to a sugar fragment and the resulting glycosides incorporates suitable properties such as a good partition coefficient, foaming, wetting and biodegradability.

The synthesis of these glycosides could be achieved enzymatically or chemically according to the O-glycoside protocols for glycosilation, particulary the Fischer, Helferich, and the Koenigs-Knorr condensation reaction (Brito-Arias, 2007).

For instance four octyl D-glucosides were synthesized in a few steps by using a Lewis acid under microwave conditions, and tested as chelating surfactants for removal from waste water.

Their main interfacial properties were determined, and their flotation properties were evaluated on laboratory scale using Fe(III) as a model contaminant metal. The performance on metal extraction was mainly dependent on the complexing functional group (Ferlin et al., 2008).

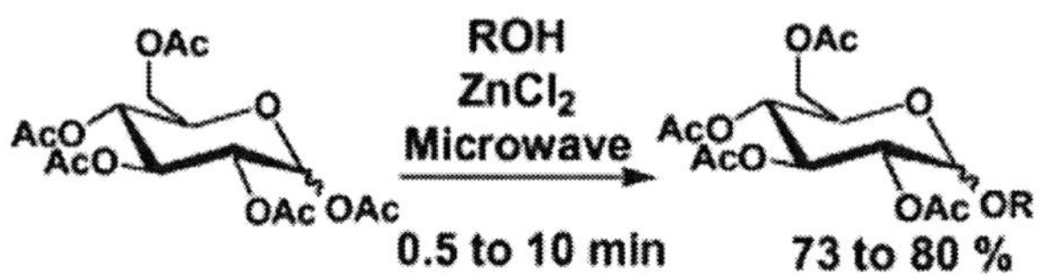

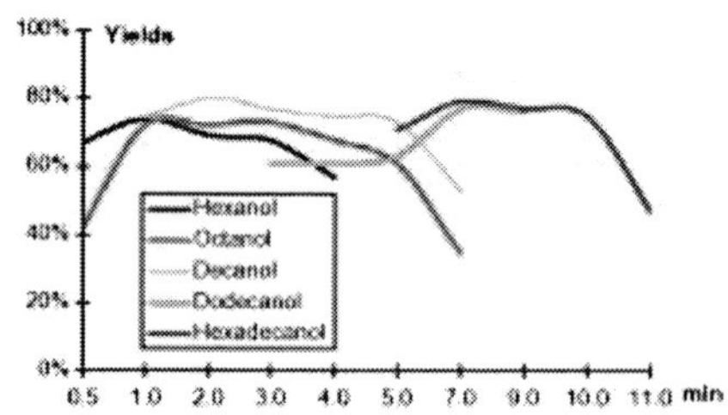

Another report describes the syntheses of dodecyl N-acetyglucosamine, and N-acetylactosamine, prepared from peracetylated sugar and conjugates with 1-dodecanol under TMS-OTf conditions. The resulting alkyl glycosides were used as saccharide primers, incorporated into B16 melanoma cells and glycosylated by glycosyltransferase (Sato et al., 2008).

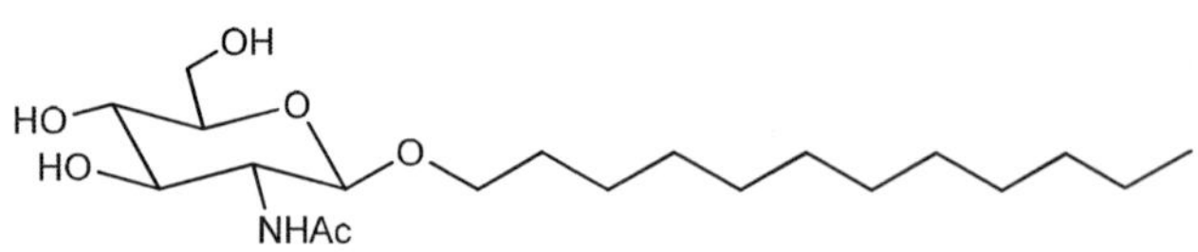

GlcNAc-C12

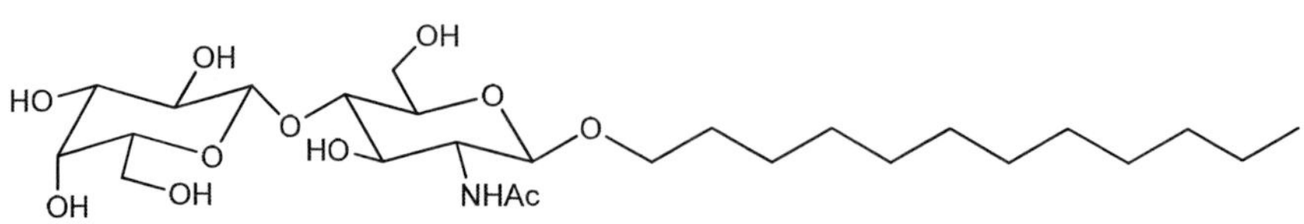

LacNAc-C12

Natural biosurfactants produced by the yeast strains of the genus Pseudozyma have been isolated and characterized.

The structural analysis reveals the structure as Mannosylerythritol lipids (MELs) which show excellent surface-active properties, also versatile biochemical actions including cell-differentiation induction of different mammalian cells as well as affinity binding toward different immunoglobulins (Fukuoka et al., 2008).

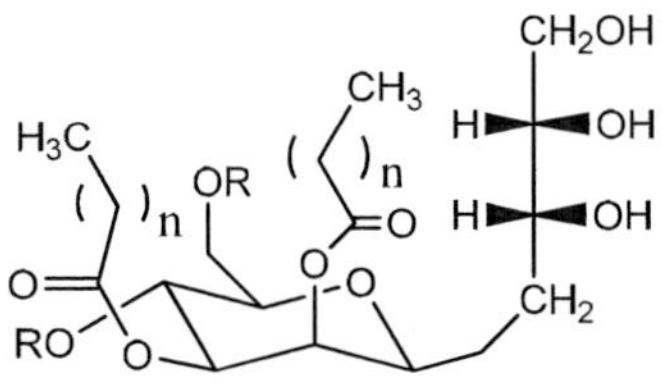

R = H or Ac

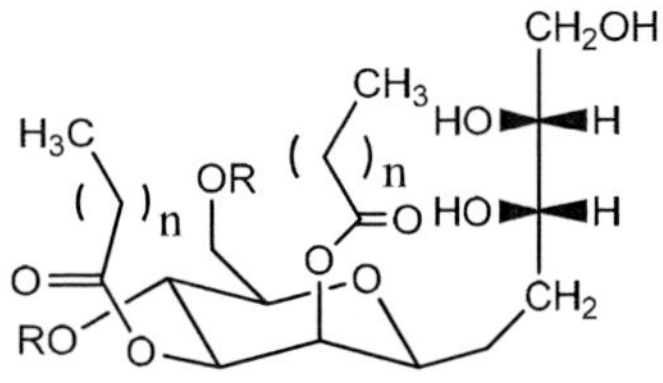

R = H or Ac

Likewise sucrose fatty acid esters, useful as mild surfactants, non toxic emulsifiers in food, cosmetics, pharmaceutics, agrochemicals or polymers have been synthesized in good yields by reaction of sucrose with fatty acid esters in a solvent-free medium, in DMSO or DMF, with a base as catalyst. High contents of monoesters and low saponification rate have been observed. The co-melting of sucrose and a multivalent cation soap such as magnesium or zinc soaps, under basic conditions, allowed the formation of a homogeneous reaction mixture, where solid sucrose was fully dissolved (Fitremann et al., 2007).

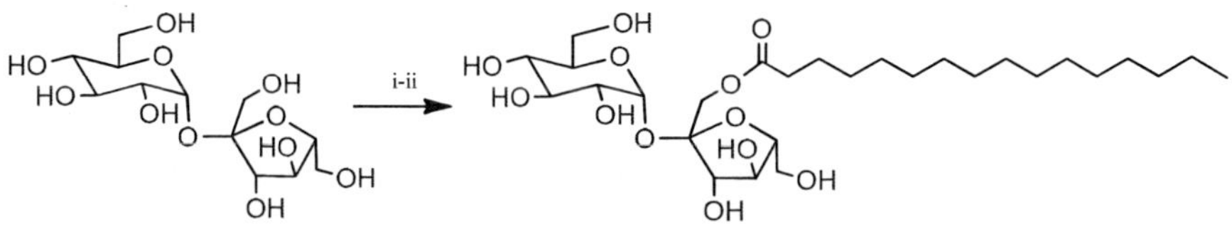

i) KOH, Magnesium Stereate. ii) Methyl palmitate 125-135°C, 3-4 h.

Glycosides Involved in Food and Metabolism

Glycosides have been recognized in their beneficial implications in the treatment and prevention of cancer and cardiovascular disease, as well as antioxidants or in self defense processes. A representative example of antioxidants are polyphenols glycosides such as stilbenes resveratroloside and piceid which constitute an important example of plant metabolites, because they are recognized for their beneficial implications in human health, such as in the treatment and prevention of cancer, cardiovascular, and anti aging due its anti oxidative properties. Due their importance a number of compounds have been also synthesized under the Koenigs-Knorr conditions and used for commercial purposes (Regev-Shoshani et al. 2003).

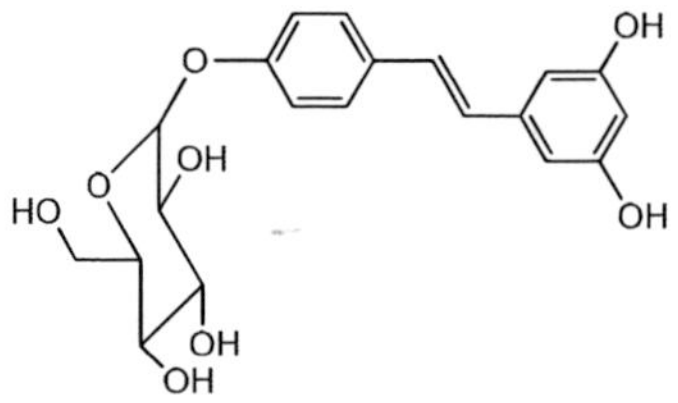

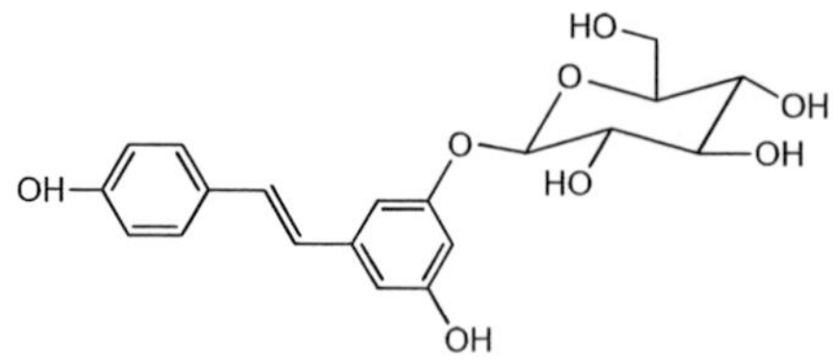

On the other hand, the steviol glycosides are 8 diterpenic glycosides being the main components the estevioside (50%) and rebaudioside A (30%) found in the leaves of S*tevia* plant having a sweet taste and low calorie content, ranging in sweetness from 40 to 300 times sweeter than sucrose (Brandle, 2006).

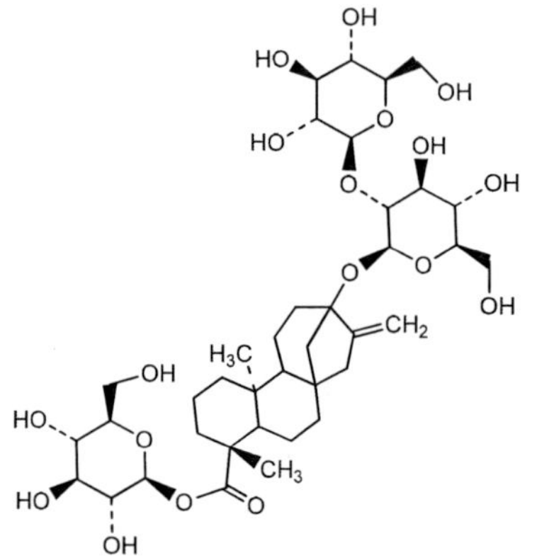

Alkaloid glycosides are a specific group of water-soluble terrestrial and marine natural compounds. Such is the case of benzoxazinone glycosides obtained from monocotyledonous plants such as maize which have been reported to act as weight loss agents, appetite suppressants, mood enhancers, and adjunctive therapy for arthritis, sleep apnea, fibromyalgia, diabetes, and hyperglycemia (Dembitsky, 2005).

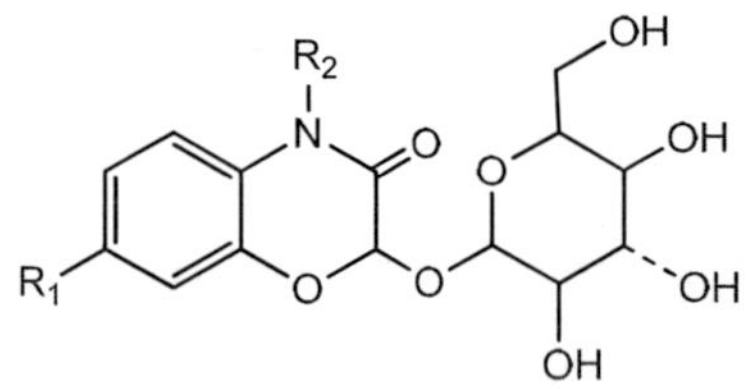

R_1 = -H, -OMe
R_2 = -H, -OMe

Alterations in metabolism leading to diabetes or obesity have been an intensive area of study. The C-aryl glucoside (dapagliflozin) was identified as potent and selective hSGLT2 inhibitors which reduce blood glucose levels in a dose-dependent manner by as much as 55% in hyperglycemic streptozotocin (STZ) rats.

These findings, combined with a favorable ADME profile, have prompted clinical evaluation of dapagliflozin for the treatment of type 2 diabetes (Meng et al., 2008).

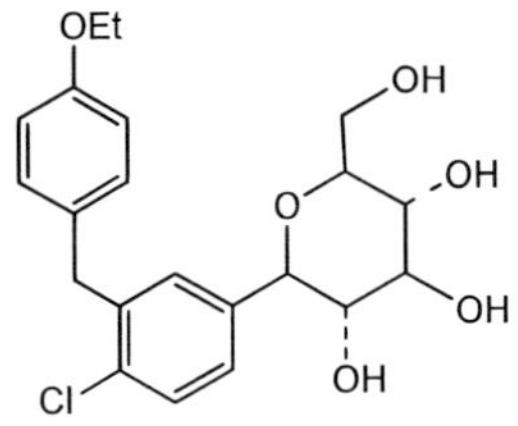

REFERENCES

Aparna P, Tiwari AK, Srinivas PV, Ali AZ, Anuradha V, Rao JM. (2009) *Phytother Res.*, 23: 591-6.

Brandle J., *Agriculture and Agri-Food* Canada, 2006. http: //res2 .agr. ca/ London/faq/stevia_e.htm.

Brito-Arias M. (2007) *Synthesis and Characterization of Glycosides*, *Springer*, 68-137.

Brito-Arias M., Pereda-Miranda R., and Heathcock H. C. (2004) *Journal of Organic Chemistry*, 69: 4567-4570.

Campbell C.T., Aich U., Weier C.A., Wang J.J., Choi S.S., Wen M.M., Maisel K., Sampathkumar S-G. and Yarema K.J. (2008) *J. Med. Chem.*, *51:* 8135–8147.

Chai X.Y., Song Y.L., Xu Z.R., Shi H.M., Bai C.C., Bi D., Wen J., Li F.F., and Tu P.F. (2008). *Journal of Natural Products*, 71: 814-819.

Choi M.K., Kim H.N., Choi H.J., Yoon J., Hyun M.H. (2008) *Tetrahedron Letters*, 49, *4522-4525.*

Collot M., Sendid B., Fievez A., Savaux C., Standaert-Vitse A., Tabouret M., Drucbert A.S., Danzé P.M., Poulain D., and Mallet J-M (2008) *Journal of Medicinal Chemistry*, 51: 6201 – 6210.

David O., Maisonneuve S., Xie J. (2007) *Tetrahedron Letters,* 48: 6527-6530.

Delaine T., Cumpstey I., Ingrassia L., Le Mercier M., Okechukwu P., Leffler H., Kiss R. and Nilsson U.J. (2008) *J. Med. Chem.*, *51*: 8109–8114.

Dembitsky V.M. (2005) *Lipids*, 40: 1081–1105.

Ferlin N., Duchet L., Kovensky J., Grand E. (2008) *Carbohydrate Research*, 343: *2819-2821.*

Fitremann J., Queneau Y., Maître J-P., Bouchu A. (2007) *Tetrahedron Letters*, 48: 4111-4114.

Fukuoka T., Kawamura M., Morita T., Imura T., Sakai H., Abe M., Kitamoto D. (2008) *Carbohydrate Research*, 343: 2845-3008.

Honcharenko D., Zhou C., and Chattopadhyaya J. (2008) *Journal of Organic Chemistry*, 73: 2829 – 2842.

Hsieh Y-C., Chir J-L., Wu H-H., Chang P-S., Wu A-T. (2009) *Carbohydrate Research*, 344: 2236-2239.

Jones G.B, Lin Y., Xiao Z., Kappen L. and Goldberg I.H. (2007) *Bioorganic and Medicinal Chemistry*, 15: 784-790.

Leu Y-L, Chen C-S, Wu Y-J., and Chern J-W (2008) *Journal of Medicinal Chemistry*, 51: 1740-1746.

Li B., Song H., Hauser S., and Wang L-X. (2006) *Organic Letters*, 8: 3081–3084.

Liu D-Z, Sinchaikul S, Reddy P.V.G., Chang M-Y, and Chen S-T (2007) *Bioorganic and Medicinal Chemistry Letters*, 17: 617-620.

McGuigan C., Serpi M., Bibbo R., Roberts H., Hughes C., Caterson B., Torrent Gibert A., and Alaez Verson C.R. (2008) *Journal of Medicinal Chemistry*, 51: 5807 – 5812.

Meng W. *et al*, (2008) *Journal of Medicinal Chemistry*, 51: 1145 – 1149.

Miki K., Nagai T., Suzuki K., Tsujimura R., Koyama K., Kinoshita K., Furuhata K., Yamada H. and Takahashi K. (2007) *Bioorganic and Medicinal Chemistry Letters*, 17: 772-775.

Otsuka H., Kuwabara H., and Hoshiyama H. (2008) *Journal of Natural Products*, 71: 1178-1181.

Patnaik S., Kumar P., Garg B.S., Gandhi R.P. and Gupta K.C. (2007) *Bioorganic and Medicinal Chemistry*, 15: 7840-7849.

Phifer S.S., Lee D., Seo E-K., Kim N-C., Graf T.N., Kroll D.J., Navarro H.A., Izydore R.A., Jiménez F., Garcia R., Rose W.C., Fairchild C.R., Wild R., Soejarto D.D., Farnsworth N.R., Kinghorn A.D., Oberlies N.H., Wall M.E., and Wani M.C (2007). *Journal of Natural Products*, 954 – 961.

Piontek C., Silva D., Heinlein C., Pöhner C., Mezzato S., Ring P., Martin A., Schmid F.X., Unverzagt C. (2009) *Angewandte Chemie International Edition*, 48: 1941-1945.

Preobrazhenskaya M.N., Olsufyeva E.N., Solovieva S.E., Tevyashova A.N., Reznikova M.I., Luzikov Y.N., Terekhova L.P., Trenin A.S., Galatenko O.A., Treshalin I.D, Mirchink E.P., Bukhman V.M., Sletta H. and Zotchev S.B. (2009) *J. Med. Chem., 52:* 189–196.

Regev-Shoshani G., Shoseyov O., Bilkis I., and Kerem Z. (2003) Biochem. J., 374: 157–163.

Rodriguez Romero V., Brito-Arias M. (2009) *Rev Soc Quim Peru*, 75 (3), 300-302.

Sato T., Takashiba M., Hayashi R., Zhu X., Yamagata T. (2008) *Carbohydrate Research*, 343: 831–838.

Shelke S.V., Gao G-P., Mesch S., Gäthje H., Kelm S., Schwardt O. and Ernst B. (2007) *Bioorganic and Medicinal Chemistry*, 15. 4951-4965.

Shoji A., Hasegawa T., Kuwahara M., Ozaki H. and Sawai H. (2007) *Bioorganic and Medicinal Chemistry Letters*, 17: 776-779.

Simerska P., Abdel-Aal A-B. M., Fujita Y., Moyle P-M., McGeary R.P., Batzloff M.R., Olive C., Good M.F., and Istvan Toth I. (2008) *Journal of Medicinal Chemistry*, 51: 1447-1452.

Tietze L.F, Schuster H.J, Krewer B. and Schuberth I. (2009) *Journal of Medicinal Chemistry, 52:* 537–543.

Ukiya M., Akihisa T., Yasukawa K., Koike K., Takahashi A., Suzuki T, and Kimura Y. (2007) *Journal of Natural Products*, 70: 783 – 788.

Urbain A., Marston A., Sintra Grilo L., Bravo J., Purev O., Purevsuren B, Batsuren D., Reist M., Carrupt P-A., and Hostettmann K. (2008) *Journal of Natural Products*, 71: 895 – 897.

Van der Eycken E., Terryn N., Goemans J.L., Carlens G., Nerinckx W., Claeyssens M., Van der Eycken J., Van Montagu M., Brito-Arias M., Engler G. (2000) *Plant Cell Reports*, 19: 966-970.

Wang D., Bhagat L., Yu D., Zhu F-G., Tang J.X., Kandimalla E.R. and Agrawal S. (2009) *J. Med. Chem., 52:* 551–558.

Wang X., Wei L. and Kotra L.P. (2007) *Bioorganic and Medicinal Chemistry*, 15: 1780-1787.

Yoon J.S., Sung S.H., Kim Y.C. (2008) *Journal of Natural Products*, 71: 208-211.

Yoon S-H, Fulton D.B.,Robyt J.F. (2009) *Carbohydrate Research*, 344: 2279-2424.

In: Biotechnology
Editors: L. G. Torres Bustillos and I. Garcia Peña

ISBN 978-1-62081-071-2

Chapter 4

State Estimation Based on Projection Observers Applied to Sequential Treatment Based on Ozonation and Aerobic Biodegradation: Application for Phenolic Wastewaters

Pamela Guerra[a], Alan García[b], Tatyana Poznyak[a], Ines García[c] and Isaac Chairez**[c]

[a]Sección de Estudios de Postgrado de la ESIQIE-IPN, Mexico
[b]Biomedical Engineering Department, ITESM-Guadalajara, Mexico
[c]Bioprocesses Department, UPIBI-IPN, Mexico

Abstract

The adaptive linearization of dynamic nonlinear systems remains, in general, as an open problem due the complexities associated to the method required to derive the linear or quasi-linear model. The problem is even more difficult if the system is uncertain, that is, when the formal description of the plant is almost unknown considering that number of states is available. This chapter discusses an adaptive linearization method for perturbed nonlinear uncertain systems based on the application of special artificial neural networks. The proposal is based on no-parametric identifier and its convergence is analyzed using the second method of Lyapunov. The suggested structure preserves some inherited structural properties like controllability. The scheme was tested using three different set of activation functions: sigmoid, wavelets and Chevyshev polynomials. The proposed method shows a good transient performance and the identification goals are fulfilled. A distillation column was used to show how the identifier works.

* isaac_chairez@yahoo.com

INTRODUCTION

Phenols and chlorophenols are toxic compounds, which can contaminate water. Sources of these compounds have different nature such as waste incineration, uncontrolled use of wood preservatives, pesticides, fungicides and herbicides, as well as via bleaching of pulp with chlorine and the chlorination of drinking water (Mangat and Elefsiniotis,1999, Contreras et al., 2003, Tarighian et al., 2003). Discharging these compounds into the environment is thus of great concern, because of the compounds toxicity and suspected carcinogenicity (Sahinkaya and Dilek, 2007).

Nowadays, treatment of these compounds has been an important area of research because the number of industries that produce wastewater with high degree of phenolic contaminants. The number of different methods proposed to eliminate or reduce the concentration of such organics has included biological and chemical principles. Among others, aerobic and anaerobic microbiological reactions and advanced oxidation process have shown remarkable efficiencies to decompose phenolics compounds. Recently, mixed schemes including biological and chemical methods have also been designed. These proposals used the advantages of both schemes: the high velocity to decompose complex and recalcitrant compounds offered by the chemical method and the low cost obtained when microorganisms are used to decompose simple and easily intake organics.

When xenobiotics biotreatment is applied, mixed cultures (usually called consortia) are particularly efficient to achieve complete mineralization (complete transformation of contaminants into carbon dioxide). Besides, some single cultures have demonstrated the accumulation of toxic intermediates that inhibit the process. This obstruction arises because the inability of a single organism to manage the presence of toxic byproducts produced by the microorganism metabolism (Buitron et al., 1998). When different species are working together to eliminate phenolic compounds and their byproducts, usually, it is accepted that one byproduct inhibiting the metabolism of one microorganism can be used to by other participating into the consortia.

Also, the elimination of such capacities of a microbial consortium can be enhanced by acclimation (Buitron and Gonzalez, 1996, Kim et al., 2002). Buitron et al., 1998 reported that acclimated activated sludge degraded the phenols and chlorophenols mixture by 1 to 2 orders of magnitude faster than pure strains obtained from the acclimated consortium.

However, chlorinated and simple phenols (CPhs and Phs) constitute a class of toxic, refractory and no biodegradable compounds that may be eliminated with a high degree of difficulty using biological treatments. The main troubles to be overcome in such situation are the longtime of elimination and the limited maximum initial concentration of contaminants. It is widely reported that chemical oxidation treatments has become into an efficient removal treatment for chlorophenols as well as their mixtures (Jogleker et al 1991, Stowell et al., 1992).

In particular, ozonation was used to eliminate a complex mixture of Phs and CPhs and producing simple organic acids. Nevertheless, complete transformation to carbon dioxide was not achieved (Poznyak et al., 2005). Complete mineralization may be achieved but after a very long treatment period (hours) with a high cost. On the other hand, ozone has many properties that can be used to design an effective wastewater treatment: it is a powerful oxidant and the reaction byproducts are more biodegradable than initial contaminants (Poznyak and Vivero, 2005).

Recently, the combination of biodegradation with preliminary or post ozonation has drawn a great attention. Some studies depict the improvement of the contaminants elimination efficiency and the avoiding of natural deficiencies of each individual process [11]. Even when some important improvements have been obtained after the sequential application of biological and chemical treatments, there is an important pending assignment to design a better process: to find an optimal time to change the water volume between the biological and chemical reactors. This issue can be solved using automatic control methods. However, most of the controllers that can solve this problem demands the complete knowledge of all variables involved in the mixed process.

An important difficult regarding the control of the combined process is the restriction to determine the values of some variables on-line. These constrain come from the lack of sensors and for the inherent complexity of the analytic techniques used to determine some important organics. In this situation, the possible optimization of the combined process cannot be carry out. The automatic control theory offers a solution by using a special class of structures called state observers or software sensors. These algorithms can be used to reconstruct variables that cannot be measured on-line using the measurable information. Most of these observers use a validated mathematical model representing the treatment process dynamics.

Contribution on the observer construction problem for nonlinear systems in the presence of *complete information* about the nonlinear dynamics, was performed by Williamson (1977), Krener and Isidori (1983), Krener and Respondek (1985), and Xia and Gao (1989). Most of these results deal with the situation where it is possible to obtain a set of rather restrictive conditions when the dynamics of the observation errors is linear and there is no observation noise. In Walcott and Zak (1990) a class of observers for nonlinear systems subjected to bounded nonlinearities or uncertainties was suggested. A canonical form and a necessary and sufficient observability condition for a class of nonlinear systems which are linear with respect to the inputs were established by Gauthier and Bornard (1981). The extended Luenberger observer for a class of SISO nonlinear systems was designed by Zeitz (1987). These results were extended in Birk and Zeitz (1988) for a class of MIMO nonlinear systems. An exponentially convergent observer was derived in Gauthier et al., 1992 for nonlinear systems that are observable for any input signal. More advanced results were obtained in Ciccarella et al. 1993 where based on simple assumptions of regularity; global asymptotic convergence of the estimated states to the true states was shown. Afurther line of investigation relates to the observation problem subjected to bounded nonlinearities or *uncertainties* (see Walcott, and H. Zak, 1987, Walcott et al, 1987).

In the situation when the plant model is incomplete or uncertain, the implementation of high-gain observers seems to be convenient (Tornambé, 1989, Dabroom, and Khalil, 1997). In Yaz and Azemi, 1994 a robust/adaptive observer is presented for state reconstruction of nonlinear systems with uncertainty having unknown bounds. More general robust nonlinear observer is considered in Shields (1996) for a class of singular nonlinear descriptor systems subject to unknown inputs. This class is partly characterized by globally Lipschitz nonlinearities. A suboptimal robust filtering of states for finite dimensional linear systems with time-varying parameters under nonrandom disturbances was considered in Poznyak and Osorio-Cordero (1994). A comprehensive survey on these problems can be found in Poznyak (2004). All these methods require the complete knowledge of the mathematical model describing the system under analysis.

This constraint can be relaxed if an artificial modeling technique based on some numerical data obtained from the process is applied. In this sense, artificial neural networks (NN), offer an excellent alternative, exploiting the fact of being universal approximations of continuous functions. In this study, Differential Neural Networks (DNN), a special kind of NN, is considered to reconstruct the behavior of the mixed process using a class of biological reaction combined with a pre-ozonation. A different robust adaptive observer for a class of nonlinear systems is proposed in Ruijun et al. (1997) based on generalized dynamic recurrent neural networks. However all designs describing observers based on DNNdo not consider state space constrains. In general, just a few results have been proposed to consider the called positive systems. Here, an important issue must be considered: all variables used to describe the treatment process never can be negative values. This is a special constraint that must be taking into account when the observer will be designed.

In this chapter, a class of adaptive observer based on DNN structures for systems where the states never can take negative values is proposed. The chapter is organized as follows: section II, includes the discussion about the class of systems represented by Ordinary Differential Equations (ODE) which will be considered, as well as presenting the main structure of the DNN observer. In section III, the main theorem about the convergence of the DNN observer is presented. Section IV shows the application of the proposed algorithm to obtain the immeasurable information of the mixed treatment using ozonation and bioprocess. Finally section V introduces final comments and conclusions.

STATE ESTIMATION WITH STATE CONSTRAINTS

Let us consider the nonlinear continuous-time system with measured output which is given by the following ODE.

$$\dot{x}_t = f(x_t,t) + \xi_t,\ x_0 \text{ is fixed}$$
$$y_t = Cx_t + \eta_t \tag{1}$$

where $x_t \in \mathrm{R}^n$ is the state-vector at time $t \geq 0$, $y_t \in \mathrm{R}^m$ is the corresponding output, available for a designer at any time, $C \in \mathrm{R}^{m\times n}$ is the state-output transformation (matrix), ξ_t and η_t are noise in the state dynamics and in the output, respectively. This system represents the behavior of biological and chemical reactions used to eliminate the Phs and CPhs compounds. State vector includes biomass, substrate and products concentrations for the biological reaction while ozonation reaction includes the ozone in the gas phase, ozone in the liquid phase and organics concentrations.

In many practical problems there is known *a* priory that the state-vector x_t always belongs to a given *compact set* X (even in the presence of noise) which has a concrete physical sense. For example, the dynamic behavior of some reagents participating in chemical reactions always keeps nonnegative the current values of these components which, in fact, are the corresponding concentrations of those reagents and, hence, cannot be negative. The same comment seems to be true for another physical variables and there dynamics such as temperature, pressure, heat and etc.

The filtering (state estimation) problem consists in designing a vector-function $\hat{x}_t = \hat{x}_t\left(y_{\tau\in[0,t]}\right) \in R^n$ *depending only on the available data* $y_{\tau\in[0,t]}$*available up to the time* t *in such a way that it would be respectively close to its real (but non-measurable) value* x_t. The measure of that *closeness* depends on the accepted assumptions concerning the state dynamics as well as the noise effects. The most of filters solving this problem are presented also as an ODE, namely, they are given by

$$\frac{d}{dt}\hat{x}_t = F\left(t, \hat{x}_t, y_{\tau\in[0,t]}\right), \ \hat{x}_0 \text{ is a fixed vector} \tag{2}$$

It seems to be very important to keep the generated state estimates $\hat{x}_t$ always remaining within a given compact set X , that is,

$$\hat{x}_t \in X \tag{3}$$

Indeed, if the obtained state estimated are supposed to be used in some (in fact, feed-back) control construction, for example, $u_t = K\hat{x}_t$, then any changing of a sign in $\hat{x}_t$ may provide significant instability effect of the corresponding close-loop dynamics. One of (seemed to be evident) feasible solutions consists in the introduction of some projection operator $\pi_S\{\cdot\}$ acting to the right-hand side of (filter) keeping the property $\hat{x}_t \in X$, i.e.,

$$\frac{d}{dt}\hat{x}_t = \pi_S\left\{F\left(t, \hat{x}_t, y_{\tau\in[0,t]}\right)\right\} \tag{4}$$

But such *evident* solution immediately implies many constructive problems since the set S , where one needs to project the right-hand side of (2), should be significantly nonlinear and non-stationary, that is, there should be $S = S\left(t, \hat{x}_t, \frac{d}{dt}\hat{x}_t\right)$ since the scheme (4) realizes the projection of the estimates of the state-derivative, but not the state-vector $\hat{x}_t$. That's why the problem of designing of other filtering structures, verifying (belonging-compact), presents a real challenge for the applied electrical engineering community. This observer using the state constrains should consider some assumptions that will be described in the following section.

BASIC ASSUMPTIONS

Hereafter, we will assume that

- A1. The function $F(x,t)$ is uniformly (on $t \geq 0$) A -Lipschitz continuous in $x \in X$, that is, for all $t \geq 0$ and all $x,x' \in X$ there exist a matrix $A \in \mathrm{R}^{n \times n}$ and a constant $L_f < \infty$ such that

$$\|f(x,t) - f(x',t) - A(x - x')\| \leq L_f \|x - x'\| \tag{5}$$

and there exists a constant $L_\delta < \infty$ such that for all $t \geq h$

$$\|\delta_t - \delta_{t-h}\| \leq L_\delta h \tag{6}$$

where $\delta_t := \hat{x}_t - x_t$ is the filtering error at time t . $x_t \in X \subset \mathrm{R}^n$ and

- A2. X is convex.
- A3. The pair (C, A) is observable, that is, there exists the gain matrix K of the projectional filter (proj-filter) - (right-hand) such that

$$\widetilde{A} := A - KC = -\lambda I,\ \lambda > 0 \tag{7}$$

- A4. The noises ξ_t and η_t acting to the system (1) are uniformly (on t) bounded such that

$$\|\xi_t\|^2_{\Lambda_\xi} := \xi_t^T \Lambda_\xi \xi \leq 1$$

$$\|\eta_t\|^2_{\Lambda_\eta} := \eta_t^T \Lambda_\eta \eta_t \leq 1 \tag{8}$$

where Λ_ξ and Λ_η are known "normalizing" non-negative definite matrices which permits to operate with vectors having components of different physical nature (for, example, meters, $mole/l$, voltage and etc.).

PROJECTIONAL FILTERS

Instead of (4) let us consider the following filter called by the projectional filter ($t \geq h$):

$$\hat{x}_t = \pi_X \left\{ \hat{x}_{t-h} + \int_{\tau=t-h}^{t} F\left(\tau, \hat{x}_\tau, y_{s\in[0,\tau]}\right) d\tau \right\} \tag{9}$$

Here π_X is the projector to the given convex compact X satisfying the condition

$$\left\| \pi_X \{x\} - z \right\| \leq \left\| x - z \right\| \tag{10}$$

for any $x \in \mathrm{R}^n$ and any $z \in X$.

RemarkNotice that the trajectories $\{\hat{x}_t\}$ *are not differentiable for any* $t \geq h > 0$.

DIFFERENTIAL NEURAL NETWORK OBSERVERS WITH STATE CONSTRAINS

Notice that (system) always could be represented as

$$\begin{aligned} F\left(t, \hat{x}_t, y_{s\in[0,t]}\right) &= f_0\left(\hat{x}_t, t \mid \Theta\right) + \widetilde{f}_t \\ \widetilde{f}_t &:= F\left(t, \hat{x}_t, y_{s\in[0,t]}\right) - f_0\left(\hat{x}_t, t \mid \Theta\right) \end{aligned} \tag{11}$$

where $f_0\left(\hat{x}, u, t \mid \Theta\right)$ is considered as a possible *nominal dynamics* which can be selected according to a designer desires and $\widetilde{f}_t$ is a vector called the *no modeled dynamics*. Here the parameters Θ are subjected to adjustment in order to obtain the complete matching between the nominal and the nonlinear dynamics. In view of (Lipschitz) and the corresponding boundness property, the following upper bound for the no modeled dynamics $\widetilde{f}_t$ takes place:

$$\left\| \widetilde{f}_t \right\|_{\Lambda_f}^2 \leq \widetilde{f}_0 + \widetilde{f}_1 \left\| x_t \right\|_{\Lambda_{\widetilde{f}}^1}^2, \ \Lambda_f, \Lambda_{\widetilde{f}}^1 > 0 \tag{12}$$

According to DNN approach (Poznyak et al., 2001), we will define the nominal dynamics as

$$f_0(\hat{x},t \mid \Theta) = A\hat{x}_t + W^0\sigma(\hat{x}_t)$$
$$A \in \Re^{n\times n},\quad W^0, \in \Re^{n\times n},\quad \sigma(\cdot) \in \Re^{l\times 1} \tag{13}$$

The activation vector-function $\sigma_i(\cdot)$ is usually constructed with sigmoid function components

$$\sigma_j(\hat{x}_t) := a_j\left(1 + b_j \exp\left(-c^T \hat{x}_t\right)\right)^{-1} \; j = \overline{1,n} \tag{14}$$

It is easy to prove that each component in the activation functions satisfies the following sector conditions

$$\left\|\sigma(x) - \sigma(x')\right\|^2_{\Lambda_\sigma} \le l_\sigma \left\|x - x'\right\|^2_{\Lambda_\sigma} \tag{15}$$

Lets introduce the adaptive state estimator based on DNN (Poznyak et al., 2005) as follows:

$$\hat{x}_t = \pi_X\left\{\hat{x}_{t-h} + \int_{\tau=t-h}^{t} \left(A\hat{x}_\tau + W_\tau\sigma(\hat{x}_\tau) + K\left[y_\tau - C\hat{x}_\tau\right]\right)d\tau\right\}$$
$$\hat{y}_t = C\hat{x}_t \tag{16}$$

Here the weights matrices ($W_{1,t}$) supply the adaptive behavior to this class of observers. The nonlinear weight *updating* (learning) law is described by the following differential equations

$$\dot{W}_t = -\frac{k_t}{2}\Omega_t\sigma^T(\hat{x}_t) \qquad \widetilde{W}_t := W_t - W^0$$
$$\Omega_t := \Pi\widetilde{W}_\tau\sigma(\hat{x}_\tau) + 2N_\delta C^T e_{t-h}$$
$$\Pi = \left(I + N_\delta C^T C \Lambda_\eta^{-1} C C^T N_\delta + \delta N_\delta \Lambda^{-1} N_\delta\right)$$
$$N_\delta = \left(C^T C + \delta I\right)^{-1},\ \delta > 0 \tag{17}$$

The matrix P is the positive definite solution for the following Linear Matrix inequality:

$$\begin{bmatrix} -\Gamma(K) & P & \mathbf{0} & \mathbf{0} & \mathbf{0} & \mathbf{0} & \mathbf{0} & \mathbf{0} \\ P & R & \mathbf{0} & \mathbf{0} & \mathbf{0} & \mathbf{0} & \mathbf{0} & \mathbf{0} \\ \mathbf{0} & \mathbf{0} & \Theta_1 & \widetilde{A}^T P & \mathbf{0} & \mathbf{0} & \mathbf{0} & \mathbf{0} \\ \mathbf{0} & \mathbf{0} & P\widetilde{A} & \mu_1 P & \mathbf{0} & \mathbf{0} & \mathbf{0} & \mathbf{0} \\ \mathbf{0} & \mathbf{0} & \mathbf{0} & \mathbf{0} & \Theta_2 & \left[W_1^0\right]^T P & \mathbf{0} & \mathbf{0} \\ \mathbf{0} & \mathbf{0} & \mathbf{0} & \mathbf{0} & PW_1^0 & \mu_2 P & \mathbf{0} & \mathbf{0} \\ \mathbf{0} & \mathbf{0} & \mathbf{0} & \mathbf{0} & \mathbf{0} & \mathbf{0} & \Theta_3 & \left[W_2^0\right]^T P \\ \mathbf{0} & \mathbf{0} & \mathbf{0} & \mathbf{0} & \mathbf{0} & \mathbf{0} & PW_2^0 & \mu_3 P \end{bmatrix} > 0$$

$$\Gamma(K) := P\widetilde{A} + \widetilde{A}^T P + Q$$

$$\widetilde{A} := A - KC$$

$$Q := \left[\|\Lambda_\gamma\| L_\sigma + \mu_1 + \mu_2 L_\sigma\right] I_{n\times n} + \omega\left(\Lambda_\alpha^{-1} + \Lambda_\beta^{-1}\right) + Q_0$$

$$R := W_1^0 \Lambda_\sigma^{-1} \left[W_1^0\right]^T + \Lambda_{\widetilde{f}} + \Lambda_\xi$$

$$\Lambda_\alpha, \Lambda_\beta, \Lambda_\gamma > 0,\ \Lambda_\alpha, \Lambda_\beta, \Lambda_\gamma \in \Re^{n\times n}$$

Main Contributions

In this chapter, the contributions may be stated as follows

- To demonstrate the workability of the projectional filter (9), when

$$F\left(t, \hat{x}_t, y_{\tau\in[0,t]}\right) = A\hat{x}_t + W^0\sigma(\hat{x}_t) + K\left(y_t - C\hat{x}_t\right),\ K \in \mathrm{R}^{n\times m} \tag{18}$$

(that corresponds to the standard Luenberger filter with a linear correction term), we have suggested to implement the "nonstandard" Lyapunov function

$$V_t := \int_{\tau=t-h}^{t} \|\hat{x}_\tau - x_\tau\|^2 d\tau \tag{19}$$

(which is, in fact, the Lyapunov -Krasovski functional) instead of the standard Lyapunov function $V_t = \frac{1}{2}\|\hat{x}_t - x_t\|^2$ which is not differentiable on the trajectories of the projectional filter (9).

- The upper bound for the averaged filtering error corresponding to (9) with (right-hand) has been found.
- The application of the suggested observer using projection method has been tested for the wastewater mixed treatment (biological and chemical ones).

Main Result

Theorem. If there exist positive real scalars μ_1, μ_2 and ω and positive solution of the Linear Matrix Inequality for the problem defined in (17), under the assumptions A1-A4, for any $\lambda_0 > 0$, and a sufficiently small positive h such that $\lambda_0 - 3h(L_f + \lambda_0)^2 > 0$ with $\lambda := 2L_f + \lambda_0$ in (7), the projectional filter (9)-(18) provides the following upper bound for the averaged filtering error:

$$\limsup_{t\to\infty} \frac{1}{t} \int_{\tau=0}^{t} \|\hat{x}_\tau - x_\tau\| d\tau \leq 2\frac{b}{a} + \sqrt{\frac{c}{a}} \tag{20}$$

where

$$a = \lambda_0 - 3h(L_f + \lambda_0)^2, \quad b = \varsigma + \frac{h}{2}\left[L_\delta(4L_f + \lambda_0)\right]$$
$$c = 3L_\delta^2 h^3 + 3\varsigma^2 h, \quad \varsigma = \|K\|\left\|\Lambda_\eta^{-1}\right\|^{1/2} + \left\|\Lambda_\xi^{-1}\right\|^{1/2}$$

Proof. Using the functional (19), in view of the property (10) and taking into account that $x_t \in X$ by A2, we have

$$\dot{V}_t = \|\delta_t\|^2 - \|\delta_{t-h}\|^2 = -\|\delta_{t-h}\|^2 +$$
$$\left\| \pi_X \left\{ \hat{x}_{t-h} + \int_{\tau=t-h}^{t} F(\tau, \hat{x}_\tau, y_{s\in[0,\tau]}) d\tau \right\} - x_t \right\|^2 \leq \alpha_t + \beta_t \tag{21}$$

where

$$\alpha_t = \left\| \int_{\tau=t-h}^{t} \left[f(\hat{x}_\tau, \tau) - f(x_\tau, \tau) - KC\delta_\tau + K\eta_\tau - \xi_\tau \right] d\tau \right\|^2$$
$$\beta_t = 2\left(\delta_{t-h}, \int_{t-h}^{t} \left[f(\hat{x}_\tau, \tau) - f(x_\tau, \tau) - KC\delta_\tau + K\eta_\tau - \xi_\tau \right] d\tau \right)$$

Since

$$\|\eta_\tau\| = \sqrt{\left(\Lambda_\eta^{1/2}\eta_\tau, \Lambda_\eta^{-1}\Lambda_\eta^{1/2}\eta_\tau\right)} \le \sqrt{\left\|\Lambda_\eta^{-1}\right\|\left\|\eta_\tau\right\|_{\Lambda_\eta}^2} \le \left\|\Lambda_\eta^{-1}\right\|^{1/2}$$

the following upper estimate holds:

$$\beta_t \le 2\left(\delta_{t-h}, \widetilde{A}\int_{\tau=t-h}^{t}\delta_\tau d\tau\right) + 2h\|\delta_{t-h}\|\varsigma + 2L_f\|\delta_{t-h}\|\int_{\tau=t-h}^{t}\|\delta_\tau\|d\tau$$

By (6) the right-hand side in the last inequality may be estimated as

$$\beta_t \le h\left[(2L_f - \lambda)\|\delta_{t-h}\|^2 + \|\delta_{t-h}\|\left(2\varsigma + h\left[L_\delta(\lambda + 2L_f)\right]\right)\right]$$

Analogously,

$$\alpha_t \le \left((L_f + \lambda)\int_{\tau=t-h}^{t}\|\delta_\tau\|d\tau + \zeta h\right)^2 \le 3h^2\left((L_f + \lambda)^2\|\delta_{t-h}\|^2 + L_\delta^2 h^2 + \zeta^2\right)$$

That finally implies

$$\dot{V}_t \le h\left[-a\|\delta_{t-h}\|^2 + 2\|\delta_{t-h}\|b + c\right] = \left[\left(-a\left[\|\delta_{t-h}\| - \frac{b}{a}\right]^2 + \frac{b^2}{a}\right) + c\right]h \tag{22}$$

Rearranging and integrating (22), by the Jensen inequality

$$\frac{1}{t}\int_{t=h}^{t}\left[\|\delta_{t-h}\| - \frac{b}{a}\right]^2 dt \ge \left(\frac{1}{t}\int_{t=h}^{t}\left[\|\delta_{t-h}\| - \frac{b}{a}\right]dt\right)^2$$

We get

$$\frac{1}{t}\int_{t=h}^{t}\|\delta_{t-h}\|dt - \frac{b}{a}\frac{t-h}{t} \le \left|\frac{1}{t}\int_{t=h}^{t}\left[\|\delta_{t-h}\| - \frac{b}{a}\right]dt\right|$$
$$\le \frac{1}{t}\int_{t=h}^{t}\left[\|\delta_{t-h}\| - \frac{b}{a}\right]^2 dt \le \sqrt{\frac{V_0}{ah\cdot t} + \frac{b^2}{a^2} + \frac{c}{a}}$$

that leads to (20).

Remark For small enough h the upper bound (20) is

$$\limsup_{t\to\infty}\frac{1}{t}\int_{\tau=0}^{t}\left\|\hat{x}_\tau - x_\tau\right\| d\tau \le dh + 2\lambda_0^{-1}\left(\left\|K\right\|\left\|\Lambda_\eta^{-1}\right\|^{1/2} + \left\|\Lambda_\xi^{-1}\right\|^{1/2}\right) + o(h)$$

Where d is a scalar positive constant. The following section will show how the proposed observer was tested using real treatment process using the combination of ozonation and biological aerobic reaction.

Materials and Methods

This section describes the set of materials and methods used in both, the biological and chemical treatments. Besides, the analytical methods are also described.

Materials

A model solution was artificially contaminated with a mixture of $4-CPh$ (60 mg/L) and $2,4-DCPh$ (60 mg/L). This one was treated by biodegradation and the combination ozonation followed by biodegradation. For biological treatment, two microbial consortia were cultivated and acclimated (by a fill-and-draw procedure Vázquez et al., 2006) to the specific carbon source (model solution with or without previous ozonation). The mineral media contains (g/L): $(NH_4)_2SO_4, 3$; $KH_2PO_4, 0.6$; $K_2HPO_4, 2.4$; $MgSO_4\ 7H_2O, 1.5$; $CaSO_4, 0.15$; $FeSO_4, 0.03$. Inoculum of the corresponding microbial consortium was added into batch reactors containing the contaminated water (model solution with or without previous ozonation). Reactors were kept at ambient temperature and shaken in an orbital shaker at 200rpm. The biomass amount and CPhs as well as ozonation products concentration of samples in reactors were periodically measured in triplicate. This section of the wastewater treatment was carried out in a batch reactor (Figure 1).

The chemical oxidation of the model solution was carried out in a semi batch reactor ($0.5L$) using an ozone generator AZCO with an ozone concentration of 30 mg/L with the gas flow of 0.5 mL /min at the pH 7. The measurement of ozone in gas phase in the outlet of the reactor was done with an ozone analyzer BMT 930, connected with a PC.

The ozone analyzer provides the ozone detection in the gas phase for the ozone monitoring to measure the ozonation degree as well as the ozone consuming and the ozone decomposition. The reactor has a by-pass gas flow. This special reactor construction allows controlling the gas flow into or out of the reactor using the same ozone analyzer (Figure. 2).

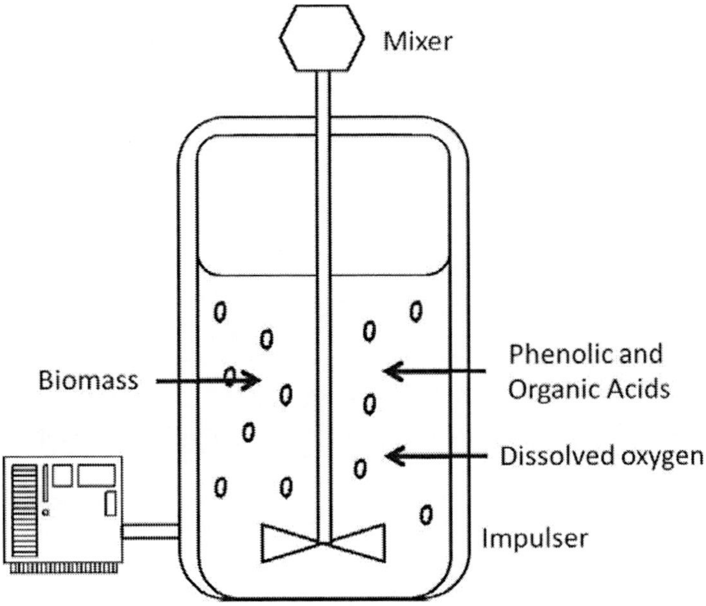

Figure 1. Schematic diagram of bioreactor. This setup includes the mixer and propeller, external heater. The variables used to describe the reaction is biomass, phenolic and organic acids used as substrate and dissolved oxygen.

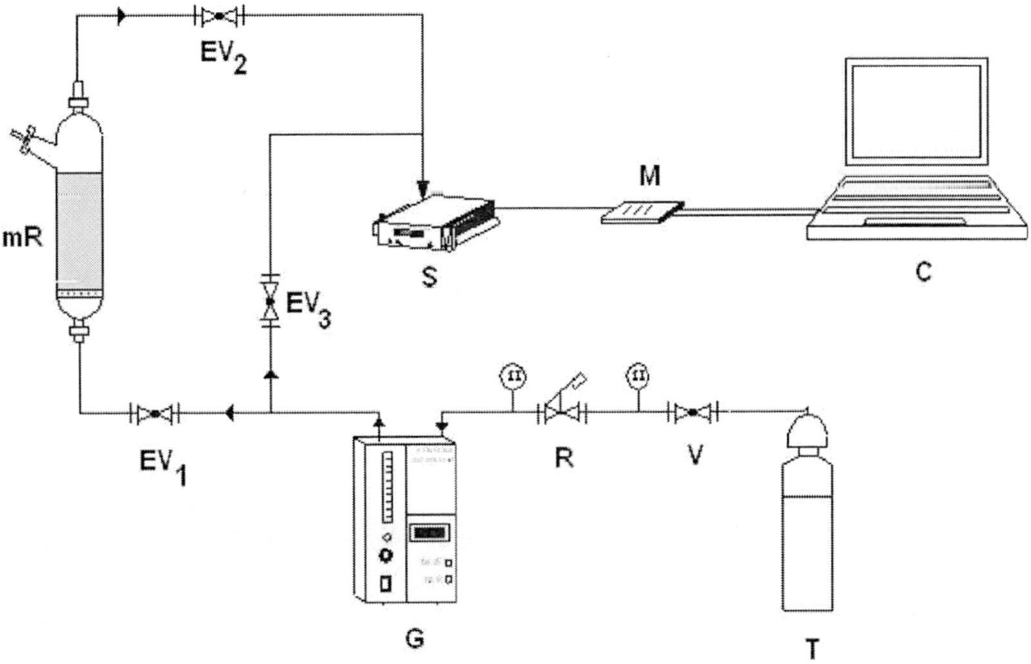

Figure 2. Schematic diagram of the ozonation device: (mR) reactor, (T) oxygen tank, (V) valve, (R) manifold, (G) ozone generator , (EV1, EV2, EV3) bypass system with electro-valves, (S) ozone analyzer , (M) Data acquisition system, (C) PC.

ANALYTICAL METHODS

The control of chlorophenols decomposition as well as the intermediates and final products formed in the ozonation step was made by high performance liquid chromatography (HPLC), (Perkin Elmer) equipped with UV-VIS detector series 200 (190 -- 460 nm) analytical details are shown in Table 1.

UV Spectroscopy (Lambda 2S, Perkin Elmer) was used for the bioprocess control (190 - 320 nm). Two wavelengths were periodically monitored (222 nm and 282 nm), each one representing the major groups of compounds formed during ozonation; organic acids and phenolic compounds, respectively. This technique was also used for monitoring the microbial growth (OD600).

Table 1. Experimental conditions for Analytical Methods.

	Compounds	
Condition	Phenolics	Organic Acids
Column	*Platinum C18* (*Alltech*) 250x4.6 *mm*	*Prevail Organic Acid* (*Grace*) 150x4.6*mm*
Carrier Phase	60:40 (water : methanol)	KH_2PO_4 25 mMol in water
λ(nm)	210	210
Flow Rate	1 ml/min	1 ml/min
Sample volume	30 μL	30 μL

RESULTS

This section describes the application of the projection observer to the mixed water treatment method. This observer uses the dissolved oxygen concentration and the output ozone concentration to reconstruct the remainder variables involved in the combined treatment process. These variables are:

- Dissolved ozone in the chemical reactor and phenols or/and chlorophenols concentrations for the ozonation method
- Ozonation by-products considered as the substrate in the biological reaction and the corresponding by-products produced by the microorganism metabolism.

Experimental procedure used to test the observer based on projectional DNN considered three different stages: the first one uses pure ozonation to decompose the phenolics and chlorophenolic compounds, the second stage only uses the biological reaction to demonstrate how the microorganisms performed a low efficiency to decompose the organics and finally the mixed algorithm where the ozonation was applied prior to the biological treatment.

Simple Biological Treatment

When the biological treatment was applied to decompose no ozonated CPhs, the organics degradation was not satisfactory because after three weeks of treatment, only 40% of the original contaminants concentration was eliminated. Here, CPh's toxicity was the key aspect to obtain such a poor degradation, the long treatment times and the corresponding inhibition troubles. Similar conclusions were reported by Sahinkaya, and Dilek (2007) where a batch reaction to treat a model mixture conformed by 4-CP and 2,4-DCP inhibited their own degradation at 53 and 25 mg/L, respectively. Moreover, this inhibition kinetics was characterized using the Haldane equation.

On the other hand, ozonation was successfully applied to decompose initial mixture of Phs and CPhs completely. However, two main drawbacks appeared when pure ozonation was used: treatment period was too long (almost one hour) and the number of oxidative by-products was high including a set of organics acids which can be eliminated with ozone after very long time periods. Therefore, complete mineralization of initial model mixture is really costly.

The following sections show the results obtained when the mixed treatment scheme was applied.

Pure Ozonation of Model Solution

Pre-ozonation was used to reduce the phenolic compounds. It has been reported that ozone may decompose Phs and CPhs after few minutes (less than 10) without taking into account the compounds concentrations but with high enough ozone concentration. The initial concentration of organics considered in this study leads to use less than 30 mg/L as initial ozone concentration. Besides, initial model solution was ozonated just 3 minutes. This period of time was enough to eliminate 53% of 4-CPh and 70% of 2,4DCPh.

Some by-products were formed: catechol, hydroquinone and oxalic acid, which are some of the most important products identified in CPhs ozonation (Poznyak and Vivero, 2005). Besides, some other by-products (phenolic compounds and organic acids) were observed without complete identification. All of these compounds constitute the carbon source for bioprocess applied in the next treatment stage.

In this work is wanted to observe the decomposition dynamics of all the compounds, which are present after ozonation, but previously was explained that some of them were not identified, so UV absorption was taking account for measure their decomposition. The two wavelengths used to build these dynamics were 222 nm for organic acids and 282 nm for phenolic and aromatic compounds.

Biological Treatment of Pre Ozonated Model Solution

Biodegradation treatment of pre ozonated phenolics compounds was tested after a short ozonation period (3 minutes). Ozonation was developed using the operation conditions described above (Table 1). The microorganism consortia was acclimated with a sequential batch process All these batch processes used a mixture of organic acids. This acclimatization was needed to improve the contaminant degradation. This degradation dynamics of organic acids and phenolic compounds was characterized by the variation of the absorbance measured at 222 nm and 282 nm, respectively.

In this study, organics concentration was obtained using a previous calibration based on equivalents of phenols and oxalic acid. Besides, microbial growth was obtained using an indirect measure of optical absorbency (Spectroscopy method using 600 nm as main wavelength). Even when the true concentrations are available, all of them are presented in a normalized way to clarify their comparison.

All biodegradation treatments showed three stages: the first one was observed between the first and third day; the second one started on the third day and finalizes on the next day and finally the third stage ended on the ninth day. This characterization was determined using the different substrate intake dynamics. Actually, organic acids were preferred by the microbial population as it was shown by the degradation of this class of contaminants (since the first stage, acids were completely eliminated). Phenolics compounds were decomposed slower than organic acids.

Therefore, microbial growth during the first stage can be associated to the organic acids decomposition. These compounds were used as carbon source during the first stage when these organics were depleted with higher velocity.

Second stage shows an important fact to take into account: when organic acids becoming scarce. In this period, microbial growth is almost stopped. Therefore, the substrates consumption is reduced because the unavailability of organic acids. Here microbial population has to use others carbon sources available in the medium (phenolic compounds). A remarkable characteristic of this consortium is associated to the capacity to produce enzymes required to use this new carbon source within the day where the second stage is occurring.

In the third stage, phenolic compounds are the main carbon and energy source for the microbial population. Therefore, the biomass and metabolites are produced using these organics. However, its assimilation is slower than the one obtained during the first stage. One may conclude that organic acids are acting as co-substrates for phenolic compounds. These compounds are easier to assimilate allowing biomass growing. Once the biomass is accumulated, this microbial population is capable to degrade more toxic substrates like the phenolic compounds. This condition will eventually reduce the toxicity and inhibition effect. This mixed behavior is similar to those results obtained in different studies where chlorophenols have been biodegraded (using mixed cultures) by co-metabolism with sugars (Basu, and Oleszkiewicz,1995, Wang, and Loh, 1999), peptone (Sahinkaya, and Dilek, 2006), or even phenol (Chiavola et al., 2004).

All of these works suggest a better behavior when the main compounds to be biodegrade are chlorophenols. Therefore, the consortium is capable to eliminate higher initial concentrations. Besides, the treatment time was reduced when co-substrates are used within the reactor.

Ozonation treatment can be considered as the preliminary process to produce co-substrates for biological process. In this sense, more biodegradable substrates are produced when pre-ozonation is used Biodegradation enhancement of lower biodegradable compounds is actually obtained with the preliminary oxidation method. It is important to mention that phenolic compounds are eliminated faster than organic acids when ozone is used as pre-treatment.

Another important fact for biodegradation is the previous acclimation of the biomass, which enhances the process, and makes possible the utilization of both types of compounds available in the medium, despite organic acids are easier and faster degraded, just one day is needed to prepare conditions that make possible to use phenolic compounds as carbon source.

NUMERICAL RESULTS OF STATE ESTIMATION USING THE PROJECTION SOFTWARE SENSOR

As it follows from the presentation above, to carry out the suggested approach one needs to fulfill the following steps:

- Define the projector.
- Select Matrices A and K .
- Select $\hat{W}$ (some hints are given by [21] and [26].
- Find P as the solution of the LMI problem (LMI), it can be done using the corresponding Matlab-Toolbox.
- Introduce P into the adapting weight law and solve it on-line.

In this study all simulations were carried out by the Matlab-Simulink software, considering the 4th order Runge-Kutta integration method, with a fixed step of 0.001 seconds. Measured information was obtained using the corresponding real sensors. These devices were connected with a personal computer using a data acquisition board (National Instruments USB-6008). Acquired information was feed to the numerical algorithm that implements the DNN based observer. This algorithm was successfully used to reconstruct the concentration of the unmeasurable compounds. The reconstruction was completed using the real concentrations for those compounds that were completely identified. On the other hand, when the organics were not identified, the identification was obtained using normalized relationships using the absorbency obtained at each treatment stage and the initial one.

Observer performance for the pure chemical reaction with ozone was tested to reconstruct the Phs and CPhs concentrations. This reconstruction was developed using the gas phase ozone concentration. Phenol and chlorophenol reconstruction were successfully obtained using the projection scheme and avoiding negative reconstructions (Figures 3 and 4).

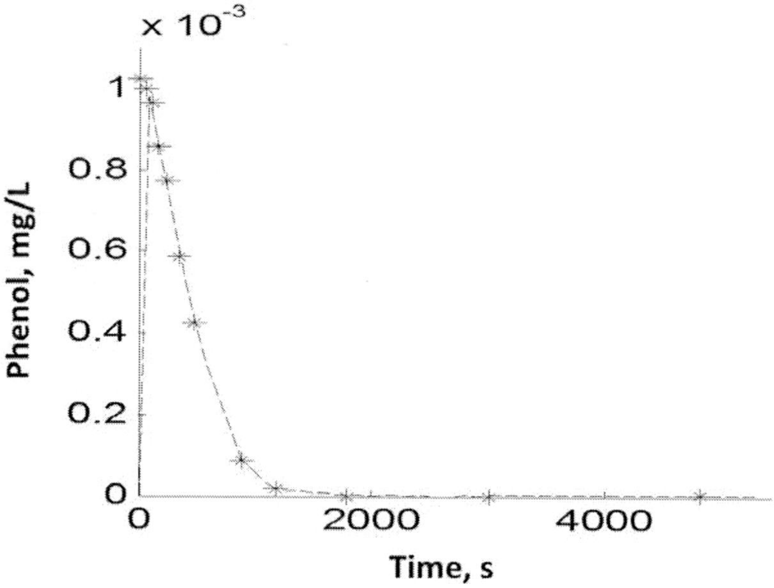

Figure 3. DNN observer performance to estimate the phenol concentration when the ozone is used to decompose this organic.

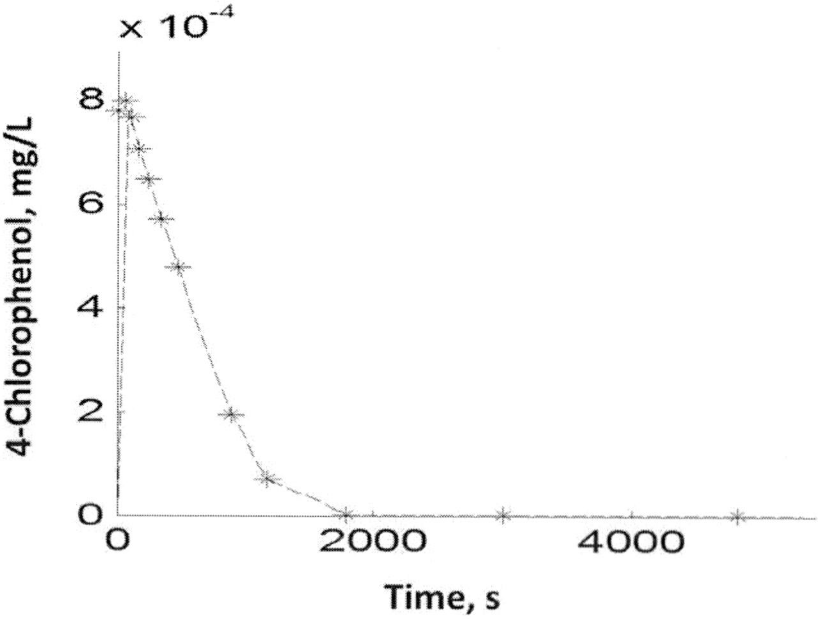

Figure 4. DNN observer performance to estimate the chloro-phenol concentration when the ozone is used to decompose this organic.

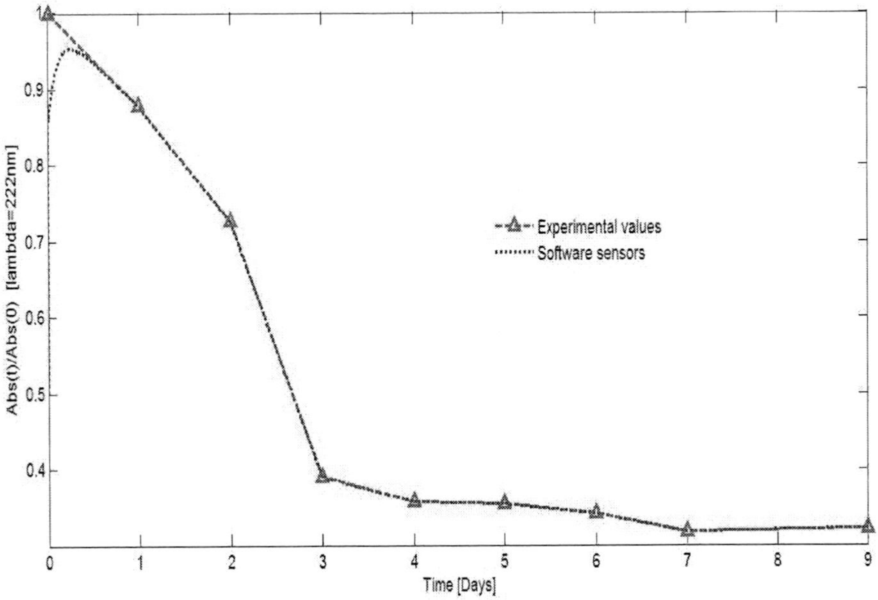

Figure 5. Software Sensor reconstruction for absorbance at λ=222nm.

Figures 5 and 6 show the Projectional Differential Neural Network estimation for those absorbency changes observed when the wavelengths λ =222nm and λ =282nm, respectively were used. Such variation represents the decomposition of the initial contaminants as well as the dynamics of their by-products. This software sensors, allow us to estimate such dynamics by the simple measure of the optical density, without an extra analytical process.

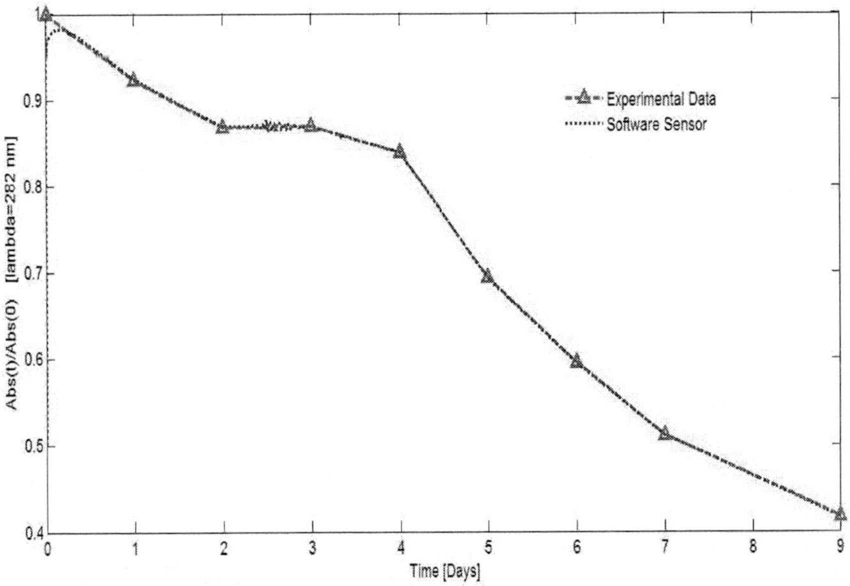

Figure 6. Software sensor reconstruction for absorbance at λ =282nm.

Conclusion

Chlorophenols toxicity is very high, so trying to eliminate them with biodegradation is complicated even when a consortium was previously acclimated to them, because inhibitory effects are presented. On the other hand, ozonation can easily decompose them and form more biodegradable species like organic acids, which can eventually improve biodegradation of lower biodegradable ones (phenolic compounds) formed during ozonation, too. It seems like organic acids act like co-substrates on phenolic compounds biodegradation. So, taking into account decomposition and accumulation dynamics of phenolic compounds and organic acids, respectively, contaminated water could be ozonated longer times and eventually enhance the next treatment step (biodegradation). The previous acclimation of the consortium also showed to improve the complete treatment scheme. On the other hand, the designed software sensor, achieve an excellent convergence with the experimental values, when the optical density is considered as a measurable output of the process. From this information it is possible to estimate the decomposition behavior of the initial compounds. Further analysis considering dissolved oxygen as a measurable output should be considered to improve the behavior and utility of the sensor.

References

Basu, S.K. and J.A.Oleszkiewicz (1995). Factors affecting aerobic biodegradation of 2-chlorophenol in sequencing batch reactors. *Environmental Technology* 16: 1135-1143.

Birk, J. and M. Zeitz (1988). Extended luenberger observer for nonlinear multi-variable systems. *Int. J. Control* 47: 1823-1836.

Buitron, G., A. González and L. M. López-Marín (1998). Biodegradation of phenolic compounds by an acclimated activated sludge and isolated bacteria.*Water Science and Technology* 37: 371.378.

Buitron, G. and A. Gonzalez (1996). Characterization of the microorganisms froman acclimated activated sludge degrading phenolic compounds. *Water ScienceTechnolology* 34: 289-294.

Chiavola, A., R. Baciocchi, R.L. Irvine, Gavasci and P. R., Sirini (2004). Aerobic biodegradation of 3-chlorophenol in a sequencing batch reactor: e¤ect ofcometabolism. *Water Science and Technology* 50: 235-242.

Ciccarella, G., M. Dalla Mora and A. Germani (1993). A luenberger-like observer for nonlinear systems. *Int. J. Control* 57: 537-556.

Contreras, S., M. Rodriguez, F. AlMomani, C. Sans and S. Esplugas (2003). Contribution of the ozonation pre-treatment to the biodegradation of aqueous solutions of 2,4-dichlorophenol. *Water Research* 37: 3164-3171.

Dabroom, A. and H. Khalil (1997). Numerical differentiation using high-gain observers. In: *Procc. IEEE Conf. on Dec. and Control* (CDC97). San Diego CA.Pp. 4790-4794.

Gauthier, J. P. and G. Bornard (1981). Observability for any u(t) of a class ofnonlinear systems. *IEEE Trans. Aut. Control* 26: 922-926.

Gauthier, J. P., H. Hammouri and S. Othman (1992). A simple observer for non-linear systems applications to bioreactors. *IEEE Trans. Aut. Control* 37: 875-880.

Gricni, O., M. Bao, K. Bariere, S. D. Burrini and F. Pantani (1999). Formation and removal of biodegradable ozonation by-products during ozonationbio.ltration treatment: pilot scale evaluation. *Ozone Science & Engineering* 21: 79-98.

Jogleker, H.S., S.D. Sament and J.B. Joshi (1991). Kinetics of wet air oxidationof phenol and substituted phenols. *Water Research* 25: 135-145.

Kim, Jung-Hwa, Kyung-Keun Oh, Sung-Taik Lee, Seung-WookKim and Suk In Hong (2002). Biodegradation of phenol and chlorophenols with definedmixed culture in shake-.asks and a packed bed reactor. *Process Biochemistry* 37: 1367-1373.

Krener, A. J. and A. Isidori (1983). Linearization by output injection and nonlinear observers. *Syst. Control Lett.* 3: 47-52.

Krener, A. J. and W. Respondek (1985). Nonlinear observers with linearizableerror dynamics. *SIAM. J. Control and Optim.* 23: 197-216.

Mangat, S. and P. Elefsiniotis (1999). Biodegradation of the herbicide 2,4-dichlorophenoxyacetic acid (2,4-d) in sequencing batch reactors. *Water Research* 33: 861-867.

Poznyak, A., E. Sánchez and W. Yu (2001). *Differential Neural Networks for Robust Nonlinear Control* (Identiffication, state Estimation and trajectory Tracking). World Scientific.

Poznyak, A. S. and A. Osorio-Cordero (1994). Suboptimal robust filtering of statesof .finite dimensional linear systems with time-varying parameters under non-random disturbances. In: Proc. *IEEE Conf. on Dec. and Control. Lake BuenaVista, FL.* Pp. 3931-3936.

Poznyak, Alex S. (2004). In Variable structure systems: from principles to implementation. Chap. Deterministic output noise effects in sliding mode observation, Pp. 45.80. *IEE Control Engineering Series. London, England.*

Poznyak, T. I. and J. L. Vivero (2005). Degradation of aqueous phenol and chlorinated phenols by ozone. *Ozone Science & Engineering* 27: 447-458.

Poznyak, T., I. Chairez and A. Poznyak (2005). Application of the model–freeneural observer to the phenols ozonation in water: Simulation and kincticparameters identi.cation. *Water Research* 39: 2611-2620.

Ruijun, Z., C. Tianyou and S. Cheng (1997). Robust nonlinear adaptive observerdesign using dynamic recurrent neural networks. In: *Proceedings of the 1997American Control Conference*. IL, USA. Pp. 1096-1100.

Sahinkaya, E. and F. Dilek (2006). Effect of biogenic substrate concentrationon the performance of sequencing bath reactor treating 4-CP and 2,4-DCPmixtures. *Journal of Hazardous Materials* 128: 258-264.

Sahinkaya, E. and F. Dilek (2007). Effect of feeding time on the performance of asequencing batch reactor treating a mixture of 4-CP and 2,4-DCP. *Journal of Environmental Management* 83: 427-436.

Shields, D. N. (1996). Observer design for nonlinear descriptor systems for controland fault detection. *International Journal of Systems Science* 22: 5-20.

Stepanyan, V. and N. Hovakimyan (2007). Robust adaptive observer design foruncertain systems with bounded disturbances. *IEEE Transactions on NeuralNetworks* 18: 1392-1403.

Stowell, J. P., J. N. Jensen and A.S. Weber (1992). Sequential chemical/biologicaloxidation of 2-chlorophenol. *Water Science and Technology* 29: 2085-2087.

Tarighian, A., G. Hill, J. Headley and S. Pedras (2003). Enhancement of 4-chlorophenol biodegradation using glucose. *Clean Technologies and Experimental Policy* 5: 61-65.

Tornambé, A. (1989). Use of asymptotic observers having high-gains in thestate and parameter estimation. In: *Proc. 28th Conf. Dec. Control.* Tampa,Florida. Pp. 1791-1794.

Vázquez, G., C. Youssef and Waissman-Vilanova (2006). Two-step modelingbiodegradation of phenol by an acclimated activated sludge. *Chemical Engineering Journal* 117: 245-252.

Walcott, B. L. and H. Zak (1987). State observation of nonlinear uncertain dynamical systems. *IEEE Transactions on Automatic Control* 32: 166-170.

Walcott, B. L., M. J. Corless and S. Zak (1987). Comparative study of non-linearstate-observation techniques. *Int. J. Control* 45: 2109-2132.

Wang, S.-J. and K.-C. Loh (1999). Facilitation of cometabolic degradation of 4-chlorophenol using glucose as an added growth substrate. *Biodegradation,*10: 261-269.

Williamson, D. (1977). Observation of bilinear systems with application to biological control. Automatica 13: 243-254.

Xia, X. U. and W. B. Gao (1989). Nonlinear observer design by observer errorlinearization. SIAM J. Control Optim. 27: 199-216.

Yaz, E. and A. Azemi (1994). Robust/adaptive observers for systems having uncertain functions with unknown bounds. In: *Proceedings of Amer.Contr.Conf.NY,* USA. Pp. 73-76.

Zak, H. and B. L. Walcott (1990). Deterministic Control of Uncertain Systems.Chap. *State observation of nonlinear control systems via the method of Lyapunov,* Pp. 333-350. Peter Peregrinus. Stevenage UK.

Zeitz, M. (1987). The extended luenberger observer for nonlinear systems. *Syst.Control Lett.* 9: 149-156.

In: Biotechnology
Editors: L. G. Torres Bustillos and I. Garcia Peña
ISBN 978-1-62081-071-2

Chapter 5

METAL ADSORPTION AND NANOPARTICLES FORMATION BY MICROORGANISMS

Luis C. Fernández Linares**, *Dorian A. Bautista Hernández and Isaías Martínez Bautista
Bioprocess Department UPIBI-IPN, Mexico

SUMMARY

Recovery of valuable metals from industrial effluents and environmental remediation are currently performed by using biomaterials. Removal of metals from solutions using biomasses offers advantages over physicochemical methods, especially when they are found at low concentrations. Biosorption can be defined as the removal of metal or metalloid species, compounds, and particulates from solution by biological material. Metals can be accumulated in biomass by a variety of processes dependent and independent from metabolism. Both living and dead biomass, as well as cellular products, can be used for metal removal. Bacterial biomass has shown important potential for bioadsorption. Microorganisms are being studied in order to produce metallic nanoparticles that can have potential applications in advanced materials used in electronics, pharmaceutics, and energy.

Nanoparticles are substances smaller than 100 nm in dimension, that possess more than one dimension and its physicochemical properties are different from those presented by material in the size range of centimeters. They can be spherical, tubular, or irregularly shaped and can be given in the form of alloys, aggregates, or agglomerates. According to their origin, nanoparticles can be divided into natural and anthropogenic.

In this chapter, an overview of metal sorption and metal nanoparticles formation by microorganisms is given. In addition, part of the research being performed at the Professional Interdisciplinary Unit of Biotechnology, National Polytechnical Institute (IPN) is presented. Research focused essentially on the isolation of microorganisms resistant to metals, and to study the ability of these microorganisms to accumulate metals and form metal nanoparticles.

* lfernandezl@ipn.mx

INTRODUCTION

Mining Industry

The mining industry has been key to the development of civilization. In 2001, the mining industry produced over six billion tons of raw product valued at several trillion dollars. Downstream beneficiation and minerals processing of these raw materials adds further value as raw materials and products are created to serve all aspects of industry and commerce worldwide.

Traditional mining countries, such as USA, Canada, Australia, South Africa, and Chile, dominate the global mining scene. These countries have become the traditional leaders in mining and exploration methods and technology. Exploration and development funding has changed over the past few years with emphasis shifting to areas that have been poorly explored or have had poor access for reasons of politics, infrastructure or legislation. Gold, base metal, diamonds, and platinum group elements (PGEs) are the most important commodities explored for and developed globally.

North America is the major producer of gold and silver. Mining production in Mexico represents 2.4% of the worldwide mining production, occupying the 9th place worldwide and the 4th in Latin America (Asali, 2004). Mexico produces ca. 30 million tons of metals and minerals per year (INEGI, 2009).

Mining activities have generated several potentially toxic materials, to which the population, flora and fauna are exposed through the soil, air or contaminated waterbodies. Likewise, the risk of metal bioaccumulation phenomena is present along the diverse stages of the foodweb (Ross, 2002). On the other hand soil erosion has a negative impact on the earth and its inhabitants, giving rise to dramatic changes in its topography, soil properties, productivity, underground water infiltration, as well as negative effects on public health, mainly through contaminating particles such as heavy metals and radionucleotides (Larney *et al.*, 1999).

Since no concentration process reaches a 100% recovery, mining wastes (tailings) always contain mineral residues, allowing for its eventual future recovery. However, it is quite common that many firms discharge their wastes into rivers or in large piles. The composition of these wastes can be highly diverse in its chemical and mineralogical nature (Sánchez, 1995).

Tailings

Tailing are wastes from the flotation processes that, generally, contain minerals in non-soluble forms, alkali or acidic salts, and soluble traces of elements derived from the oxidation of tailings, like As and Se among others. The most common example of soil contamination due to mining activities in Mexico is that generated during gold and silver beneficiation processes, usually performed through Hg amalgamation and cyanidation. In none of these two cases is it possible to totally recover the added compounds and/or elements; therefore, it is common to find them in a soluble form in wastes of the process (Sepúlveda, 2005).

From the geotechnical perspective, these deposits represent also a high risk as they are non-consolidated materials whose particles can be moved readily by action of the wind or by hitting rainwater, dispersing easily. Besides, the lack of hydraulic infrastructure to deviate the water from the runoff into the capture basin poses a high risk of hydric erosion. Once rain has fallen for several days, pores reach a saturation limit close to 100% and the tailings can flow due to their weight, resulting in losses of stability and sliding, mainly due to the inclination of the slopes (Terlien, 1998).

Environmental Problems Associated with Metals

All human activities, in which metals are involved from their collection until their disposal, are potential contamination sources. These activities are of highly different nature and, in general, can be classified as:

1) Those related to obtaining gold by mining industries, in processes such as mine excavation, metal removal, and the extraction process. Mining wastes (tailings) always contain wastes of the metal or metals.
2) Wastewaters and solid wastes from the process. After extracting the metals from the ores, the metals are used in a large variety of industrial processes. Contamination can come from the wastewaters of these processes or from the generation of dangerous wastes, which, without treatment or adequate disposal, respectively, can cause environmental contamination.
3) Domestic solid wastes and disposal of products containing heavy metals. Other underestimated sources that generate metal or metalloid contamination are: rainwater and domestic sewages, sanitary landfills, which can contain important amounts of metals, according to their specific characteristics (Csuros and Csuros, 2002).

Mexico is one of the Latin American countries located in a volcanic region rich in minerals, and its mining tradition is well known. The potentially toxic elements most commonly derived from these processes, in the case of Mexico, are Pb, Cd, Zn, As, Se, and Hg (Volke *et al.* 2005).

Biotechnology and Metals

Biotechnology is the technology based on biology, it develops within a multidisciplinary approach that involves diverse disciplines and sciences like biology, biochemistry, genetics, virology, agronomy, engineering, physics, chemistry, medicine, and veterinary, among others.

Recovery of minerals from the environment either for remediation purposes or because of their high economic value is a widely studied process. The use of microorganisms is a viable alternative to contaminating physicochemical processes, allowing for environmental quality and a sustainable development (Atlas and Bartha, 2006).

Heterotrophic fungi have been used as bioleaching agents of minerals rich in iron, such as geothite, limonite, or hematite. Most species studied for bioleaching are members of the *Penicillium* and *Aspergillus* genera (Wainwright, 1995). Heterotrophic bioleaching represents an option to recover metals from minerals that do not contain sulphur, i.e., carbonates and metalic silicates. Bioleaching of metals is mainly achieved with fungi, it is a process mediated by the production of organic acids (citric and oxalic acids).

Another area for fungal application is adsorption of metal ions in solution. The goal of this process is to eliminate heavy metals present in effluents using the adsorption capacity of live or dead mycelia. Alternatively, the metals of the solution can be eliminated by passing the effluent through a column containing components of the fungal wall, such as chitin. These metal ions recovering biological systems can be used to clean contaminated effluents or to recover precious metal ions from solutions. In both cases, it will necessary to demonstrate that the fungal biomass can compete with the physicochemical methods, because it has not yet been proven definitively that the fungal system is economically superior (Gadd, 1990). Mycelia from *Rhizopus* and *Absidia* have been reported as excellent biosorbents for Pb, Cd, Cu, and Zn, and for other heavy metals up to 25% of the dry weight of the biomass (Volesky and Holan, 1995).

Miocroorganisms, such as bacteria, also have a wide surface:volume relation and intensive metabolic activity, which makes them very efficient vectors for the introduction of heavy metals into the trophic chains. Microbial production of acids and chelating agents affects the adsorption and mobilization processes of toxic metals in the soil, Once the heavy metal has been mobilized it can be incorporated and accumulated intracellularly by microorganisms (Atlas and Bartha, 2006).

Specific defense mechanisms allow bacteria to diminish the stress caused by the toxicity due to the high concentration of metal ions or metals in the environment. These mechanisms involve alteration in solubility and toxicity due to changes in the redox potential of metal ions, formation of extracellular complexes, intracellular precipitation, and the lack of specific mechanisms for metal transport. This constitutes the basis for important applications of microorganisms, such as bioleaching, microbial corrosion, biomineralization, and synthesis of nanoparticles. To this respect a wide range of bacteria, yeasts, and algae have been studied, which are able to accumulate metal ions up to several orders of magnitude above the concentration of the metal in the environment (Narayanan and Sakthivel, 2011, Bhattacharya and Gupta, 2005).

Undoubtedly, biosorption and bioleaching by microorganisms represent the most practical applications of the interactions of microorganisms with metals. Through bioleaching it has been possible to recover metals from sulfur or iron minerals, based on the activity of chemolithotrophic bacteria that oxidize iron and sulphur (iron and sulpho-oxidants), among them *Thiobacillus ferrooxidans, T. thiooxidans,* and *Leptospirillum ferrooxidans*, which turn insoluble metallic sulphurs (SO) to soluble sulfates and sulphuric acid (Bosecker, 2001). A disadvantage of conventional bioleaching occurs when it is used to recover metal from industrial scraps, because most of the metals are oxides instead of sulfides. In this case, the metal oxides can be leached by the microbial production of acids (as the sulfuric acid generated by *T. thiooxidans*).

Metals adsorbed by the biomass can be separated through treatment with sulfuric acid, hydrochloric acid, sodium hydroxide, or complexing agents, notwithstanding that live or dead biomass had been used. Besides, the system can be regenerated through a treatment (alkali) and reused in further adsorption-desorption cycles. Synthetic ion exchange resins with well-established methodologies present a similar efficiency to those obtained with biomaterials, and constitute the main competitors of biomaterials. Despite of this, bioadsorption methods have been demonstrated to be more efficient for the removal of metals at low concentrations (below 2 to 10 mg/L). In addition, the specificity of bioadsorbents can avoid the oversaturation with alkali metals present in contaminated effluents, which occurs when using ionic exchange resins (Valls and De Lorenzo, 2002).

INTERACTIONS MICROORGANISMS-METAL

Microorganisms have coexisted with metals since the begining of life. This is reflected in the wide range of divalent or transition cations as active centers of many enzymes. Microorganisms have used the chemical properties of metals to catalyze key reactions or to maintain the protein structure. In consequence, metals are required in minimal amounts for cell metabolism, and absorption of the metal is subject to homeostatic mechanisms that ensure that only the required metal is acquired. Many other metals do not seem to have a relevant biological role. To the contrary, they harm the cell, mainly due to their affinity to the sulfhydryl groups of proteins which they block and inactivate. In this way, from a physiological point of view, metals are grouped in three main categories: (i) essential and non-toxic (Ca, Mg), (ii) essential but harmful at high concentrations (typically Fe, Mn, Zn, Cu, Co, Ni y Mo), and (iii) toxic (Hg, Cd). Interactions with metals do not depend only on a given element, but also on the chemical speciation (Valls and De Lorenzo, 2002). In nature, metals exist principally as cations, oxyanions, or in both forms in solution, and as salts or oxides in crystalline form (mineral) or as amorphous precipitates in an insoluble form (Ehrlich, 1997).

FUNCTIONS OF THE CELL WALL IN THE INTERACTION WITH METALS

The interaction mechanisms between metals and microorganisms take place in the cell wall (Shumate and Strandberg, 1985), which plays a fundamental role in the capacities of microorganisms for the sorption of metals from the environment. This includes the absorption of the metal by the whole biomass (alive or dead) through physicochemical mechanisms such as adsorption, surface complexing, ionic exchange or superficial precipitation. These mechanisms occur in the cell wall which is a rigid layer covering the cell.

A significant amount of chemical microenvironments exist in the cell wall (Table 1). These include phosphate, carboxyl, hydroxyl, and amino functional groups. Many Gram-negative bacteria release polysaccharides that cover the cell surface granting it a negative charge that attracts (adsorbs) cations. Bioadsorption also occurs in bacteria that naturally form an extracellular surrounding polysaccharide, which provides sites for the binding of metal cations (Patrick, 2004).

In general, eukaryotes are more sensitive to toxicity by metals and the typical regulation mechanisms of metal ions concentration is the production of metallothionein (TMs), a group of metal-chelating proteins. Metallothioneins have also been reported in the cyanobacteria *Synechococcus,* which confer resistance to Zn^{2+} and Cd^{2+}. Production of metallothionein as the main resistance mechanisms is exceptional in the bacterial world. In contrast, a specific number of resistance mechanisms (including active efflux and cleavage or transformation to other chemical species) become activated at concentrations higher than homeostasis or of non-toxic levels. These mechanisms are usually ligated to plasmids, which eases transfer from cell to cell (Valls and De Lorenzo, 2002).

Table 1. Functional groups in the surface of the cell wall of microorganisms.

Basic		Acid	
$-NH_2$	Amine	$-CO_2H$	Carboxylic
$=NH$	Imine	$-SO_3H$	Sulfonic
-N=	Heterocyclic nitrogen	$-PO(OH)_2$	Phosphonic
-CO	Carbonyl	-OH	enolic, phenolic
-O-	Ether	=N-OH	Oxim
-OH	Alcohol	-SH	thioenolic and thiophenolic
-S-	Thioether		
$-PR_2$	phosphine substituted		

Source: Birch and Bachofen, 1990.

In bacteria, plasmid-mediated metal resistance is highly specific for a given metal, resistance systems to metals have been found in plasmids of different groups of bacteria (Silver and Phung, 1996). Resistance to metals, via plasmids, can move rapidly withing the cells of a population (Bruins *et al.*, 2000). Notwithstanding, certain resistance mechanisms, specifically for essential metals, are determined by the chromosome.

METAL TOLERANCE AND RESISTANCE MECHANISMS IN MICROORGANISMS

There are several resistance and tolerance mechanisms to metals, which can be enzymatic, such as: oxidation, reduction, methylation, and alkylation; or non-enzymatic by passive processes.

Bioadsorption (Microbial Immobilization of Metals)

The term bioadsorption is employed to describe the adsorption or passive binding of heavy metals, radionucleotides, and metalloids to cellular components of live or dead biological material. Bioadsorption can be considered as the first step in the microorganism-metal interaction (Valls and De Lorenzo, 2002). Bioadsorption implies physicochemical mechanisms by which metal species are sorbed, complexed, or precipitated in the biomass or microbial products; it can also imply ionic exchange mechanisms. These mechanisms are accomplished in the cell wall; hence, its composition is determinant in adsorption processes. In general, it depends on organisms either in suspension or immobilized, the presence of different cations and anions, the pH of the medium and on the chemical characteristics of the metal (Chojnacka, 2010; Katarzyna, 2010; Eccles, 1999).

Precipitation of Metals (Microbial Immobilization)

Microbiological precipitation is a widely diffused process that can be due to the dissimilar reduction or to a secondary mechanism of metabolic processes not related with the transformation of metals.

Reducing precipitation allows organisms to reduce the mobility and toxicity of a metal or metalloid by turning into a lower redox state. Many of the microorganisms that catalyze these reaction use metals or metaloids as electron acceptors during their anaerobic respiration, reducing highly soluble oxidized forms to insoluble elemental forms (reduced), resulting in detoxification and/or precipitation of the contaminant (Lovely and Coates, 1997).

Production of sulfhydric acid by sulfur-reducing bacteria (SRB) is one of the best known natural precipitation mechanisms. This phenomenon has been attributed to the deposition of metal sulfurs on the surface of biofilms or in the liquid phase, followed by capture of the precipated sulfurs by exopolymers (Valls and De Lorenzo, 2002).

Precipitation by phosphates is given by the release of inorganic phosphate from organic phosphate donor molecules. The largest research has been performed in *Citrobacter sp.*, a strain isolated from a metal-contaminated soil. This bacterium accumulated high levels of uranium, nickel, and zirconium through the formation of metallic phosphates.

In dissimilar reduction, transformation of the metal is not related with the synthesis of cellular matter; therefore, chemical species resulting from the biological activity usually end up in the extracellular medium. Diverse organisms that accomplish dissimilar reduction of uranium, selenium, chrome, gold, and other metal, under diverse environmental conditions have been reported. Dissimilar reduction in bacteria is catalyzed by enzymes associated to the membrane, which can be cytochrome type c (Valls and De Lorenzo, 2002). It has been reported that *Desulfovibrio desulfuricans* combines the oxidation of a variety of electron donors for the reduction of Tc(VII) by means of a periplasmic hydrogenase.

Enzymatic Transformation of Metals and Metalloids

Enzymatic transformation leads to precipitation and mobilization of metals through the following mechanisms: capture of trace metals for their incorporation to metallo-enzymes or their utilization in enzyme activation, this interaction occurs in all microorganisms and metals must be in ionic form; utilization of metals or metalloids as electron donors or acceptors in energy metabolism (it occurs in eubacteria and archeobacteria, their utilization in eukaryotes is unknown); enzymatic detoxification of toxic metal species, the toxic species are converted into less toxic or nil toxicity through oxidation or enzymatic reduction. Another enzymatic activity that transforms certain metal species is methylation, this is a resistance mechanism that is activated by the presence of metals; transformation by alkylation is a form of chelation that involves transfer of organic ligands (methyl groups) to the metal, thereby giving rise to stable organometallic compounds.

Metal Adsorption by Microorganisms

The metal adsorption process by microbial biomass includes the phenomena of adsorption and absorption. Adsorption involves a solid phase (sorbent) and a liquid phase (solvent, commonly water) which contains the dissolved species that are to be adsorbed (sorbate, metallic ions). The affinity between the sorbent and the sorbate leads to the binding of the sorbate, through diverse mechanisms, to the solid phase (sorbent). This process is exerted until equilibrium is established between the amount of sorbate bound to the solid phase and the amount of sorbate as residue in solution Cf (final concentration or equilibrium concentration). The degree of affinity of the sorbent for the sorbate determines the distribution of the latter in the solid and liquid phase.

The performance of the sorbent material is established considering the amount of sorbate it can retain. For this, the adsorption (q) of the biosorbent is determined, expressed as the amount of sorbate adsorbed (mg) per unit of solid phase (g). Calculation of the adsorbed metal (mg/g) is based on the balance of the system's materials: where the sorbate lacking in the solution must be in the solid phase.

Adsorption Isotherm

Adsorption isotherms are points defined by the amount of adsorbed sorbate (adsorption [q]) and the final or equilibrium concentration in the supernatant (Cf). Given that adsorption is mostly an exothermic process, it is influenced by temperature, therefore a constant temperature is required during the adsorption process. The temperature range for bioadsorption applications goes from 10 to 70°C; therefore, temperature has a lesser effect in large scale applications.

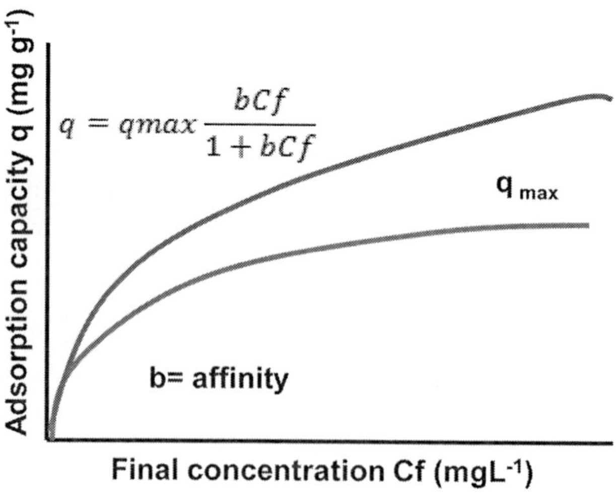

Figure 1. Langmuir Isotherm.

The relation between *q* and Cf was determined mathematically by Langmuir and Freundlich (Volesky and Holan, 1995), who studied the adsorption process with activated charcoal. The Lagmuir isotherm is a hyperbole (Fig. 1).
where:

qmax is the maximum adsorption capacity of the sorbent under determined conditions; for example, the concentration of the sorbate in the solution (mg/L). ***b*** is a coefficient related with the affinity between the sorbent and the sorbate and indicates the initial slope of the isotherm. The greater the value of ***b***, the greater is the affinity of the sorbent for the sorbate.

To obtain the ***qmax*** and ***b*** parameters, the Langmuir isotherm must be linearized by plotting $1/q$ vs. $1/Cf$ or Cf/q vs. Cf (Fig. 2).

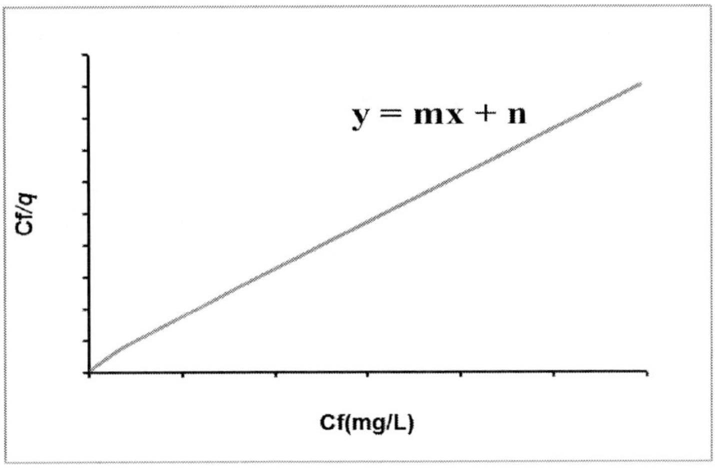

Figure 2. Linearization of the Langmuir isotherm.

where: $\mathbf{qmax} = \frac{1}{m}$ and $\mathbf{b} = \frac{1}{n} / \mathbf{qmax}$

Nanoparticles Formation by Microorganisms

Interest on the synthesis of nanoparticles from diverse materials through the use of the physical properties and the biosynthetic activity of microorganisms has risen recently. Studies have been performed with diverse microorganisms that produce nanostructures of mineral crystals and metal nanoparticles with similar features to those materials synthetized chemically (Klaus, *et al.*, 2001). However, the interest in the synthesis of nanoparticles by microorganisms recently emerged (Zhang *et al.*, 2011, Zhang and Yan, 2011)

The main studied microorganisms are prokaryotes (Table 2), one of the best advantages to use prokaryotes to synthetize nanoparticles is because they can be modified with genetic engineering techniques for a higher expression of the enzymes responsible for the reduction and capture of metals, aside from being easier to handle (Bhattacharya y Gupta, 2005). However, the use of eukaryote microorganisms, specifically fungi, is quite promising and has the following advantages: they excrete higher amounts of proteins as compared to bacteria; therefore a higher productivity in biosynthesis can be achieved as well as the possibility to cover large surfaces through an adequate growth of the mycelium. For these reasons, the biosynthesis process can be escalated and is economically feasible (Mukherjee, *et al.*, 2001; Sastry, *et al.*, 2003).

Aside from the formation of gold and silver nanoparticles, great attention has been given to the development of protocols for the synthesis of semiconductors (Quantum dots, QDs), such as CdS, ZnS, and PbS. These luminescent semiconductors have risen as new materials that can be employed for biological detection and cellular bioimaging (Mandal *et al.,* 2006). Some of these examples are: formation of ZnS nanospheres of 2 to 5 nm in size within natural biofilms of sulfate reducing bacteria of the *Desulfobacteriaceae* family (Labrenz, *et al.,* 2000); formation of CdS nanocrystals by incubating *E. coli* cells with cadmium and sodium sulfide (Sweeney, *et al.,* 2004).

Table 2. Synthesis of nanoparticles by microorganisms.

Microorganism	Type of nanoparticle
Bacteria	
Bacilus subtilis	Gold
Shewanella algae	Gold
Pseudomonas stutzeri	Silver
Lactobacillus	Gold, silver, gold-silver alloys
Clostridium thermoaceticum	Cadmium sulfide
Klebsiella aerogenes	Cadmium sulfide
Escherichia coli	Cadmium sulfide
Desulfobacteriaceae	Zinc sulfate
Thermoanaerobacter ethanolicus	Magnetite
Magnetospirillium magnetotacticum	Magnetite
Thermonospora sp.	Gold
Rhodococcus	Gold
Chlorella vulgaris	Gold
Phaeodactylum tricornutum	Cadmium sulfide
Yeasts	
Candida glabrata	Cadmium sulfide
Torulopsis sp.	Lead sulfide
Shizosacharomyces pombe MKY3	Cadmium sulfide
Fungi	
Veticillium	Gold, silver
Fusarium oxysporum	Gold, silver, gold-silver alloys, Cadmium and Zirconium sulfide
Colletotrichum spp.	Gold

Source: Mandal, *et al.* (2006)

Mandal, *et al.* (2006) reported the formation of CdS from $CdCl_2$ in the presence of cysteine chlorhydrate. This precipitation occurred in both the cell surface of *Clostridium thermoaceticum* and the culture medium (the source of sulfate was attributed to cysteine). Likewise, the formation of CdS with sizes of 20 to 200 nm has been reported on the cell surface of *Klebsiella aerogenes*, through exposing it Cd^{2+} ions in the culture medium. Ahmad, *et al.* (2002) attained extracellular formation of CdS nanoparticles through enzymatic reduction of sulfate ions by the fungal strain *Fusarium oxysporum*. Kowshick, *et al.* (2002) reported the intracellular synthesis of CdS nanocrystals with semiconductor properties in the yeasts of *Schizosaccharomyces pombe* and *Candida glabrata,* as well as the intracellular synthesis of PbS nanocrystals with semiconductor properties by the yeast *Toulopsis sp.* Other reports on nanoparticles formation by microorganisms are: extracellular production of selenium nanospheres by anaerobic bacteria that use selenium in their metabolic process, *Sulfurospirillum barnesii, Bacillus selenitireducens, Selenihalanaerobacter shriftii*; extracellular synthesis of zirconium nanoparticles (ZrF_6^{2-}) mediated by proteins of the fungus *Fusarium oxysporum* (Bansal *et al.*, 2004); formation of magnetic particles by magnetotactic bacteria (Roh, *et al.*, 2001); and the formation of paladium nanoparticles by using sulfate reducing bacteria (SRB) in the presence of an external electron donor (Yon, *et al.*, 2002).

Biological Applications of Nanoparticles

Detection of biomolecules: Nanoparticles are used as biomolecule detectors in DNA assays, immunoassays, and cellular bioimaging. Nanoparticles are released with different functional groups such as nucleic acids, oligonucleotide probes, antibodies, and proteins. Specifically, gold nanoparticles have been used for the identification of a pathogenic bacterium through a DNA microarray technology (Taton, *et al.,* 2000). One of the main advantages that have risen from the use of the gold nanoparticles detection system is that visible light can be used. Alternatively, nanoparticles can be used as fluorophores in *in situ* fluorescent hybridization (FISH) applications. The Quantum dots annexed to a specific oligonucleotide probe or to immunoglobulin G (IgG) have been successfully used for the detection of the human Y chromosome, as well as for the localization of cancer markers in cellular imaging studies (Pathak, *et al.,* 2001; Wu, *et al.,* 2002).

Separation, purification, and concentration of samples: Magnetic nanoparticles are used mainly in the separation, purification, and concentration of diverse biomolecules. For this, diverse molecules such as antibodies and oligonucleotides are immobilized on the surface of magnetic nanoparticles. On the other side, magnetic nanoparticles can be adhered to the inside of microspheres such as polystyrene spheres, which are then superficially immobilized. After target molecules react in solutions that contain up-take probes, these are bound or captured on the surface of magnetic microspheres. Thereafter, by applying a magnetic field, these particles can be attracted and concentrated in a specific site and thereby separated from the solution. The use of magnetic nanoparticles or nanospheres has been reported mainly in biological studies. Some examples are: screeining and recovery of stem cells, antigens concentration from solutions, use of nanoparticles adhered with enzymes as nanocathalysis agents, separation and concentration of microbial cells from water samples, purification of ultrapure supercoloidal DNA plasmids from bacterial cultures (Liu, 2006).

Coding of substrates. One of the unique qualities of nanoparticles is their capacity to code substrates with different colors during multiplex assays. One technique consists of embedding a mixture of Quantum dots (QDs) that emit red, green, and blue colored light at different radii. Another technique to code substrates is the production of submicrometric metal bars similar to a bar code. Each metal bar contains differente bimetal bands, such as gold and silver bands. By conjugating different metal bars with oligonucleotide or antibody probes a multiplex analysis can be achieved in the same solution (Nicewarner-Peña, *et al.*, 2001).

Signal transduction and amplification: Nanoparticles can function as signal sensors or amplifiers for fluorescence assays in electrical detection systems (Park, *et al.* 2002; Nam, *et al.*, 2003).

Monitoring and microbial detection: Despite that there are more biomedical applications of nanoparticles as compared to environmental ones; the use of QDs as fluorescence labeling systems for the detection of microorganisms has been successfully demonstrated. QDs can be conjugated with specific antibodies for the detection of pathogenic microorganisms such as *Cryptosporidium parvum* and *Giardia lambli* (Zhu *et al.*, 2004). Detection and monitoring of microorganisms can be accelerated with the use of nanoparticles for the fluorescence labeling systems in microfluid devices; an example is the development of a microfluid device for fast detection of viruses present in aqueous environments. Likewise, fluorescence systems based on nanoparticles have been developed for fast detection of bacteria even to the level of individual cells (Zhao *et al.*, 2004; Zhang *et al.*, 2005).

Chemical degradation and environmental remediation: The use of bimetal nanoparticles as oxidants has led to a greater efficiency in contaminants removal from the environment when compared with the use of zero valence metals; this efficiency is attributed to the properties of the nanoparticles that diffuse or penetrate contaminated zones that cannot be reached by microparticles, besides nanoparticles can reach a high reactivity towards redox potential-sensitive contaminants (Nurmi *et al.*, 2005).

Removal of hydrophobic organic contaminants during bioremediation can be improved with the use of nanoparticles, and this constitutes one of the key challenges in soil and aquifer bioremediation. Mycelia-based surfactants have been used commonly to improve the mobility and bioavailability of these hydrophobic contaminants. As an alternative, the use of polymeric nanoparticles prepared from polyethylene glycol-modified urethane-acrylite (MPUA) precursor has been proposed to achieve availability of polyaromatic hydrocarbons (PAHs) in soils and in aqueous phase (Tungittiplakorn *et al.*, 2005; Tungittiplakorn *et al.*, 2004). Nanoparticles can also be used to immobilize microbial cells that can degrade or recover specific chemical substances (Shan *et al.*, 2005).

Risk of Nanoparticles to Human Health and the Environment

Wastes generated during production and use of nanoparticles represents a latent risk because if not handled adequately they can become incorporated into the environment. Likewise, it is highly possible for humans to become exposed to these nanoparticles through inhalation, dermal absorption, and digestion. The highest risk of nanoparticles exposure lies in that their effects on health are still unknown, and it is a difficult task to determine the risks. To this respect, several studies have been conducted that demonstrate the cytotoxicity of QDs (nucleus CdSe) in cell envelopes (Kirchner *et al.*, 2005; Derfus *et al.*, 2004). Nanoparticles can simply be absorbed in the surface of cell membranes, be digested and degraded in the cells, deriving in toxicity effects. Nanocables and nanotubes can act similarly to a microneedle, damaging the cell wall and cell growth (Liu, 2006).

The potential effects of nanoparticles on organisms in natural environments is still more difficult to determine, however, with the current practices of nanoparticles-containing wastes discharge, the risk of transporting them to the aquatic enviroments is evident (Liu, 2006). For these reasons, and in parallel with the production and application of nanoparticles in scientific research, a controlled handling of nanoparticles must be sought and their negative effects on human health and organisms in the natural environments must be investigated.

Therefore, our group worked in the study of metals adsorption and nanoparticles formation with bacteria isolated from Mexican´s tailings. This study was made mainly with four metals Ag, Au, Zn and Pb. The methodology and results are presented below.

METHODOLOGY

Isolation and Resistance Tests

Samples from different mine tailings were collected: Oaxaca, El Oro de Hidalgo, and Zacualpan, State of Mexico, Mexico. Enrichment was achieved in a nutritive broth additioned with $ZnSO_4 \cdot 7H_2O$, $NiCl_2 \cdot 6H_2O$, $CuCl_2 \cdot 2H_2O$, $Na_2SeO_3 \cdot 5H_2O$, $Pb(NO_3)_2$ or $Al_2(SO_4)_3 \cdot 18H_2O$, AgNO3 to which 1 g of the mine tailings sample was inoculated and incubated at 35°C for 48 h.

We isolated the microorganisms from the enriched media by dilution in a nutritive agar plate enriched with the aforementioned metals at a 0.5 mM concentration. Cultures were incubated at 27°C. The minimum inhibitory concentration (MIC) of the metal at which the bacteria showed growth was determined by reseeding by puncture in nutritive agar with different metal concentrations (1, 5, 10, 25, and 50 mM). The strain with the highest MIC was selected.

Biosorption Tests

Sorption tests were conducted with biomass grown in a nutritive broth centrifuged at 5000 rpm for 15 min (Cole-Parmer Model 752400) and washed with sterile saline solution; the cellular pellet was dried at 80°C for 24 h. Dry biomass (0.015 g) was placed in contact with 10 mL of a metal solution at different time intervals (kinetics) and different initial concentrations (isotherm). The pH at initial contact was measured. The biomass was subsequently separated by centrifugation (3500 rpm for 15 min, Cole-Parmer Model 752400) and the non-absorbed metal was determined in the supernatant by atomic absorption (AA) spectrometer GBC 932 plus. The procedure was done in triplicate. The indicator q, which represents the metallic capture per unit of weight of the biosorbent, was calculated according to the following equation:

$$q = \frac{Vi\ (Ci - Cf)}{S} \tag{1}$$

Where: q = Sorption capacity (mg g^{-1}); V_i = Initial volume (mg L^{-1}); C_i = Initial concentration (mg L^{-1}); C_f = Final or equilibrium concentration (mg L^{-1}); S = Biomass weight (mg)

Using the median and standard deviation of the three series, the pseudo second-degree model was adjusted to the sorption kinetics:

$$q = \frac{t}{(1/K_s^{2} \cdot q_{eq}^{2} + t/q_{eq})} \tag{2}$$

Where: t = Time (min); K_s^2 = Sorption constant; q_{eq} = Sorption capacity in equilibrium. Linearization of t/q vs. t was used to estimate the value of the sorption constant (K_s^2) and the sorption capacity in equilibrium (q_{eq}).

Both commonly used models in literature, Langmuir and Freundlich, were used to represent the sorption phenomenon. The first is mathematically represented as:

$$q = \text{qmax} \frac{bCf}{(1 + \text{bCf})} \tag{3}$$

Where: q_{max} = Maximal sorption capacity (mgg^{-1}); b = Affinity related coefficient. Linearization of C_f/q vs. C_f was used for estimating parameters.

Freundlich´s isotherm is represented as:

$$q = \text{k}\ C_f^{(1/n)} \tag{4}$$

Where: k and n are constants that indicate sorption capacity and sorption intensity, respectively. Linearization of Ln q vs. Ln Cf was used in this case.

Silver and Gold Biosorption

Experiments were done in a bacterial strain labeled OC_4. Gold biosorption was performed in Eppendorf with 2 mg biomass/2 mL $HAuCl_4$. Variable gold concentration was from 5 to 30 mg Au/L. Silver biosorption was performed using 2 mg biomass/2 ml $AgNO_3$. Variable silver concentration was from 100 to 800 mg Ag/L. The reaction was done in an Eppendorf AG shaker, 800 rpm, 31° C. Reaction time was 1 h. At the end of the experiment the microbial biomass was separated by centrifugation (Hermile Z233 MK-2) 10000 rpm, 15 min. Remaining metals in the solution were determined by atomic absorption spectrometry. Metals biosorption results were analyzed using the mathematical Langmuir isotherm model. The model gives the maximum metal biosorption value ($q_{max:}$ mg metals/g of biomass) and the affinity value of the biomass to the metal (b) (Volesky and Holan, 1995).

Silver and Gold Nanoparticle Formations

Silver and gold nanoparticles formation was determined in a bacterial strain OC4. An Eppendorf tube was supplemented with 2 mg of bacterial biomass with $HAuCl_4$ and $AgNO_3$, respectively, were reacted for 72 hours in an Eppendorf AG shaker, 800 rpm, 31° C. At the end of the reaction time, the biomass was separated from the metal solution by centrifugation 10 000 rpm,15 min. Cellular samples for the TEM microscope (transmission electron microscope) were prepared using established methodologies (Gericke and Pinches, 2006; Sastry *et al.*, 2003; Klaus, *et al.*, 1999). Microscope observation was done in a JEOL 1210 TEM, 80 kV.

To determine the 16S rRNA gene sequence of the strain, cells were lysed according to The 16S rDNA fragment was amplified by PCR using the following universal primers: forward, 59-AGAGTTTGATCATGGCTCGA-39; and reverse, 59-GGCTACCTTGTTACGACTT-39 (positions 1510–1492). The sequence of the amplified 16S rDNA fragment (1400 bp) was analyzed using the Codon Code Aligner software (V.3.6.1) and compared with the National Center for Biotechnology Information (NCBI) database.

RESULTS

Thirty bacterial strains, mostly Gram negative, and 7 fungal strains were isolated. Strains were resistant to different concentrations and different metals. Two bacterial strains (Zac4 and OC4) and one fungi *Penicillium spp.* were selected. Strain Zac4 corresponded to Gram-negative bacillus, which form irregular, cream-colored colonies, of viscous consistency and soft brilliant elevated surface. Compared to the NCBI GenBank sequences, the strain's sequence showed 99% homology to *Delftia tsuruhantensis*. The strain OC4 was a Gram negative bacillus was identified *as Microbacterium oxydans* with 100% homology

The most silver resistant microorganisms were bacteria *Microbacterium oxydans*, which grew in 5.0 mM $AgNO_3$, bacteria NC1 (Gram positive), which grew in 0.5 mM $AgNO_3$, and fungal strains *Penicillium spp.* grew in 0.1 mM $AgNO_3$. *Microbacterium oxydans* had the highest specific silver biosorption (mg Ag/g of biomass), but exhibited low affinity for silver ions (biosorption of silver in low concentrations). Fungal strains had the highest specific biosorption of gold (mg Au/g of biomass) (Table 3).

Table 3. Biosorption of gold and silver by microbial biomass.

Metal	Microbial strain	
Silver	*Penicillium spp.*	Bacteria OC4
r^2	0.96	0.95
q max	3.3 mg Ag/g	68 mg Ag/g
Affinity (*b*)	0.06	0.01
Gold	*Penicillium spp.*	Bacteria OC4
r^2	0.98	0.95
q max	91 mg Au/g	26 mg Au/g
Affinity (*b*)	0.01	0.45

M. oxydans in presence of $CoCl_2$ or $Cucl_2$ presented a MIC value of 1 mM for both metals, in solid medium. No nanopraticles of Co nor Cu were formed with this strain. Nevertheless, gold and silver nanoparticles were observed; bacterial cells reacted with the $HAuCl_4$ solution, leading to the precipitation of gold ions and the formation of nanoparticles. Gold nanoparticles were circle- and triangle-shaped, with sizes ranging from 10 to 100 nm and an average size of 50 nm (Fig. 4).

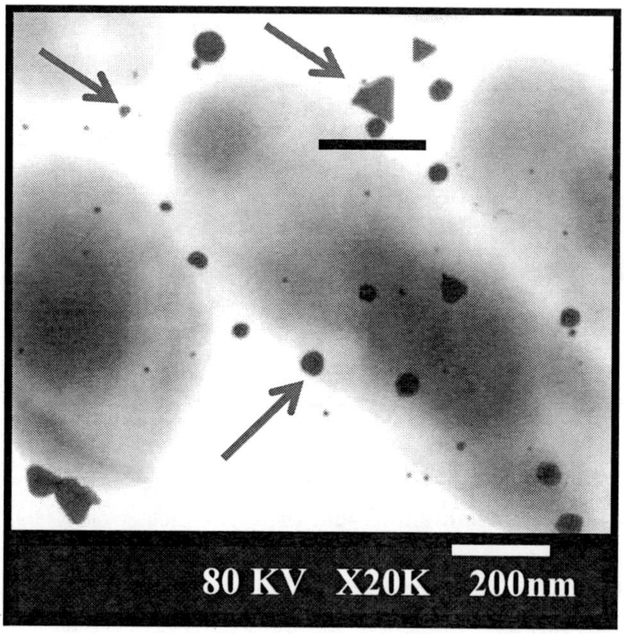

Figure 4. Gold nanoparticles in bacterial strain *M. oxydans*. Microscope observation was done with a JEOL 1210 TEM, 80 kV.

Reaction of bacterial *M. oxydans* cells with the $AgN0_3$ solution led to the precipitation of silver ions and the formation of nanoparticles. Silver nanoparticles were circle-shaped, sizes ranged from 25 to 100 nm and the average size was of 50 nm (Fig. 5).

Although, it is generally acknowledged that eukaryotes are more sensitive to metal toxicity than prokaryotic microorganisms, there are reports where fungal strains have shown higher resistance compared to the resistance of bacterial strains (Pümpel and Schinner, 1986). In this study, bacterial strains, compared to fungal strains, exhibited higher resistance to silver ions. Taking into account the silver concentration in which microbes were able to grow, metal toxicity in fungal strains might be regulated through metallothionein enzyme production, whereas in bacterial strains this might be through resistance mechanisms such as active efflux, transport, enzymatic transformation, and precipitation. These mechanisms might be plasmid mediated (Valls and De Lorenz, 2002; Gadd, 1992). Silver ions resistance in bacterial NCl (Gram positive) is comparable to *E. coli* R1 and *Pseudomonas stutzeri* AG259 resistance (Gadd, *et al.,* 1989). Bacterial *M. oxydans* (Gram negative) exhibited the highest resistance compared to the previous microorganisms. Hence, resistance of silver ions by bacterial *M. oxydans* can be explained because Gram negative bacteria have a more complex cell wall and polysaccharides that can immobilize effectively metal ions (Patrick, 2004).

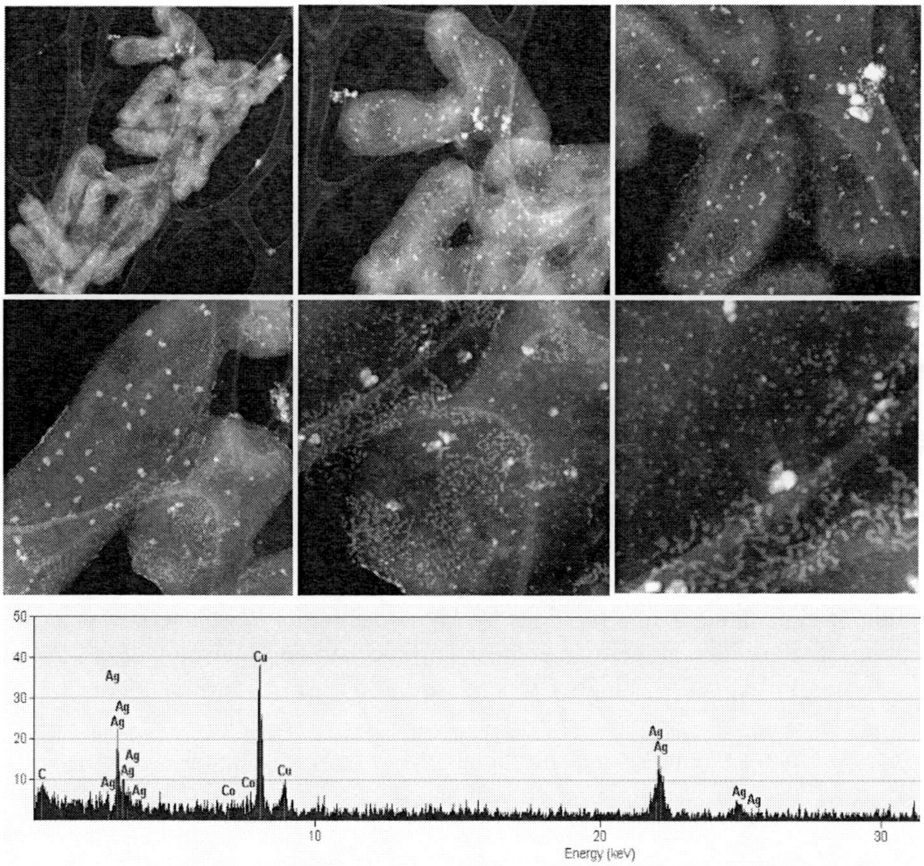

Figure 5. Silver nanoparticles in bacterial strain *M. oxydans and* Energy-dispersive X-ray spectroscopy. Analysis made at the *Centro de Nanociencias y Micro y Nanotecnologías del Instituto Politécnico Nacional* [Center of Nanoscience and Micro and Nanotechnologies National Polytechnic Institute], by Mayahuel Ortega, PhD.

Results of silver biosorption kinetics in *Penicillium spp.* biomass were similar to several biosorption reports in microbial biomass, not only for silver biosorption (Merroun, *et al.*, 2001), but also for biosorption of gold ions (Gericke and Pinches 2006; Tsuruta, 2004; Pethkar and Paknikar, 1998). These results supported our decision to set one hour as the standard time to develop biosorption experiments. Specific biosorption of silver ions by bacteria *M. oxydans* was higher compared to several microorganisms (Pümpel and Schinner, 1986; Merroun, *et al.*, 2001; Shakibaie, *et al.*, 1999) and it was comparable to *Escherichia coli* K12, which has been reported with high silver biosorption capability (67 mg Ag/g of biomass) (Ghandour, *et al.,* 1988). Since *M. oxydans* had the ability to grow in agar containing silver and the biomass exhibited high levels of silver biosorption, we consider that the biosorption capability is highly related to the silver resistance (Pümpel and Schinner, 1986; Merroun, *et al.*, 2001). Our results also suggest that this strain can be used for the recovery of silver from effluents containing at least 800 mg Ag/L. On the other hand, biosorption of gold in *M. oxydans* was lower than results reported for other microorganisms (Tsuruta, 2004).

Specific biosorption of silver by *Penicillium spp.* and *Mucor spp.* was lower than biosorption by bacterial strains (Merroun, *et al.*, 2001), nevertheless the biosorption was similar compared to other fungal strains (Pümpel and Schinner, 1986). *Penicillium spp.* exhibited high affinity for the biosorption of silver ions at low concentrations. These results suggest the potential use of the biomass for the recovery of silver ions from effluents with concentrations lower than 100 mg Ag/L.

TEM images of *M. oxydans* showing the formation of gold and silver nanoparticles suggest that in the uptake of ion metals there were not only ion biosorption but also precipitation mechanisms. Such precipitation mechanisms have been hypothesized by other studies. For instance, in *Bacillus subtilis* 168 inorganic phosphate acted as complexing agents for gold, and it was established that carboxyl groups were the main sites for the deposition of metals (Beveridge and Murray, 1980). Gold nanoparticle formation has been attributed also to intracellular reduction of gold ions (Nair and Pradeep, 2002), and to the complexation in organic compounds such as amino acids and cyanides (Karamushka and Gadd, 1999). In *Pseudomonas stutzeri* AG259 precipitation of silver has been attributed to resistance mechanisms such as accumulation of precipitates outside the citoplasmic membrane, efflux, biosorption (Klaus, *et al.,* 1999), and the immobilization via sulphate (Slawson, *et al.,* 1994). Biosorption of high amounts of silver ions by *Myxococcus xanthus* bacteria was attributed to precipitation of silver ions in the extracellular polysaccharides (Merroun, *et al.*, 2001). Such mechanisms might explain the high silver biosorption capacity we found in the bacterial strain *M. oxydans*.

Delftia tsuruhantensis presented the highest MIC in zinc solution (25 mM), zinc being the metal in which it showed the best growth and represented the enrichment from which it was isolated. The MICs observed with other metals were 6 mM, 3 mM, 1 mM, 1 mM, 0.5 mM for lead, selenium, copper, aluminum, and nickel, respectively.

The zinc MIC for *Delftia tsuruhatensis* was slightly higher than the reported MIC for *Bacillus circulans* (22 mM) (Yilmaz, 2003) and for the unidentified strain reported by Ansari and Malik (2006), (3 mM). These authors also report the MIC of lead for bacterial growth as 5.57 mM, which is similar to the value shown by *Delftia tsuruhatensis* (6 mM).

Sorption kinetics of Zn and Pb with *D. tsuruhatensis* show that equilibrium was reached at 40 and 20 min, respectively (Fig. 1). Puranik and Paknikar (1997) reported sorption kinetics for Zn and Pb by *Streptoverticillium cinnamoneum*, reaching equilibrium at 15 and 30 min, respectively. This would suggest sorption affinities of the biosorbent towards certain metallic ions.

It is assumed that as soon as the adsorption process proceeds, the adsorbed sorbate tends to be released and eventually the rates of adsorption and desorption adjust in a state of equilibrium. Thus, a static equilibrium is reached in a relatively longer period than a dynamic equilibrium (Mustafiz *et al.,* 2002). Once this equilibrium has been reached, adsorption becomes irreversible since very little metal will be released even if the solution remains under constant agitation (Tsezos, 1995).

The experimental data on Pb and Zn sorption fitted the Langmuir model better than the Freundlich model. The affinity coefficients in both models (b Langmuir and n Freundlich) were greater for Pb than for Zn. This means that the resistance to mass transfer in the process of adsorption is overcome at lower initial concentrations (Liu *et al.*, 2003). The maximal sorption capacity coefficients were also greater for Pb in both models (q_m Langmuir and K_F

Freundlich), indicating that *Delftia tsuruhatensis* possesses more sites for Pb than for Zn adsorption (Fig. 6) (Table 3).

Table 3 Parameters and adjustment of adsorption isotherms, Langmuir, Freundlich, and pseudo second order model, in zinc and lead adsorption by *Delftia tsuruhatensis* biomass.

Langmuir isotherms model				Freundlich isotherms model			Kinetics model		
	Q_m mg g^{-1}	b L mg^{-1}	R^2	K_F mg g^{-1}	n	R^2	Q_m mg g^{-1}	K_s^2	R^2
Pb(II)	44.8	0.147	0.988	18.66	5.54	0.803	28.8	0.018	0.99
Zn(II)	13.56	0.045	0.904	3.97	4.61	0.711	9.37	0.036	0.99

Two main variables have been suggested to explain the difference in adsorption for different metal ions by the same bioadsorbent, greater absorption of the metallic ion with lower charge/mass ratio (Esposito *et al.,* 2001) and smaller ionic radius (Ansari and Malik, 2006). However, these criteria do not explain the absorption of Zn and Pb by *Delftia tsuruhatensis*, since, in this case, greater adsorption of Pb than of Zn was achieved, i.e., greater adsorption of the element with larger ionic radius and lower charge/mass ratio (Table 4).

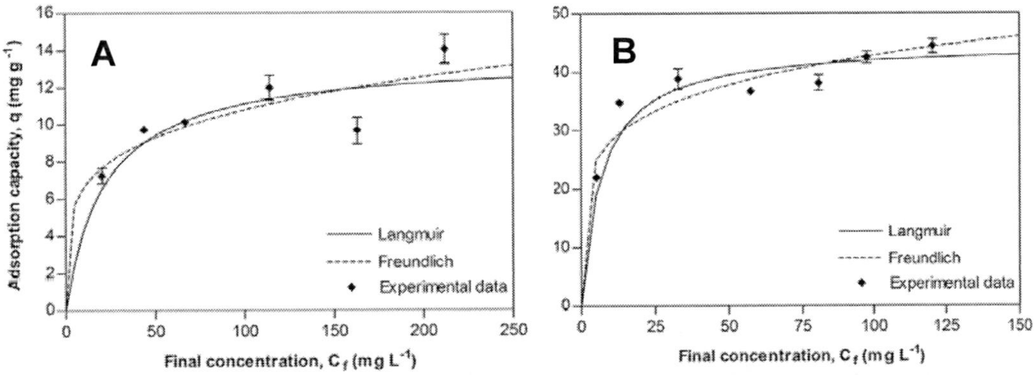

Figure 6. Metallic adsorption capacity (q) of Delftia tsuruhatensis at different final concentrations of zinc, A and lead, B. Adjustment with the Langmuir and Freundlich adsorption isotherms.

Table 4 . Atomic characteristics of Zn and Pb.

	Radius Covalent Å	Radius Ionic Å	Radius Atomic Å	Electronic Configuration	Atomic weight
Lead	1.47	1.2	1.75	[Xe]4f145d106s26p2	207.2
Zinc	1.31	0.74	1.38	[Ar] 3d104s2	65.39

A greater Pb capture could be attributed to its ability to form more stable compounds, compared to Zn, which is reflected in its covalent index (Can and Jianglong, 2007; Chandrasekhar *et al.,* 2003; Brady and Tobin, 1995).

Conclusions

The most silver resistant microorganisms were *Microbacterium oxydans* which grew in 5.0 mM $AgNO_3$, bacteria NC1 (Gram positive) which grew in 0.5 mM $AgNO_3$, and fungal strains *Penicillium spp.* and *Mucor spp.* which grew in 0.1 mM $AgNO_3$. *Penicillium spp.* biomass showed 93% of biosorption after 1 hour of reaction. *Microbacterium oxydans* had the highest specific silver biosorption (mg Ag/g of biomass), but exhibited low affinity for silver ions and could form Gold and silver nanoparticles. Gold nanoparticles were circle and triangle shapes, with sizes ranged from 10 to 100 nm and average size of 50.

The Zn and Pb sorption capacities shown by *Delftia tsuruhatensis* can be considered of median magnitude compared to other reported biosorbents, the strain had a maximal sorption capacity (mgg^{-1}) for Pb and Zn of 44.8 and 13.5; and affinity of 0.147 and 0.045, respectively. Although *Delftia tsuruhatensis* shows notable resistance to Zn and Pb, the sorption capacity for these metals did not prove superior that of other reported biological materials.

The use of microorganisms, whether bacteria or fungi, as biomass for bio-accumulation and valorization of metal residues and for the formation of nanoparticles, represents an important biotechnological potential. There are still technological and scientific challenges to make these technologies feasible and economically competitive with physical-chemical processes.

References

Ahmad A., Mukherjee P., Mandal D., Senapati S., Khan M. I., Kumar R. y Sastry M. (2002). Enzyme mediated extracellular synthesis of CdS nanoparticles by the fungus *Fusarium oxysporum*. *Journal of the American Chemical Society*. 124: 12108-12109.

Ansari, M.I. and Malik A. (2006). Biosorption of nikel and cadmium by metal resistant bacterial isolates from agricultural soil irrigated with industrial wastewater. *Bioresour. Technol*. 98: 3149-3153.

Asalil, G. A. M. (2004). El sector minero en México; diagnóstico,propectiva y estrategia. *El sector minero en México*. México: Centro de Estudios de Competitividad del ITAM.

Atlas R. and Bartha R. (2006). *Ecología microbiana y microbiología ambiental.* Pearson Addison Wesley. 4ª edición. Madrid.

Beveridge T. J. y Murray R. G. E. (1980). Sites of metal deposition in the cell wall of *Bacillus subtilis*. *Journal of Bacteriology*. 141: 876- 887

Brady, J.M.and Tobin, J.M. (1995). Binding of hard and soft metals ions to *Rhizopus arrhizus* biomass. *Enz. Microb. Technol*. 17: 791-796.

Bhattacharya B. and Gupta R. (2005). Nanotechnology and potential of microorganisms. *Critical Reviews in Biotechnology*. 25: 199-204.

Can, C., Jianlong, W. (2007). Influence of metal ionic characteristic on their biosorption capacity by *Saccharomyces cerevisiae*. *Appl. Microbiol. Biotechnol*. 74: 911-

Chandrasekhar, K., Kamala, C.T., Chary, N.S., Anjanuyulu, Y. (2003). Removal of heavy metal using plant biomass with reference to environmental control. *Int. J. Miner. Process*. 68: 37-45.

Chojnacka, K. (2010) Biosorption and bioaccumulation – the prospects for practical applications. *Environment International*. 36: 299–307.

Bansal V., Rautary D., Ahmad A., Sastry M. (2004). Biosynthesis of zirconia nanoparticles using the fungus *Fusarium oxysporum. Journal of Materials Chemistry.* 14: 3303 – 3305.

Bhattacharya B. and Gupta R. (2005). Nanotechnology and potential of microorganisms. *Critical Reviews in Biotechnology.* 25: 199-204.

Birch L. and Bachofen R. (1990). Complexing agents from microorganisms. *Experientia.* 46: 827-834.

Bosecker K. (2001). Microbial leaching in environmental clean-up programs. *Hydrometallurgy.* 59: 245-248.

Bruins M. R., Kapil S., Oehme F. W. (2000). Microbial resistance to metals in the environment. *Ecotoxicology Environment Safety.* 45: 198-207.

Csuros M. and Csuros C. (2002*). Environmental sampling and analysis for metals.* Lewis Publisher. 372 pp. Cunningham, S.D. y D.W. Ow. 1996.

Derfus A. M., Chan W. C. W., Bhatia S. N. (2004). Probing the cytotoxicity of semiconductor quantum dots. *Nano Letters.* 4: 11-18.

Eccles H. (1999). Treatment of metal-contaminated wastes: why select a biological process? *Trends in Biotechnology.* 17: 462-465.

Ehrlich H. L. (1997). Microbes and metals. *Applied Microbiology and Biotechnology.* 48: 687-692

EPA. 2007. Nanotechnology White Paper. U. S. Environmental Protection Agency Report EPA 100/B-07/001. Washington DC 20460, USA. *http://www.epa.gov/ OSA/nanotech.htm*

Esposito, A., Pagnanelli, F., Lodi, A., Solisio, C., Vegliò, F. (2001). Biosorption of heavy metals by *Sphaerotilus natans*: an equilibrium study at different pH and biomass concentrations. Hydrometallurgy. 60: 129-141.

Gadd G. M. (1990). Heavy metal accumulation by bacteria and other microorganisms. *Cellular and Molecular Life Sciences.* 46: 834-840.

Gericke M. and Pinches A. (2006). Biological synthesis of metal nanoparticles. *Hydrometallrgy.* 83: 32-140.

Ghandour W., Hubbard J. A., Deistung J., Hughes M. N. y Poole R. K. (1988). The uptake of silver ions by *Escherichia coli* KI2: toxic effects and interaction with copper ions. *Applied Microbiology Biotechnology.* 28: 559-565

INEGI 2009. La minería en México 2009. *Serie estadistícas sectoriales* Aguascalientes Instituto Nacional de Estadítica y Geografía

Karamushka V. I., Gadd G. M. (1999). Interaction of *Saccharomyces cerevisiae* with gold: toxicity and accumulation. *Biology of Metals.* 12: 289-294.

Katarzyna, C. (2010). Biosorption and bioaccumulation – the prospects for practical applications. *Environ. International***.** 299-307.

Kirchner C., Liedl T., Kudera S., Pellegrino T., Javier A. M., Gaub H. E., Stölzle S., Ferting N., Parak W. J. (2005). Cytotoxicity of colloidal CdSe and CdSe/ZnS nanoparticles. *Nano Letters.* 5: 331-338.

Klaus J. T., Joerger R., Olsson E. y Granqvist C. (1999). Silver based crystalline nanoparticles, microbially fabricated. *Proceedings of the National Academy of Sciences.* 96: 13611-13614.

Klaus J. T., Joerger R., Olsson E., Claes G. G. (2001). Bacteria as workers in the living factory: metal-accumulating bacteria and their potential for material science. *Trends in Biotechnology.* 19: 15-20.

Kowshick M., Vogel W., Urban J., Kulkarni K., Paknikar M. (2002). Microbial synthesis of semiconductor PbS nanocrystallites. *Advanced Materials.* 14: 815-818.

Labrenz M., Druschel G. K., Thomsen-Ebert T., Welch S. S., Bond P. L. y Banfield J. F. (2000). *Biomineralization of nanocrystalline sphalerite (ZnS) from dilute solutions by natural communities of sulfate-reducing bacteria.*

Larney, F. J., Cessa, A. J., Bullock, M. S. (1999). Herbicide transport on wind-eroded sediment. *Environ. Qual.,* 28: 1412–1421.

Liu Wen-Tso. (2006). Nanoparticles and their biological and environmental applications. *Journal of Bioscience and Bioengineering.* 102: 1-7.

Liu, Y., Xu, H., Yang, S.F., Tay, J.H. (2003). A general model for biosorption of Cd^{2+}, Cu^{2+} and Zn^{2+} by aerobic granules. *J. Biotechnol.* 102: 233-239.

Lovely, D. R., J. D. Coates. (1997). Bioremediation of metal contamination. *Current Opinion in Biotechnology* 285-289.

Mandal D., Bolander M. E., Mukhopadyay D., Sarkar G. y Mukherjee P. (2006). The use of microorganisms for the formation of metal nanoparticles and their application. *Applied Microbial Biotechnology.* 69: 485-492.

Merroun M. L., Omar N. B., Alonso E., Arias J. M., Gonzalez-Muñoz M. T. (2001). Silver sorption to *Myxococcus xanthus* biomass. *Geomicrobiology Journal.* 18:183-192.

Merroun M. L., Omar N. B., Alonso E., Arias J. M., Gonzalez-Muñoz M. T. (2001). Silver sorption to *Myxococcus xanthus* biomass. *Geomicrobiology Journal.* 18:183-192.

Mukherjee P., Ahmad A., Mandal. D., Senapati S., Sainkar S. R., Khan M. I., Parishcha R., Ajayakumar M. A., Kumar R. y Sastry M. (2001). Fungus-Mediated synthesis of silver nanoparticles and their inmobilization in the mycelial matrix: a novel biological approach to nanoparticle synthesis. *Nano Letters.* 1: 515-519.

Nair B., Pradeep T. (2002). Coalescence of nanoclusters and formation of submicron crystallites assisted by *Lactobacillus strains. Cristal Growth & Design.* 2: 293-298.

Narayanan, k. B. & Sakthivel, N. (2011). Facile green synthesis of gold nanostructures by NADPH-dependent enzyme from the extract of Sclerotium rolfsii. *Colloids and Surfaces A: Physicochemical and Engineering Aspects.* 380: 156-161.

Nam J. M., Thaxton C. S., Mirkin C. A. (2003). Nanoparticle-based bio-bar codes for the ultrasensitive detection or proteins. *Science.* 301: 1884-1886.

Nicewarner-Peña S. R., Freeman R. G., Reiss B. D., He L. Peña D. J., Walton I. D., Cromer R., Keating C. D., Natan M. J. (2001). Submicrometer metallic barcodes. *Science.* 295: 137-141.

Nurmi J. T., Tratnyeck P. G., Sarathy V., Baer D. R., Amonette J. E., Pecher K., Wang C., Linehan J. C., Matson D. W., Penn R. L., Driesssen M. D. (2005). Characterization and properties of metallic iron nanoparticles: spectroscopy, electrochemistry, and kinetics. *Environmental Science Technology.* 39: 1221-1230.

Park S. J., Taton T. A. and Mirkin C. A. (2002). Array-based electrical detection of DNA with nanoparticles probes. *Science.* 295: 1503-1506.

Pathak S., Choi S. K., Arnheim N., Thompson M. E. (2001). Hydroxilated quantum dots as luminescent probes for in situ hybridization. *Journal of the American Chemical Society.* 123: 4103-4104.

Patrick K. J. Environmental microbiology. (2004). Principles and applications. Science publishers. Inc. USA. India. 204 p.

Pethkar A. V. and Paknikar K. M. (1998). Recovery of gold from solutions using *Cladosporium cladosporioides* biomass beads. *Journal of Biotechnology.* 63: 121-136.

Philip, D. (2010). Green synthesis of gold and silver nanoparticles using Hibiscus rosa sinensis. *Physica E.* 42: 1417-1424.

Puranik, P. and Paknikar, K.M., (1997). Biosorption of lead and zinc from solutions using *Streptoverticillium cinnamoneum* waste biomass. J. Biotechnol. 55, 113- 118.

Pümpel T. y Schinner F. (1986). Silver tolerance and silver accumulation of microorganisms from soil materials of a silver mine. *Applied and Environmental Microbiology.* 24: 244-247.

Roh Y., Lauf R. J., McMillan A. D., Zhang C. Rawn C. J., Bai J. Phelps T. J. (2001). Microbial synthesis and the characterization of metal-substituted magnetites. *Solid State Communications.* 118: 529-534.

Ross, J. A. M. (2002). Minería. *Elementos a considerar para integrar la base politica para la prevenciónde la contaminación del suelo y su remedición.* México: Secretaría de Medio Ambiente y Recursos Naturales.

Sastry M., Ahmad A., Khan M. I. y Kumar R. (2003). Biosynthesis of nanoparticles using fungi and actinomycete. *Current Science.* 85: 162-170.

Sánchez (1995). *Manejo de residuos sólidos en minería. Aspectos Geológicos.* UNESCO, 239-250.

Sepúlveda, T. (2005). *Suelos contaminados por metales y metaloides: muestreo y alternativas para su remediación,* México ISBN: 968-817-492-0, Instituto Nacional de Ecología.

Shakibaie M. R., Kapadnis B. P., Dhakephalker P. y Chopade B. A. (1999). Removal of silver from photographic wastewater effluent using *Acinetobacter baumannii* BL54. *Canadian Journal of Microbiology.* 45: 995-1000

Shan G. B., Xing J. M., Zhang H. Y., Liu H. Z. (2005). Biodesulfurization of dibenzothiophene by microbial cells coated with magnetite nanoparticles. *Applied and Environmental Microbiology.* 71: 4497-4502.

Shumate S. E. y Strandberg G. W. (1985). Accumulation of metals by microbial cells. En: Moo-Young M., Robinson C. N. y Howell J. A. (eds). *Comprehensive Biotechnology.* Vol. 4. Pergamon Press. New York. pp 235–247

Sweeney R. Y., Mao C., Gao X., Burt J. L., Belcher A. M., Georgiou G., Iverson B. L. (2004). Bacterial biosynthesis of cadmium sulfide nanocrystals. *Chemical Biology.* 11: 1553 – 1559.

Silver S. y Phung L. T. (1996). Bacterial heavy metal resistance: new surprises. *Annual Review of Microbiology.* 50: 753-789.

Taton T. A., Mirkin C. A., Letsinger R. L. (2000). Scanometric DNA array detection with nanoparticle probes. *Science.* 289: 1757-1760.

Terlien, M. T. (1998). The determination of statistical and deterministic hydrological landslide-triggering thresholds. *Environmental Geology,* 35, 124-130

Tsezos, M., Remoudaki, E., Angelatou, V., (1995). A systematic Study on Equlibrium and Kinetics of Biosorptive Accumulation. *The case of Ag and Ni. Int. Biodet.*

Tsuruta T. (2004). Biosorption and recycling of gold using various microorganisms. *Journal of General and Applied Microbiology.* 50: 221-228.

Tungittiplakorn W., Lion L. W., Cohen C., Kin J. Y. (2004). Engineered polymeric nanoparticles for soil remediation. *Environmental Science Technology.* 38: 1605-1610.

Tungittiplakorn W., Cohen C., Lion L. W. (2005). Engineered polymeric nanoparticles for bioremediation of hydrophobic contaminants. *Environmental Science Technology.* 39: 1354-1358.

Valls M. and De Lorenzo V. (2002). Exploiting the genetic and biochemical capacities of bacteria for the remediation of heavy metal pollution. *FEMS Microbiology Reviews.* 26: 327-338.

Volke S. T., Velasco T. J. and De la Rosa P. D. (2005*). Suelos contaminados por metales y metaloides: muestreo y alternativas para su remediación.* INE. México. pp. 144.

Volesky B. and Holan Z. R. (1995). Biosorption of heavy metals. *Biotechnology Progress.* 11: 235-250.

Wainwright M. (1995). *Introducción a la biotecnología de los hongos.* Acribia. Madrid.

Wu X., Liu H., Liu J., Haley K. N., Treadway J. A., Larson J. P., Ge N., Peale F., Bruchez M. P. (2002). Immunofluorescent labeling of cancer marker Her2 and other cellular targets with semiconductor quantum dots. *Nature Biotechnology.* 21: 41-46.

Yilmaz, I.E., (2003). Metal tolerance and biosorption capacity of *Bacillus circulans* strain EB1. *Res. Microbiol.* 154, 409-415.

Yon P., Rowsen N. A., Farr J. P. G., Harris J. R., Macaskie L. E. (2002). Bioreduction and biocrystallization of palladium by *Desulfovibrio desulfuricans* NCIMB 8307. *Biotechnology and Bioengineering.* 80: 369-379.

Zhang Q., Zhu L., Feng H. H., Ang S., Chau F. S., Liu W.-T. (2005). Microbial detection in microfluidic devices trough dual staining of quantum dots-labeled immunoassay and RNA hybridization. *Analytica Chimica Acta.* 556: 171-177.

Zhang, X., Yan, S., Tyagi, R. D. & Surampalli, R. Y. (2011). Synthesis of nanoparticles by microorganisms and their application in enhancing microbiological reaction rates. *Chemosphere.* 82: 489-494.

Zhao X. J., Hilliard L. R., Mechery S. J., Wang Y. P., Bagwe R. P., Jin S. G., Tan W. H. (2004). A rapid bioassay for single bacterial cell quantitation using bioconjugated nanoparticles. *Proceedings of the National Academy of Sciences USA.* 101: 15027-15032.

Zhu L., Ang S., Liu W.-T. (2004). Quantum dots as a novel immunofluorescent detection system for *Cryptosporidium parvum* and *Giardia lamblia. Applied and Environmental Microbiology.* 70: 597-598.

In: Biotechnology
Editors: L. G. Torres Bustillos and I. Garcia Peña
ISBN 978-1-62081-071-2

Chapter 6

MICROALGAE CULTURE AS FEEDSTOCK FOR BIOFUELS

Luis Fernández Linares
Bioprocess Department UPIBI-IPN, Mexico

SUMMARY

World population and energy consumption per capita are increasing every year; they are expected to double from 2000 to 2050. This energetic demand cannot be satisfied entirely by fossil oil. Renewable energies are needed for petroleum replacement, because fossil energetics contribute to global warming and are of limited availability. In recent years, technologies for renewable energy have been developed and implemented, such as wind, solar, hydropower, geothermal or biomass. The power sector can be supplied with these renewable sources, nevertheless, the transportation sector, which represents 28% of global energy consumption, strongly depends on renewable liquid fuels. Biodiesel and bioethanol are the two potential renewable fuels that have been considered as fossil fuels substitutes. Presently, biofuels are produced from common sugar and starch crops (bioethanol) and vegetable oils or animal fats (biodiesel). This feedstock production in large quantities is not sustainable. Microalgae, photosynthetic microorganisms that convert sunlight, water and carbon dioxide to biomass, represent a remarkable alternative to produce biofuels. The algal cultures have several advantages over plants, the grow extremely rapidly and many of them show high oil yields and productivities, algae does not compete with food or feed crops or farmland; they do not have high requirements for nutrients or water, algal cultures benefit small scale farmers and energy security, and contribute to mitigate climate change.

In this chapter, an overview of biofuels production from algae is given, emphasizing on the production of algal biomass as raw material. In addition, it presents part of the research that is being performed in the Professional Interdisciplinary Unit of Biotechnology, National Polytechnical Institute (IPN, for its initials in Spanish). Research focuses mainly on the development of algal biofuel technologies. Some results on the obtained algal strains and improvement of culture media are also presented.

Introduction

The problems of increasing fuels consumption, the effect on the environment from burning fossil fuels, and the reduction of proven oil reserves, which provide energy security for many countries, are currently widely known. Between 1985 and 2008, the total worldwide petroleum consumption increased from 73,628 to 84,455 million barrels daily (Figure 1), with an average annual increase of 1.38% (EIA, 2010). For 2030, a consumption of 118,368 million barrels per day is projected. However, the world proven oil reserves were estimated at 1,184.208 to 1,342.207 billion barrels (EIA, 2011), with a reserve-to-production (R/P) ratio of 42 years. Petroleum is a finite source and rapidly becoming more expensive (Figure 2), because of difficulties in its extraction and processing, and it has become scarcer and of lower quality.

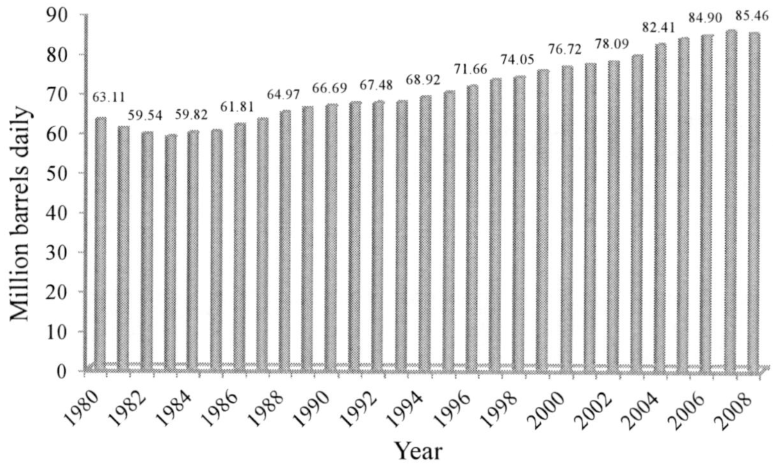

Figure 1. Total worldwide petroleum consumption (1980-2008).

Because anthropogenic emissions of carbon dioxide result primarily from the combustion of fossil fuels, world energy use continues to be at the center of the climate change debate. Petroleum-based products are one of the main causes of anthropogenic CO_2 emissions to the atmosphere. Today, the transportation sector worldwide is almost entirely dependent on petroleum-derived fuels. One-fifth of global CO_2 emissions are created by the transport sector (Balat and Balat, 2010). World energy-related carbon dioxide emissions are projected to grow from 29.7 billion metric tons in 2007 to 33.8 billion metric tons in 2020, and 42.4 billion metric tons in 2035.

The increase in anthropogenic greenhouse gas (GHG) emissions has become progressively more important to develop renewable energy sources, to minimize the dependency on fossil fuels, and to promote sustainability (Brennan and Owende, 2010; Singh *et al.*, 2011). According to the United Nations Development Program (UNDP), renewable energy sources are those energy forms that regenerate naturally, or whose rate of use does not affect the existence of the energy source. Besides, renewable energy can be defined as the energy produced by continuous and repetitive cycles that occur in the environment. Between the main renewable energies are geothermal, hydroelectricity, solar energy, tidal power, wave power, wind power, and biomass production, including biofuels. In 2008, about 19% of global final energy consumption came from renewable sources, with 13% coming from traditional biomass and only 0.6% from biofuels. Biomass is commonly used to generate both power and heat, generally through combustion, and some biomass can be converted to biofuels for transport purposes.

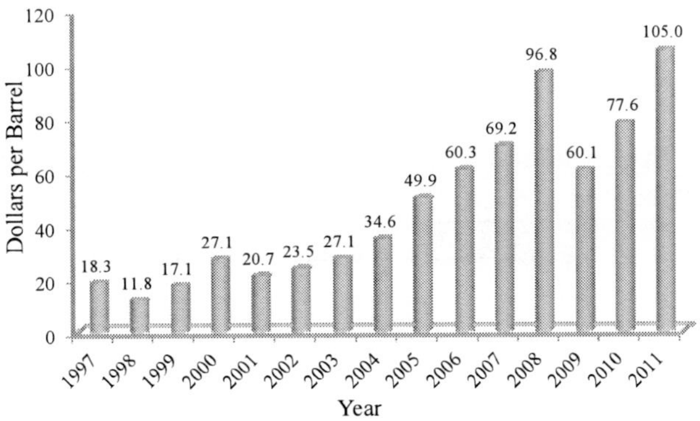

Figure 2. Annual total world spot price FOB weighted by estimated export volume (Dollars per Barrel).

BIOFUELS

Biofuels refer to renewable fuels from biological sources that can be used for heat, electricity and combustion engines. Biofuels are solid, as firewood and charcoal; liquid as bioethanol, methanol, and biodiesel; or gaseous fuels such as hydrogen, methane (Demirbas, 2008) and syngas (Syngas is a gas mixture that comprises carbon monoxide, carbon dioxide and hydrogen. The syngas is produced by gasification of a carbon-containing fuel to a gaseous product that has some heating value). Worldwide, while many countries look for alternatives to contribute to displace petroleum-derived products for economic and energy security reasons, support for biofuels is also being driven by the promise of biofuels being both environmentally and socially sustainable.

Bioethanol is a carbohydrate-derived ethyl alcohol (C_2H_5OH); a variety of common sugar crops can be used as feedstock, including sugar cane stalks, sugar beet tubers, and sweet sorghum stalks, that is, sugars that can be easily fermented into ethanol. A second group is constituted by the starch crops such as corn, wheat, and cassava that are hydrolyzed into sugars (saccharification), which can then be fermented into ethanol; nevertheless, this stage requires additional energy, adds to the cost of production, and produces a number of co-products in addition to ethanol. A third feedstock generation exists, obtained from waste and lignocellulosic materials, such as wood, straw or leaves, which are the major source of organic matter on the planet (Lange, 2008). Bioethanol production from lignocellulosic residues presents some challenges, optimization of the hydrolysis of waste (chemical or enzymatic), removal of toxic compounds produced during this stage, compounds that produce a limitation during sugar fermentation; fermentation of pentoses that are generated in addition to hexoses during the hydrolysis of lignocellulosic waste. Another problem is the collection of wastes, which are scattered and the transportation of lignocelluloses residues. Nevertheless, cellulosic ethanol has the potential to perform better in terms of energy balance, greenhouse gas emissions, and land-use requirements than starch-based biofuels.

In 2009, production of ethanol fuel reached an estimated of 478 million barrels. The United States, with starch from corn, and Brazil, which utilizes sugarcane, accounted for 88% of the global ethanol production in 2009; Followed by China, Canada, and France, in that order (Table 1) (WEO, 2010). Most of the increase in production occurred in the United States, with significant increases also in Canada, Germany, and France. Other countries that produce sizable volumes of ethanol fuel include Australia, Belgium, China, Colombia, India, Spain, and Thailand.

Table 1. The five principal bioethanol-producer countries (WEO, 2010).

Country	Production kb/d
United States	470
Brazil	287
China	24
Canada	13
France	9

Brazil's sugarcane yield averages about 82.4 tons/ha. The yield of bioethanol per hectare currently is around 6650 L/ha (Balat, 2011). About 520 Tg of dry corn is produced annually in the world, the main productive region (42%) with the highest corn yield occurs in North America, in which 7.2 Mg of dry corn per hectare is produced (Kim and Dale, 2011). In 2009, the United States' production of corn was 331 million tons and 104.1 million were used for bioethanol production. Bioethanol and bioethanol/gasoline blends have been employed as alternative transportation fuels. Bioethanol is the most commonly used biofuel for transportation worldwide. It has a high octane number (Balat, 2009), which permits both the rising of the compression ratio and having lower emissions (Çelik, 2008).

BIODIESEL

Biodiesel is the monoalkyl ester of long-chain fatty acids derived from renewable feedstocks, raw vegetable oils derived from soybean, palm oil, *Jatopha curca*, canola, sunflower, rapeseed oil, peanut oil, coconut, corn, as well as animal fats and waste oils (Kulkarni and Dalai, 2006; Felizardo *et al.*, 2006; Meher *et al.*, 2006; Karmakar *et al.*, 2010). These oils are converted to biodiesel principally by transesterification, using methanol or ethanol by alkali or catalytic acid, heterogeneous catalytic enzymes, or supercritical methanol transesterification (Karmakar *et al.,* 2010); Other possibilities are microemulsions and pyrolysis (Balat and Balat, 2010). Sometimes, vegetable oils can be used as non-transformed material, this is not recommended due to the risks of engine damage; because plant oils usually contain free fatty acids, phospholipids, sterols, water, odorants, and other impurities (Shakeel *et al.*, 2009). The biodiesel market can be broadly classified into four main end-use applications: transportation, non-road applications, marine and space heating biodiesel. Biodiesel is a particularly interesting bioenergetic because it is a non-toxic biofuel, it produces low pollutant emissions (CO_2, sulfurs, hydrocarbons, aromatics, particles) during combustion, and it is a biodegradable material (Van Gerpen, 2005; Balat and Balat, 2010). Biodiesel can be used in existing diesel engines without modification, and can be blended in at any ratio with petroleum diesel (Meher *et al.*, 2006). Biodiesel has lubricating properties, is a safe material to transport and handling due to its high flash point (150°C) and low volatility (Demirbas, 2009). In addition, the existing infrastructure for storage and distribution of petroleum diesel can be used for biodiesel. The main disadvantages of biodiesel are its higher viscosity, lower energy content, higher cloud point and pour point, higher nitrogen oxide emission, lower engine speed and power, injector coking, engine compatibility, high price and higher engine wear (Balat and Balat, 2010).

There are minimal differences between biodiesel and diesel properties (Table 2). Viscosity is the most important property of diesel since it affects the operation of the fuel injection equipment. Biodiesel has a viscosity close to that of diesel fuels; nevertheless, viscosity of biodiesel is higher than that of diesel (Table 2), this difference is reduced when it is blended with fossil diesel.

The European Union is the world leader in biodiesel industry, with a production of 8.812 million liters in 2009, and it is expected to remain as such in the next decade. Emerging countries, such as Malaysia, China, Brazil, Colombia, Argentina, and Indonesia, are developing promising technologies in biodiesel production (Li *et al.*, 2008a). World production of this biofuel will show a substantial increase with a market of 168,206 million liters by 2016 (Li *et al.*, 2008a; Timilsina *et al.*, 2010).

Table 2. ASTM standards of maximum allowed quantities in diesel and biodiesel (Balat and Balat, 2010)

Property	Diesel	Biodiesel
Standard	ASTM D975	ASTM D6751
Carbon (wt.%)	87	77
Cetane number	40–55	48–60
Cloud point (K)	258–278	270–285
Composition	C_{10}–C_{21}	C_{12}–C_{22}
Flash point (K)	333–353	373–443
Hydrogen (wt.%)	13	12
Kin. viscosity (mm^2/s) at 313 K	1.9–4.1	1.9–6.0
Oxygen (wt.%)	0	11
Pour point (K)	238–258	258–289
Specific gravity (g/mL)	0.85	0.88
Sulfur (wt.%)	0.05	0.05
Water (vol. %)	0.05	0.05

Despite the tendency of consumption and production of biodiesel, the reality is that biodiesel strongly depends on subsidies and government plans in order to be an economically competitive fuel. More than half of the production costs are dependent on the price of raw materials, so the profitability of conventional biodiesel production is also sensitive to feedstock prices (Bhattacharyya and Timilsina, 2010).

Biodiesel from oil crops and bioethanol from sugarcane are being produced in increasing amounts as renewable biofuels, but their production in large quantities is not sustainable yet. Biofuels production have been linked to rising food prices, questioning their ability to displace fossil energy, and their potential contribution to monoculture and deforestation has been analyzed (Searchinger *et al.,* 2008; Fargione *et al.*, 2008). If automotive fossil fuels are to be replaced with biofuels, new feedstock such as algae will be likely required (Chisti, 2007). Third generation biofuels from algal cells grown on non-arable land is the obvious answer to the food-fuel competition (Singh *et al.*, 2011).

MICROALGAE

Microalgae are prokaryotic cyanobacteria (Cyanophyceae) or eukaryotic photosynthetic microorganisms that are the primary synthesizers of organic matter in aquatic environments. They have high surface area to volume ratios, enabling the rapid uptake of nutrients and carbon dioxide (CO_2) and a much faster cell growth rate than land-based plants. Eukaryotic microalgae are, for example, green algae (Chlorophyta) and diatoms (Bacillariophyta) (Li *et al.,* 2008; Li *et al.*, 2008b; Mata *et al.*, 2010). Microalgae are present in all existing earth ecosystems, aquatic and terrestrial, representing a big variety of species living in a wide range of environmental conditions. It is estimated that more than 50,000 species exist, but only around 30,000 have been studied and analyzed (Richmond, 2004).

Microalgae comprise a vast group of photosynthetic organisms that have been cultivated commercially, mainly by the pharmaceutical industry and protein-producers, but has recently gained attention as a potential source of biomass for biofuels. Microalgae can provide several different types of renewable biofuels, including both liquid biofuels, biodiesel (Roessler *et al.*, 1994; Sawayama *et al.*, 1995; Dunahay *et al.*, 1996; Sheehan *et al.*, 1998; Belarbi *et al.,* 2000; Banerjee *et al.*, 2002; Gavrilescu and Chisti, 2005; Chisti, 2007) and bioethanol (Razif *et al.*, 2010; Rojan *et al.*, 2011), and gas biofuels, as methane, that are produced by anaerobic digestion of the algal biomass and production of biohydrogen (Melis, 2002; Fedorov, *et al.*, 2005; Kapdan and Kargi, 2006).

Microalgae have high potentials in biofuels production compared with other available feedstocks. Microalgae grow extremely fast and many have high intracellular oil contents. They have much higher growth rates and productivity when compared to conventional forestry and agricultural crops, they promise a potentially high productivity per hectare. Microalgae commonly double their biomass within 24 h (Chisti, 2007). They are easy to cultivate, can grow with little or even no attention, using water unsuitable for human consumption as wastewaters or brine, and there is no requirement of fertile land. Therefore, the competition for arable soil is completely reduced. Otherwise, the production of biofuels from algae as feedstock implies additional advantages such as: reduction of emissions of particulate matter, CO, hydrocarbons, and SO_x; potential recycling of CO_2 and other nutrient waste streams (Darzins *et al.*, 2010; Wang *et al.*, 2008), reducing the GHG emissions; valorization of residues after oil extraction; the resulting algal biomass can be processed into ethanol, methane, used as organic fertilizer, or simply burned for energy cogeneration (Mata *et al.*, 2010). In Mexico, the ancient Aztecs collected algae of the genus *Spirulina* from the alkaline lake of Texcoco for food consumption. Cultivation of microalgae has been used as food (the main species *Chlorella* and *Spirulina*), for the extraction of β-carotene (*Dunaliella salina*), for the extraction of astaxanthin (*Haematococcus pluvialis*), and various other species used as feed in aquaculture (Borowitzka, 1999).

Some algae species are known to produce high levels of carbohydrates instead of lipids as reserve polymers. These species are ideal candidates for the production of bioethanol. The metabolism by which the microalgae produce sucrose is through the production of a complex enzyme sucrose-phosphate synthetase/phosphatase (SPS/SPP) (Porchia and Salerno, 1996). There are few studies about the genes involved in the sugar production by microalgae (Satoh, 1998; Hagemann, 1997). The results indicate that cyanobacteria synthesize sucrose with the same route as do plants and algae (Lunn, 2002).

The osmoregulation mechanisms response is an option to increase sugar production in algae, these mechanisms are induced through saline stress. Three categories of algae have been proposed, according to their response to salt stress: low tolerance to salinity (above 0.7 M NaCl), with production of sucrose and trehalose, medium tolerance to salinity (above 1.8 M NaCl) with production of glucosylglycerol, and with high tolerance to salinity (above 2.7 M NaCl) with production of amino compounds such as glycine or glutamate-betaine. There are some exceptions within these groups, for example, some isolated marine cyanobacteria have been found capable of producing sucrose or trehalose (Page-Sharp *et al.*, 1999). The metabolism by which the cyanobacteria, as well as other microorganisms, produce sucrose is accomplished through the complex enzyme sucrose-phosphate synthetase/phosphatase (Porchia and Salerno, 1993).

However, algae cultivation faces several challenges; cultivation of algae and extraction of the oil is currently expensive. The low lipid content and low biomass that can be achieved, which leads to the high cost of biodiesel from microalgal oils, are the bottleneck for industrial production (GuanHua *et al.*, 2010). Production cost estimates for the raw oil vary between USD 0.75/L to more than USD 5.00/L (Darzins *et al.*, 2010). Commercially-viable production of biofuel from algae will depend on effective strategies to produce high volumes of biomass and oil at low-cost. The challenges to achieve this are:

1. Selection of wild microalgal species or genetic engineering of microalgae with (Amaro *et al.*, 2011; Darzins *et al.*, 2010):
    a) High growth rate and excellent biomass and oil productivity,
    b) Ability to grow under certain environmental conditions;
    c) Amount and type of fat;
    d) Weak response to environmental changes (pH, light, nutrients);
    e) Preference to nutrients and rate of incorporation especially CO_2, nitrogen, and phosphorus;
2. Possibility to obtain (valorization) value-added compounds;
3. Nutrient profile manipulation
    a) Medium composition and culture conditions
    b) Nitrogen deprivation
    c) Iron availability
4. Light supply;
5. Reactor design;
6. Scaling up production;
7. Availability of locations with sufficient sunshine and water

For the viability of biodiesel production from algae, it is also important to consider downstream manipulation, which include, cell harvesting, cell disruption, drying of biomass, oil extraction and oil transesterification (Amaro *et al.*, 2011). The principal factor to produce biodiesel successfully is finding cheaper oleaginous materials; improving transesterification technologies is another key to be considered to biodiesel production success. Biomass is easily recovered from the broth by filtration, centrifugation (Molina *et al.*, 2003); nevertheless, the cost of harvesting can be a significant proportion of the total algal production costs. This is because obtaining the algae biomass from the relatively diluted culture broths requires processing large volumes of water. Harvesting processes could be flocculation: chemical flocculation with inorganic or polyelectrolytes, or natural flocculation with organic polymers; bioflocculation (sedimentation); electroflocculation; dissolved air floatation, centrifugation, filtration, ultrafiltration and other means.

Research is mainly oriented to finding strains with high yield and high oil content or to genetic modification and to algal culture, including photobioreactors. The economic success of biodiesel production from algae should take into account the valorization of residual algal cells mass and recycling of wastewater and growth nutrients, as well as CO_2 capture; producing 100 t of algal biomass fixes roughly 183 t of carbon dioxide (Chisti, 2007). Residual biomass can be used as additional energy or as chemical products. Protein concentrations in algae vary from 15 to 71% (Spolaore *et al.*, 2006; Becker, 2007). Algal protein residues could be used as human food or supplements, livestock feed, and for aquaculture. Food additives, pigments, fibers, enzymes, sugars, fats, amino acids, vitamins, antioxidants, cosmetic products, oils, etc., are also being discussed as potential high value products from algae.

The components and contents of lipids in microalgal cells vary from species to species. The lipids are classified into neutral lipids and polar lipids. Triglycerides as neutral lipids are the main materials in the production of biodiesel. The synthesis routes of triglycerides in microalgae may consist of the following three steps: (a) formation of acetyl coenzyme A (acetyl-coA) in the cytoplasm, (b) the elongation and desaturation of carbon chain of fatty acids, and (c) the biosynthesis of triglycerides in microalgae.

The oil transesterification reaction is very simple; three alcohol molecules react with a triglyceride, under alkali, acid, or enzyme catalyzing processes (Figure 3) (Canakci and Vangerpen, 1999)

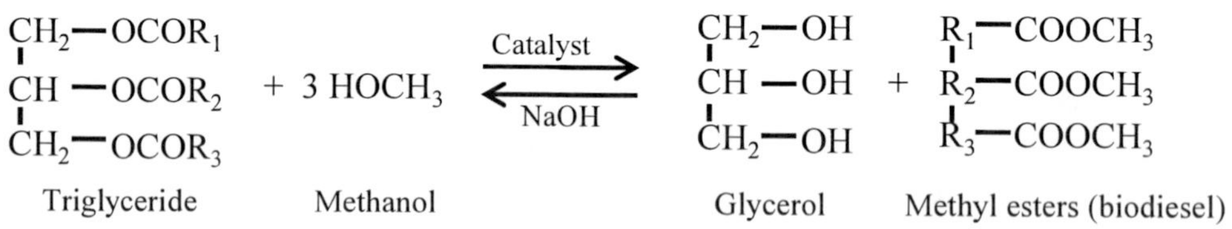

Figure 3. Transesterification of oil to biodiesel. R = hydrocarbon chains.

An excess of methanol is used to force the reaction to favor the right side of the equation. The excess methanol is later recovered and reused. The main used alcohol is methanol, but also ethanol, propanol, butanol, and amyl alcohol. Alkali-catalyzed transesterification of triglycerides is about 4000 times faster than the acid-catalyzed reaction (Fukuda, *et al.*, 2001).

PHOTOBIOREACTORS

One of the essential factors to consider for the increase in the production of biomass, algae oil yield, and cost reduction are the biomass production systems. The two methods of production more likely to be applied are open ponds or raceways ponds (Molina, 1999) and photobioreactors (Molina *et al.*, 1999; Sánchez Mirón *et al.,* 1999; Tredici, 1999). Large-scale production of microalgal biomass generally uses continuous culture, using daylight (Molina *et al.*, 1999).

Cultivation of algae in open ponds has been extensively studied (Boussiba, *et al.*, 1988; Tredici and Materassi, 1992; Hase, *et al.*, 2000). A raceway pond is made of a closed loop recirculation channel with typical dimensions: depth –20-30 cm; area, 100-250 hectare; production, 0.1 to 0.5 g dry weight per liter. Mixing and circulation are produced principally by paddlewheels but also by airlifts and pumps (Becker, 1994). One of the major advantages of open ponds is that they are less expensive and easier to construct than photobioreactors. A raceway can be built in concrete or compacted earth, and may be lined with white plastic or PVC to reduce seepage losses. Nevertheless, raceways show some limitations: significant evaporation losses, poor thermal or temperature control, poor utilization of CO_2 due to evaporation, or stripping and diffusion of CO_2 to the atmosphere, poor light utilization by the cells, contamination from unwanted algae and microorganisms, low biomass concentration due to poor mixing and optically dark zones. There are a number of mixed open pond systems such as: 1) inclined ponds, in which the culture flows from the top to the bottom of a sloping surface and needs to be pumped to the top of the slope (Richmond, 1999); 2) circular ponds with rotating agitator (central pivot) are the oldest large-scale microalgae cultivation systems and have also been used for wastewater treatment; normally, dimensions are 45 m in diameter and from 0.3 to 0.7 m depht. Circular ponds can be uncovered or covered by glass domes (Kanazawa, *et al.*, 1958); and 3) raceways.

A photobioreactor (PBR) is a reactor in which phototrophs (microbial, algal or plant cells) are grown or used to carry out a photobiological reaction under controlled conditions. PBRs, known as closed systems, permit cultivation of single species for longer duration. Microalgal broth is circulated through a degassing tank with either a mechanical pump or an airlift pump. Either horizontal or vertical array is possible. Illumination could be natural or artificial. There are different PBR designs, tubular photobioreactors, which consists of an array of straight transparent tubes, are the most used and are made of plastic or glass. Other PBR that have been used or assayed are: bubble column (Degen *et al.,* 2001), airlift column (Harker *et al.*, 1996; Kaewpintong *et al.*, 2007), stirred-tank (Ogbonna *et al.,* 1996, 1999), helical tubular (Watanabe and Hall, 1996; Ancién *et al.*, 2003), conical (Watanabe and Saiki, 1997), flat-plate photobioreactors, tubular, vertical column, internally-illuminated photobioreactors (Ugwu *et al.*, 2008).

PBR have some operational difficulties: growth of algae in tube walls (biofouling) or high biomass density blocking light; a high concentration of dissolved oxygen in combination with intense sunlight produces photooxidative damage to algal cells and can inhibit photosynthesis; problems in temperature control, PBR require cooling during daylight hours. Furthermore, temperature control at night is also useful (Chisti, 2007). Additionally, there is another disadvantage consisting of limited length and diameter of bioreactor tubes for scaling up.

For culture system selection, different parameters have to be taken into account. The closed system can guarantee increased volumetric yield and higher quality biomass, but due to its technical complexity, it requires higher energy input, and higher investment and operating costs (Table 3).

Table 3. Comparison of open vs. closed culture systems for microalgae growth (Chisti, 2007; Alabi *et al.,* 2009; Mata *et al.*, 2010).

	Culture systems for microalgae	
	Photobioreactors	**Open systems (Ponds)**
Area/volume ratio	High (20–200 m^{-1})	Low (5–10 m^{-1})
Biomass concentration	3–5 times higher	Lower
Biomass density	High	Low
Capital/operating costs ponds	Higher cost (3-10 times)	Lower cost
Comercial availability	Usually custom built	Readily available
Contamination control	Easy	Difficult
Contamination risk	Reduced	High
Control over parameters	Medium	Low
Construction cost per unit volume produced	High	Medium
Dilution rate (d^{-1})	0.384	0.25
Easiness of cleaning	Difficult	Very easy
Evaporation of growth medium	Low	High
Gas transfer control	High	Low
Hydrodynamic stress	Low–high	Very low
Investment	High	Low
Light utilization efficiency	High	Poor
Limiting factor for growth	Light	Light
Mixing	Uniform	Very poor
O_2 inhibition	Greater problem	Low porblem
Oil yield	High	Low
Operating costs	High	Medium
Operation costs	High	Low
Operation regime	Batch or semi-continuous	Batch or semi-continuous
Process control	Easy	Difficult
Scale-up	Difficult	Easy
Sterility	Possible	None
Surface area to volume ratio	Very High	High
Technology base	Under development	Readily available
Temperature control	More uniform temperature	Difficult
Volumetric porductivity	High (10 to 13 times more)	Low
Water losses	Low	High

The Professional Interdisciplinary Unit of Biotechnology, National Polytechnical Institute (IPN) together with other schools and centers of the same IPN are carrying out research and development focused on biofuels. The different subjects involved in biofuels production, such as strain selection, culture conditions, photobioreactors, biomass separation and drying, obtaining sugars, lipid extraction, and transesterification among others, including the study of the life cycle and industrial ecology of the process are being studied and integrated.

In this chapter, only strains selection for sugar and oil production, the effect of osmotic shock in the production of sugars, and the conditions of nitrogen limitation in the production of oils are discussed.

METHODOLOGY

Algae Isolation and Culture

Samples were taken from different sites in Mexico: the Lake of Texcoco, located in the Metropolitan area of Mexico City (salinity, 2000 ppm; pH, 8-10); the Lake of Guadalupe, in the State of Mexico (fresh water, pH, 7.2; salinity, 0.3 ppm); a lake in Tabasco, southeastern Mexico (pH, 7.2-7.8; salinity, 2000 ppm); Dam Madin (State of Mexico) (pH, 6.0-6.5; salinity, 177 ppm); Lagoon of Tamiahua (Veracruz) (pH, 7.5-8.1; salinity, 25 000-35 000 ppm); Lagoon of Tecuitlapa (Puebla) (pH, 7.2-8.9; salinity, 1652-2442 ppm). Samples were taken from microbial mats on rock surfaces and from water surface, 0.5 m depth, collected in bottles, and pH and salinity were determined through conductivity immediately upon arrival to the laboratory.

Samples were enriched through cultivation in different media, according to their salinity, either BBM (Bold's Basal Medium) (Ben-Amotz *et al.,* 1985), BG11, (Barsanti and Gualtieri, 2006), Guillard or Zarrouk (Zarrouk, 1966). The culture conditions were, in all experiments: fluorescent light in periods of 12:12 h (100 $\mu E/m^2s$), 25 °C, and aeration 4 vvm. After two replanting procedures, they were plated onto a solid medium. After 15 days of incubation, samples of each separate colony were taken from the plate and seeded again in a liquid medium culture under the same aforementioned conditions. Isolated strains were conserved in liquid medium with natural light at room temperature (Rippka *et al.*,1979). The strains were identified by optical microscopy using a LEICA-DMLB microscope. Each strain was observed and compared with a taxonomy library (Bourrely, 1966).

Strains

Most of the strains used in the work that is being done at the UPIBI have been isolated by the same research group; some species came from collections such as *Neochloris oleoabundans*, *Chlorella vulgaris, Nitzschia laevis, Scenedesmus obliquus, Nannochloropsis sp,* and *Botryococcus braunii.*

Growth Systems

Different systems were used according to the study. For lipid production, 500- mL culture flasks and flat cell culture dishes (Figure 4), whereas for sugars, 250-mL culture flasks were used. All PBR had three ports: aeration, vent, and sampling port and were inoculated (10% inoculum) and cultivated under the previously fixed conditions. Growth was monitored by absorbance (O.D. 600 nm), dry weight (dw) by filtering the culture through a 0.42 μm pre-weighed filter, dried at 80°C for 24 h, and weighed; and by direct counts with a Neubauer counting chamber.

Figure 4. Photrobioreactors used in experiments: flat cell culture dishes (500 mL) with 350-mL operating volume; and 500-mL glass culture flasks.

Biomass Composition

From the dry biomass, which was collected from culture broth and dried at 40°C in an oven, oil, sugars, and proteins were established. Lipids were extracted by Soxhlet in a chloroform–methanol–system according to Bligh and Dyer (1959). Using 150 mL of chloroform:methanol (2:1, v/v) for 8 h, all extractions were performed on 0.5 g of dry biomass. The extract was evaporated using a rotary evaporator to remove solvents. Then, each lipid fraction was transferred into a pre-weighed vial and residual solvents were evaporated under nitrogen and then weighed. Sugars in biomass, after lipid extraction, was determined by the Bubois method as described in intra- and extra-cellular sugars analysis. For biomass protein, 5 mg of dried biomass was hydrolyzed in 1 mL of 1 N NaOH, during 30 min in a water bath (bain-marie). Then, 0.2 mL was used to determine protein content by the Bradford method.

The effect of limiting nitrate in lipids accumulation in the cells was inferred by fluorimetry with Nile red stain. One milliliter of experimental culture was sampled and the biomass was adjusted to a concentration of 5 to 15 × 10^6 cells/Ml, then 2 μL of reagent Nile red, 20 μg/L (dissolved in acetone) were added (Wei *et al.*, 2009). After 20 min under darkness, stained cells were analyzed in a Trilogy Fluorometer. According to the pre-scan of excitation and emission characteristics of neutral lipid standards, the excitation and emission wavelengths of 486 nm and 570 nm were selected.

Effect of Intensity of Light, Nitrate Concentration, and Glucose on *Chlorella* Growth

Chlorella was grown in BG11 medium. To assess the effect of light intensity, 5 values were essayed 66, 60.5, 43.8, 35.5, and 25.8 μmol photons/m^2s. For glucose, 0, 0.5, 1, 5, and

10 g/L were added to the BG11 medium and cultivated as mentioned before (with 66 µmol photons/m^{2}s), Finally, for nitrate concentration effect, BG11 medium was formulated with four different concentrations of nitrate, three concentrations (0, 0.4, and 0.7 g/L) were lower than the original medium and the original concentration (1.5 g/L). Biomass was determined during growth.

Effect of Nitrate Limitation in Accumulation of Lipids

Neochloris oleoabundans was grown in BBM medium supplemented with five different concentrations of $NaNO_3$ (250, 200, 150, 100, and 50 mg/L), under the same conditions mentioned above. Each experiment was performed in duplicate. Representative samples were taken every 24 hours (maximum). Evaporated medium (loss by evaporation) was replenished with sterile distilled water. The analyzed variables were: biomass, lipids by spectrofluorometry, and concentration of nitrate in the medium, which was determined using a selective ion electrode, Oakton™.

Effect of Salt Stress on Growth and Osmotic Shock on Sugar Production

To assess the NaCl concentration effect on growth, isolated strains were grown in BG11 medium with 0, 0.2, 0.4, and 0.6 M NaCl under the same conditions mentioned above. The effect of osmotic shock on sugar production was completed by stressing the growth cultures (biomass at 1.5 OD_{600nm}) with the addition of NaCl, as concentrated solution, to achieve 0.1, 0.2, 0.3, 0.4, 0.5, and 0.6 M; subsequently shocked cultures were incubated during three additional days, growth and intra- and extracellular sugars were determined every 24 h during 3 days.

Intra- and Extracellular Sugars Analysis

Cell culture was centrifuged at 5000 rpm, for 20 min at 4°C. The liquid phase was separated and carbohydrates were determined. For intracellular sugars, biomass was lyophilized, the dry cells (0.01-0.05 g) were extracted with 10 mL of 70% ethanol and incubated at 65°C during 4 h, and then they were centrifuged at 10 000 rpm for 15 min. The supernatants were collected and dried at 40°C in an oven. The obtained material was re-dissolved in 10 mL of distilled water. Total carbohydrate content was established by the Dubois Method (Dubois *et al.*, 1956) and identified by HPLC as established by Müller *et al.*, (1994), the supernatants were collected and deionized by shaking with mixed-bed resin (Bio-Rad 501 X8), the solute was filtered through a microfilter (22 µm) prior to injection. A Varian 9002 HPLC with refractive index detector, using a column Phenomenex REYEX organic acid 300X7.8 75985 00H-0138-KO, mobile phase H_2O/H_2SO_4 at a flow rate 0.3 mL/min, was employed.

Results and Discussion

Microalgae Isolation and Growth (Yield)

Different types of photosynthetic microorganisms (prokaryote and eukaryote) were isolated and identified; they showed different growths and biomass productions from 0.3 to 2.5 g/L dw (Table 4).

Among the isolated strains, maximum growth (biomass) occurred in increasing order: *Lyngbya,* X1 Tam, *Nodularia, Synechocistis, Scenedesmus, Chlorella, Synechocystis, and Oscillatoria;* they showed a final biomass (dw) higher than 1 g/L (Table 5). A second group of strains: *Gloecapsa, Chlorella (T), Trebouxia, Nodularia, and Chrysosphaera* showed the lowest biomass dw (<1), i.e., 50% less than the *Lyngbya* strain.

These biomass yields can be considered as good for some strains; however, increasing the productivity of the systems should be searched for. Besides, these yields have been obtained in photobioreactors and in small volumes (fewer than 2 L), as it is escalated, less biomass is produced. Volumetric productivity (g/Ld) in photobioreactors and raceways are 1.535 and 0.117, respectively (Chisti, 2007). Other volumetric productivity may vary depending on the strain, between 0.02 and 7.7 g/Ld (Mata *et al.*, 2010).

Table 4. Identification of isolated algae and biomass produced (Yield)

Order and genus	Sampling site	Biomass dw (g/L)
Oscillattoriales *Lyngbya*	Texcoco Lake	2.56
X1 Tam	Tamiahua Lake	2.51
Nostocales *Nodularia*	Texcoco Lake	2.16
Chroococcales *Synechocystis*	Texcoco Lake	1.74
Oscillatoria spp	Tecuitlapa lagoon	1.74
Chlorococcales *Scenedesmus*	Canada	1.58
Synechococcus spp.	Tecuitlapa lagoon	1.45
Chlorella spp.	Guadalupe Lake	1.36
Chroococcales *Synechocystis*	Tabasco	1.18
Chroococcales	Tabasco	0.9
Chlorococcales *Trebouxia*	Guadalupe Lake	0.6
Chroococcales *Gloecapsa*	Tabasco	0.53
Synechocystis spp.	Madin Dam	0.52
Chysospheaceae *Chrysosphera*	Tabasco	0.32

Osmotic Shock Effect on Sugar Production in *Synechocustis* and *Synechococcus*

After three days of osmotic shock (NaCl from 0 to 1 M), increments in the production of intra- and extra-cellular sugars were observed (Figure 5). *Synechocustis* showed better production of intracellular sugar than *Synechococcus*; which increased, for both strains, with salinities of up to 0.6 M. Total intracellular sugar production was 3-fold higher after 3 days of osmotic shock using 0.6 M NaCl (Fig. 5A). Extracellular sugars were similar for both strains, and an increase, almost linear, in concentration was observed with salinity (Figure 5B).

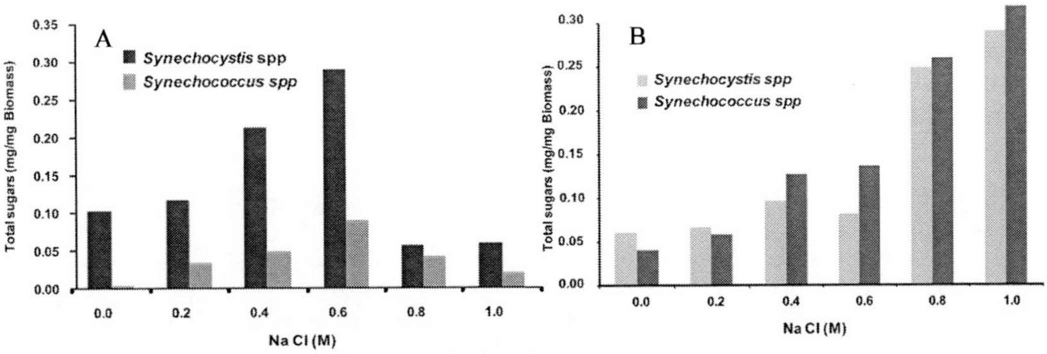

Figure 5. Osmotic shock effect on total intra-cellular, A; and extracellular, B; sugar production in *Synechocustis* and *Synechococcus* grown in BG11 medium.

Sugar Production by Isolated Eukaryotic Algae

In order to establish the sugar-production efficiency for each strain, the yield of total sugars production per gram of dried biomass per liter of algae culture was calculated for each strain. *Chlorella sp., Chysospheaceae, Chrysosphaera, Trebouxia, and Scenedesmus* were the most efficient producer *strains* (Table 5).

Table 5. Sugar efficiency production factor of eukaryotic strains

Order and Genus	Efficiency factor (mg of total sugar / L)
Chlorococcales *Chlorella*	29.11
Chlorococcales *Trebouxia*	6.22
Chysospheaceae *Chrysosphaera*	5.44
Chlorococcales *Scenedesmus*	5.37
Chlorococcales *Chlorella*	2.7
Oscillattoriales *Oscillatoria*	2.61
Chroococcales *Synechocystis*	1.81
Nostocales *Nodularia*	1.3
Oscillattoriales *Lyngbya*	1.02
Chroococcales *Synechocystis*	0.83
Chroococcales *Cistococcus*	0.7
Chroococcales	0.27

Given its high efficiency in sugar production, Chlorella sp. was selected for osmotic shock assays.

Osmotic Shock Effect on Sugar Production in *Chlorella*

When *Chlorella* was treated under osmotic shock, total sugar production was 18-fold higher after 24 h of osmotic shock at 0.4 M NaCl, (421 mg/g of biomass dw) (Figure 6). The maximum concentration of sugars was found at 24 h, after that there was a decrease, mainly of intracellular sugars. However, in all NaCl concentration and times after the shock, total sugar concentration was higher than the control. No intracellular sugars were found in media without NaCl.

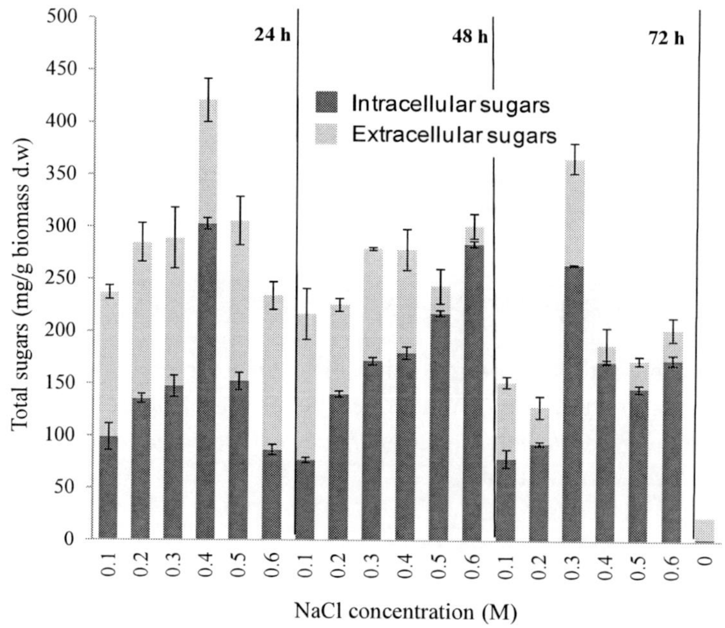

Figure 6. Effect of osmotic shock, after 24, 48, and 72 h, in total sugar production by *Chlorella sp* grown in BG medium.

Among the accumulated sugars, sucrose and trehalose were identified, which also increased with salinity. However, the concentration of these sugars was less than those quantified as total sugars. Under salt stress, *Scytonema* produced sucrose and trehalose (Page-Sharp *et al.*, 1999); perhaps because trehalose has shown superior protein stabilization capacity (Ben Amotz and Avron, 1983). *Chlorella,* under osmotic stress, produces proline, sucrose (Ben-Amotz and Grunwald, Ben Amotz and Avron, 1983; 1981; Oren, 2007;) and trehalose as osmolytes (Bremauntz *et al.*, 2011).

Effect of Intensity of Light, Nitrate Concentration, and Glucose on *Chlorella* Growth

Several studies have shown that the quantity and quality of lipids within the cell can vary as a result of changes in growth conditions (temperature and light intensity) or nutrient media characteristics (concentration of nitrogen, phosphates, and iron) (Illman, *et al.*, 2000; Liu, *et al.*, 2008).

Some species of algae depict heterotrophic metabolism. The presence of organic carbon source in the culture media can stimulate the growth of algae. In the case of *Chlorella*, the presence of sucrose in the medium, between 0.5 and 10 g/L, produced an increase in biomass. The maximum concentration of biomass was obtained in the medium supplemented with 5 g/L (Fig. 7A), getting up to three times the biomass achieved in the medium without glucose. Glycerol is a byproduct of the transesterification, which could be used as carbon source and to increase algae growth.

The presence of adequate concentrations of nitrogen in the environment is important for biomass production, but nitrogen limitation conditions are desirable for the production of lipids. In the case of *Chlorella*, it was determined whether the nitrate content in the original formulation of the culture medium (BG11, NaNO3, 1.5 g/L) was the most suitable. Lower concentrations were tested (0, 0.4 and 0.75 g/L), with half of the original concentration, greater biomass was obtained (Figure 7B). With 0.75 g/L, a cost reduction of substrate can be achieved and, with this concentration, algae will reach faster nitrogen limitation, due to the biological activity.

Increasing light intensity (μmol photons m−2 s−1) resulted in increased biomass, maximum biomass concentration was obtained with maximum intensity attained under bright fluorescent lamps (Figure 8). If the light intensity is increased further, the biomass could be augmented; however, there may be photoinhibition effects. In this work, the effect of light intensity on lipids has not been determined yet. In conditions without external CO2 supply, light intensity alone has not been found to be a major signal affecting lipids accumulation, revealing the absence of a direct regulatory link between light stress and lipids accumulation (Simionato, *et al.*, 2011).

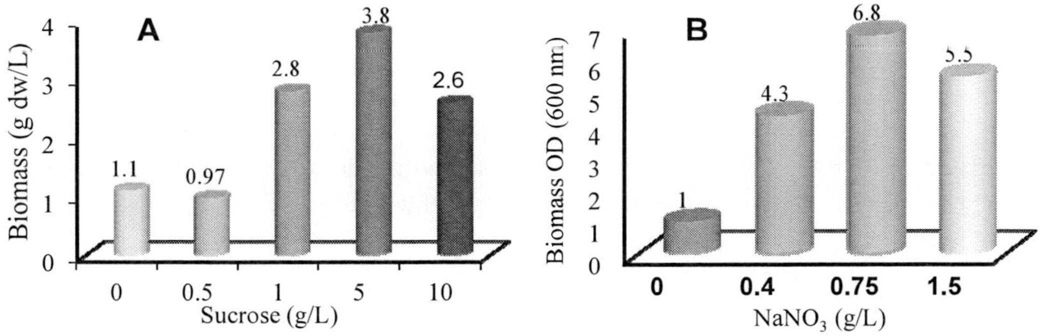

Figure 7. Sugar concentration, A; and nitrate concentration, B; effect on *Chlorella* growth.

As for all photosynthetic organisms, light is a major factor to be considered as it provides all the energy to support metabolism. Radiation, however, can also be dangerous because, if in excess, it may drive the formation of reactive oxygen species (ROS) and oxidative stress (Li *et al.*, 2009). When cells are exposed to illumination, one component of the photosynthetic apparatus, photosystem II (PSII), is continuously damaged and inactivated in a well known process called photoinhibition (Murata *et al.*, 2007). *C. vulgaris* could withstand a light intensity ranging from 150 to 350 μmol photons m^{-2} s^{-1}. Further increases in light intensity resulted in a decline of the electron transport rate (Bhola, *et al.,* 2011).

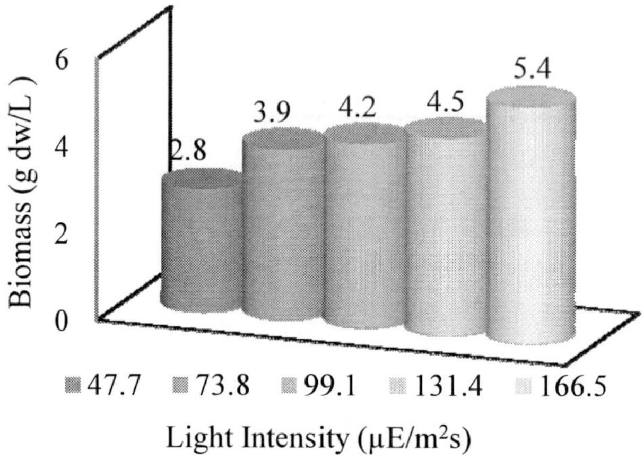

Figure 8. Light intensity effect on *Chlorella* growth.

Effect Of Nitrate Limitation in Accumulation of Lipids in *Neochloris Oleoabundans*

Recent studies have found that the lipid content of algae could be increased through changing cultivation conditions, such as CO_2 aeration fixation, temperature, salinity, and nutrient concentration. Sufficiency of nitrogen promotes high growth rates and low lipid content, whereas nitrogen deficiency reduces growth and results in high lipid content. Algae increase their lipid content from two to three times during the first 4 to 9 days of absence of nitrogen in the culture medium (Ben-Amotz *et al.,* 1985). Besides, the absence of nitrogen influences the lipid profile, increasing the proportion of triglycerides (Hu *et al.*, 2008).

In some studies of nitrogen limitation, first, the biomass is grown in suitable conditions and harvested; then the algae are left in nitrate-starved conditions. In this study, we tested different initial concentrations of nitrate. Its consumption and its effect on growth and lipid content in *N. oleoabundans* were established.

When the concentration of nitrate was increased from 50 to 250 ppm in culture media, growth and biomass of *Chlorella* augmented too. The $NaNO_3$ was consumed in parallel to cell growth; decreasing rapidly, reaching lowest concentrations between 50 and 100 hours of culture (2-20 ppm).

Nitrates up-take and nitrate concentration drop to low levels producing a decrease in the number of cells; however, the same biomass determined by optical density continued increasing over time (Figure 9). This indicates a decrease in cell division or growth, which is coupled to increased cell size, probably due to the accumulation of lipids. Increasing cell size of *N. oleoabundans* and reducing number of cells were observed by microscopy (Figure 10). Moreover, the cultures with low initial concentrations of nitrate (50 and 100 ppm) showed a color change from green to yellow-brown, indicating a change of metabolism (Figure 11).

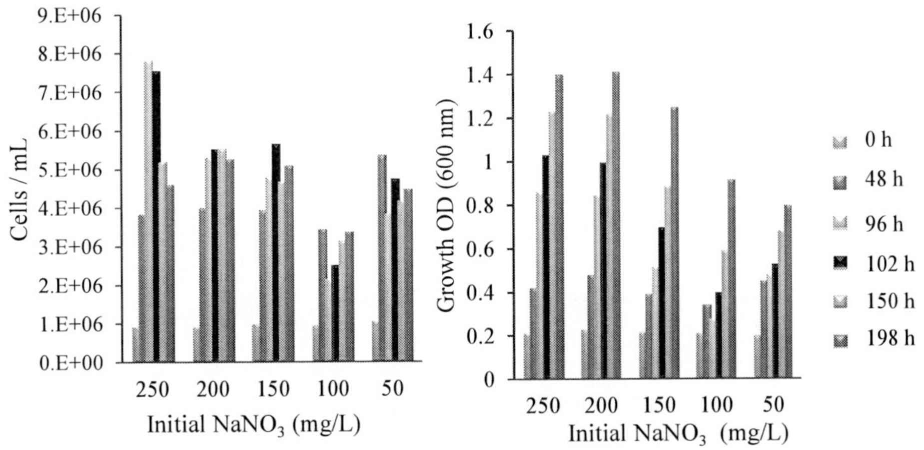

Figure 9. Effect of initial nitrate concentration on *N. Oleoabundans* growth.

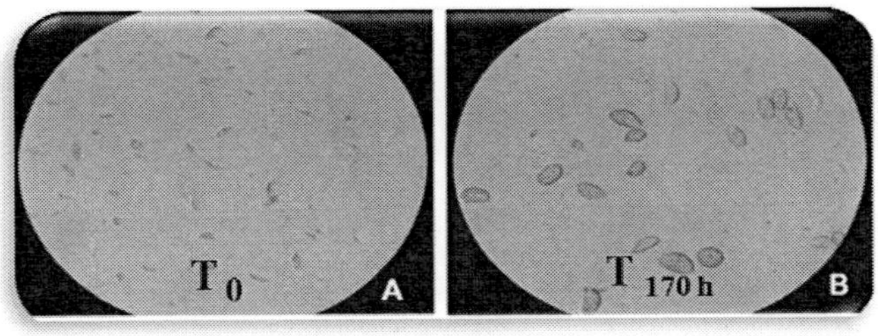

Figure 10. Effect of nitrate concentration on the number and size of cells of *N. oleoabundans*. Nitrate concentration: T_0= 250 ppm, A; T_{170} = 2-20 ppm, B.

Figure 11. Effect of initial nitrate concentration on color of cultures of *N. oleoabundans* grown in BG11 (50 to 250 ppm of $NaNO_3$).

The technique for fluorometric determination of lipids is an indirect indicator of lipid content in algae. In the cultures of *Neochloris oleoabundans*, nitrogen limitation occurred at approximately 96 hours (fourth day); at between 90 to 100 hours after nitrate limitation (eighth day), cell suspensions showed an increase in fluorescence (lipid content) in all tested initial concentrations of nitrate (Figure 12).

These results allow us to infer that, at the beginning, high nitrate conncentrations permitted the growth of *N. oleoabundans*. Subsequently, the consumption of $NaNO_3$ induced a limiting phase, addressing the metabolism to lipid accumulation. Ammonia, urea, and nitrate are often selected as the nitrogen source for the mass cultivation of microalgae (Li *et al.*, 2008). Urea and nitrate have been found to be better than ammonia for the growth and lipid accumulation in *Chlorella* sp. and *Neochloris oleoabundans* (Liu et al., 2008; Li *et al.*, 2008; Hsieh and Wu, 2009; Pruvost *et al.*, 2009). On the other hand, a 75% decrease of the nitrogen concentration in the medium, with respect to the optimal values for growth, increased the lipid fractions of *N. oculata* from 7.90 to 15.31% and of *C. vulgaris* from 5.90 to 16.41%, respectively (Converti, *et al.,* 2009). Lipid productivity can be increased by the knowledge and control of the culture, establishing times of growth, nutrient limitation and harvesting (Ben-Amotz *et al.,* 1985; Huntley and Redalje, 2007).

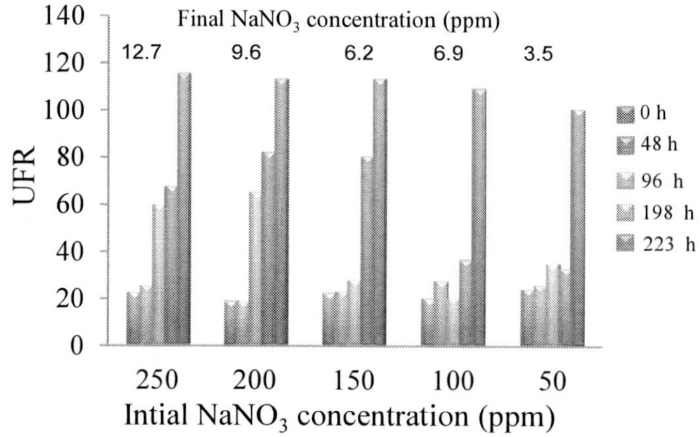

Figure 12. Effect of initial nitrate concentration and nitrate limitation on cellular lipid content that was determined by fluorometry with Nile red stain. Average of nitrate concentration after 96.5 h (2-20 ppm).

Algae have been used to obtain different products of interest for humans. Microalgae can provide several different types of renewable sources, whether liquid, gaseous or solid, fuels. Sugars and cellulose produced by algae are used to obtain bioethanol, lipids can be used for the synthesis of biodiesel, proteins as fodder, and biomass as another energy source, for direct combustion or methane and hydrogen production. Considering the results obtained in the present work regarding sugars and oil production by *Chlorella*, we estimated the annual production, considering a hectare of raceways surface of 0.3 m depth (6000 m^3), and a harvested biomass of 20% of the volume per day, under two different scenarios: 1) 50% of the maximum biomass production achieved (equivalent to 0.7 g/L) and 2) the maximum biomass (equivalent to 1.36 g/L), both with sugars yield without osmotic shock (equivalent to 0.02 kg sugar/kg biomass dw,) and after osmotic shock (equivalent to 0.25 kg sugar/kg dw). For lipids, both scenarios have the same oil yield obtained until now at the laboratory. Algal sugar production was compared to the sugar cane productivity in Colombia, where the best yield is 80 and 9.6 ton/ha of sugar cane and sugar, respectively (Córdova–Izquierdo *et al.*, 2010; Visser *et al.*, 2011), and lipids production was compared to palm oil from Africa (Carvalho *et al.*, 2006; Yusoff, 2006) (Table 6).

Lipids from *Chlorella*, proteins, and residual biomass after oil extraction (cake) were estimated in a very conservative way (based on our results), the percentage of yields were 10%, 20%, and 50%, respectively. The values of lipids with *Chlorella* are more than 5- (scenario 1) to 10- (scenario 2) folds higher than those obtained with African palm oil (*Elaeis guineensis*), and proteins are 10- to 20- times higher than those obtained with African palm oil in scenarios 1 and 2, respectively (Table 6).

As for the production of sugars, in the case of *Chlorella* in the most conservative condition (scenario 1), the production of sugars is lower (31%) than sugar cane. Nevertheless, considering maximum biomass, it is 13% higher than that of sugar cane. The two scenarios with osmotic shock are 7- and 14-fold higher than sugar cane, respectively. However, the dilution of sugars in the culture broth is a limiting factor that needs to be considered in contrast to sugar cane juice. Many technical challenges remain regarding scaling-up and adequate processes to make the use of algae feasible for the production of biofuels.

Table 6. Comparative scenarios production of sugars, lipids, and protein considering a hectare of raceways surface and 0.3 m depth; and a harvested biomass of 20% of the volume per day. Scenarios (1), 50% of the maximal biomass production achieved (0.7 g/L); (2) maximal biomass (1.36 g/L) both with sugars yield without osmotic shock (0.02 kg sugar/kg biomass dw) and after osmotic shock (0.25 kg sugar/kg biomass dw). Sugar cane from Colombia and palm oil form Africa.

Feedstock (scenario)	Biomass	Sugars	Sugars*	Protein	Biomass after extraction	Oil
	(Ton/Ha/year)					
Chlorella (1)	571	11	142.8*	114	286	57.1
Chlorella (2)	294	5.9	73.5*	59	147	29.4
Sugar cane	80	9.6	9.6	5.2[(d)]	26	-
Palm oil	22 [(a)]	2.6[(a)]	2.6[(a)]	3.5 [(b)]	5[(c)]	5.7[(a)]

*sugars yield after osmotic shock (equivalent to 0.25 kg sugar/kg dw).

These results show the potential for the use of algae to produce biofuels; however, there are still many efforts to be made before reaching a feasible and sustainable technological and economic development. In Mexico, there are no clear policies on the production of biofuels and support is limited. Nevertheless, diverse public and private educational institutions are performing research and development on this issue. In Mexico, thanks to its geographic location, agricultural and energetic situation, production of algae is highly feasible.

Chisti (2007) established the potential of microalgal biodiesel and demonstrated that it is technically feasible. Biodiesel is the only renewable biofuel that can potentially displace completely liquid fuels derived from petroleum. In addition, this author mentions that economics of producing microalgal biodiesel need to improve substantially to make it competitive with fossil diesel and that producing low-cost microalgal biodiesel requires primarily improvements to algal biology through genetic and metabolic engineering.

In the UPIBI and IPN, efforts to develop the production of liquid biofuels from algae are underway. Although, in this chapter, we mention only some efforts related to strains and culture conditions, other works in photobioreactors, harvesting systems, biomass drying, transesterification, and the genetic manipulation of strains are being developed.

Producing algal biodiesel requires large-scale cultivation and harvesting systems, with the challenge of reducing the cost. At a large scale, the algal growth conditions need to be controlled. Photobioreactors provide a controlled environment that can be tailored to the specific demands of highly productive microalgae, but a favorable assessment of the economics of production is necessary. Such processes are most economical when combined with the valorization of high value compounds, with CO_2 sequestration, with the utilization of wastewater and introducing industrial ecology. Cultivation of algae biomass in open ponds could be a very economic option. Research in advanced cultivation and downstream technologies will benefit the future development of microalgae for biodiesel production. Consequently, a significant investment in technological development and technical expertise is still needed before biofuels are economically feasible.

REFERENCES

Acién Fernández, F.G., Hall, D.O., Cañizares Guerrero, E., Rao, K., Molina Grima E. (2003). Outdoor production of *Phaeodactylum tricornutum* bioma in a helical reactor. *J. Biotech.* 103(2): 137-152.

Amaro, H., Guedes, A., Malcata., F.J. (2011). Advances and perspectives in using microalgae to produce biodiesel. *Appl. Energy*. 88: 3402–3410

Balat, M. (2009). Bioethanol as a vehicular fuel: a critical review. *Energy Source*. 31: 1242–55.

Balat, M. (2011). Production of bioethanol from lignocellulosic materials via the biochemical pathway: A review. *Energy Conversion and Management* 52: 858–875

Balat, M., Balat, H. (2010). Progress in biodiesel processing. *Appl. Energy*. 87: 1815–1835

Banerjee, A., Sharma, R., Chisti, Y., Banerjee, U. (2002). *Botryococcus braunii*: a renewable source of hydrocarbons and other chemicals. *Crit. Rev. Biotechnol*. 22: 245–79.

Barsanti, L., Gualtieri, P. (2006). *Algae – Anatomy, Biochemistry, and Biotechnology*. CRC Press, Taylor and Francis Group, Boca Raton, Florida. pp. 215–235

Becker, E. (2007). Microalgae as a source of protein. *Biotechnol. Advances*. 25: 207-210.

Becker, E. (1994). *Microalgae: Biotechnology and Microbiology*. Cambridge: Cambridge University Press.

Belarbi, E-H., Molina Grima, E., Chisti, Y. (2000). A process for high yield and scaleable recovery of high purity eicosapentaenoic acid esters from microalgae and fish oil. *Enzyme Microb. Technol*. 26: 516–529

Ben Amotz, A., Avron, M. (1983). Accumulation of metabolities by halotolerant algae and its industrial potential. *Ann. Rev. Microbiol*. 37: 95-119

Ben-Amotz A., Tornabene, T.G., Thomas, W.H. (1985). Chemical profile of selected species of microalgae with emphasis on lipids. *J. Phycol,* 21: 72-81.

Ben-Amotz, A., Grunwald, T. (1981). Osmoregulation in the halotolerant algae *Asteromonas gracilis. Plant Physiol*, 67: 613-616.

Bhola, I., Desikan, R., Kumari, S., Santosh, Subburamu, K., Sanniyasi, E., Bux, F. (2011). Effects of parameters affecting biomass yield and thermal behavior of *Chlorella vulgaris*. *J. Biosc. Bioengin*. 111(3): 377–382

Bligh, E., Dyer, W. (1959). A rapid method for total lipid extraction and purification. *Can. J. Bioche. Physiol.* 37: 911–917.

Borowitzka, M. (1999). Commercial production of microalgae: ponds, tanks, tubes and fermenters. J. *Biotechnol*. 70: 313–321

Bourrely, P. (1966). *Les Algues d'eau douce*, Vol I. Ed. N. Boubeé & Cie., París France.

Boussiba, S., Sandbank, E., Shelef, G., Cohen, Z., Vonshak, A., Ben-Amotz, A., Shoshana Arad, Richmond, A. Outdoor cultivation of the marine microalga Isochrysis galbana in open reactors. *Aquaculture. 72(3-4): 247-253.*

Bremauntz, M.P., Torres- Bustillos, L., Cañizares-Villanueva, R., Duran-Paramo, E., Fernández-Linares, L. (2011). Trehalose and sucrose osmolytes accumulated by algae as potential raw material for bioethanol. *Natural Resources. Accepted in press.*

Brennan, L., Owende, P. (2010). Biofuels from microalgae-a review of technologies for production, processing and extractions of biofuels and co-products. Renew. Sustain. *Energy Rev*. 14: 557–577

Canakci, M., Vangerpen, J. (1999). Biodiesel production via acid catalysis. *Trans ASAE*. 42: 1203–10.

Carvalho, L., Cabrita, A., Dewhurst, R., Vicente, T., Lopes, Z., Fonseca, A. (2006). Evaluation of palm kernel meal and corn distillers grains in corn silage-based diets for lactating dairy cows," *J. Dairy Sci*. (89): 2705-2715.

Çelik, M. (2008). Experimental determination of suitable ethanol–gasoline blend rate at high compression ratio for gasoline engine. *Appl Therm Eng*. 28: 396–404.

Chisti, Y. (2007). Biodiesel from microalgae. *Biotechnol. Adv*. 25: 294–306

Converti, A., Casazza, A., Ortiz, E., Perego, P., Del Borghi, M. (2009). Effect of temperature and nitrogen concentration on the growth and lipid content of *Nannochloropsis oculata* and *Chlorella vulgaris* for biodiesel production *Chem. Eng. Proc*. 48: 1146–1151.

Córdova–Izquierdo, A., Lara Torres, J.A., Amaro Gutiérrez, R., Peña Betancourt, S.D., Xolalpa Campos, V.M. (2010). "Composition of milk from Holstein cows supplemented with sugar cane-enriched comprehensive paperback saccharine,". Composición de la leche de vacas Holstein suplementadas con caña de azúcar integral y sacarina rústica enriquecida *Rev. vet*.1(1): 66–68

Darzins, A., Pienkos, P., Edye, L. (2010) *Current Status and Potential for Algal Biofuels Production, a report to IEA Bioenergy* Task 39. Transesterification. 43-44.

Degen, J., Uebele, A., Retze, A., Schmid-Staiger, U., Trosch, W. (2001). A novel airlift photobioreactor with baffles for improved light utilization through the flashing light effect. *J. Biotechnol*. 92: 89–94.

Demirbas, A. (2008). Biofuels sources, biofuel policy, biofuel economy and global biofuel projections. *Energy Convers Manage*. 49:2106–2116

Demirbas, A. (2009). Progress and recent trends in biodiesel fuels. *Energy Convers Manage*. 50:14–3

DuBois, M., Gilles, K., Hamilton, J., Rebers, P., Smith, F. (1956). Colorimetric Method for Determination of Sugars and Related Substances. *Anal. Chem*., *28* (3): 350–356

Dunahay, T., Jarvis E., Dais, S., Roessler, P. (1996). Manipulation of microalgal lipid production using genetic engineering. *Appl Biochem. Biotechnol*. 57/58: 223–231.

EIA (2010), *World Energy Projection System Plus* (2010)

EIA (2011), *The Annual Energy Outlook 2011*

Kaewpintong, K., Shotipruk, A., Powtongsook, S., Pavasant, P. (2007). Photoautotrophic high-density cultivation of vegetative cells of *Haematococcus pluvialis* in airlift bioreactor. *Bioresource Technol*. 98(2): 288-295

Fargione, J., Hill, J., Tilman, D., Polasky, S., Hawthrone, P. (2008) Land clearing and the carbon debt. *Sci Mag*. ;319: 1235e8.

Fedorov, A,. Kosourov, S., Ghirardi, M., Seibert, M. (2005). Continuous H_2 photoproduction by Chlamydomonas reinhardtii using a novel two-stage, sulfate-limited chemostat system. *Appl Biochem*. 121: 403–412.

Felizardo, P., Correia, M., Raposo, I., Mendes, J., Berkemeier, R., Bordado, J. (2006). Production of biodiesel from waste frying oil. *Waste Manag*. 26(5): 487–494.

Fukuda, H., Kondo, A., Noda, H. (2001). Biodiesel fuel production by transesterification of oils. *J. Biosci. Bioeng*. 92: 405–416.

Gavrilescu, M., Chisti, Y. (2005). Biotechnology – a sustainable alternative for chemical industry. *Biotechnol Adv*. 23: 471–99.

GuanHua, H., Feng, Ch., Dong, W., XueWu, Z., Gu, Ch. (2010). Biodiesel production by microalgal biotechnology. *Applied Energy* 87: 38–46.

Hagemann, M., Erdmann, H. (1997). In: *Environmental Stress*. Rai, A. Springer. New Delhi. 156-221

Harker, M., Tsavalos, A., Young, A.J. (1996). Autotrophic growth and carotenoid production of *Haematococcus pluvialis* in a 30 liter air-lift photobioreactor. *J. Ferment. Bioeng*. 82 (2): 113-118

Hase, R., Oikawa, H., Sasao, Ch., Morita, M., Watanabe, Y. (2000). Photosynthetic production of microalgal biomass in a raceway system under greenhouse conditions in Sendai city. J. *Biosci. Bioeng*. 89(2): 157-163

Hsieh, C.H., Wu, W. (2009). Cultivation of microalgae for oil production with a cultivation strategy of urea limitation. *Bioresource Technol. (100): 3921-3926*

Hu, Q., Sommerfeld, M., Jarvis, E., Ghirardi, M., Posewitz, M., Seibert, M., Darzins, A. (2008). Microalgal triacylglycerols as feedstocks for biofuel production: perspectives and advances. *Plant J.* 54: 621-639.

Huntley, M., Redalje, D. (2007). CO_2 mitigation and renewable oil from photosynthetic microbes: a new appraisal. *Mitig Adapt Strat Glob Change,* 12: 573-608.

Illman, A., Scragg, A., Shales, S. (2000). Increase in Chlorella strains calorific values when grown in low nitrogen medium. *Enz. Microbial Technol*. 27(8): 631-*635*

Kanazawa, Z., Fujita, C., Yuhara, T. and Sasa, T. (1958). Mass culture of unicellular algae using the "open pond circulation method". *J. Gen. Appl. Microbiol*. 4: 135-139.

Kapdan, I., Kargi, F. (2006). Bio-hydrogen production from waste materials. Enzyme. *Microb. Technol*. 38: 569–82.

Karmakar, A., Karmakar, S., Mukherjee, S. (2010). Properties of various plants and animals feedstocks for biodiesel production. *Bioresource Technol*. 101: 7201–7210.

Kim, S., Dale, B. (2011). Global potential bioethanol production from wasted crops and crop residues. *Energy Conversion and Management* 52: 858–875.

Kulkarni, M., Dalai, A. (2006). Waste cooking oil - an economical source for biodiesel: A review. *Ind Eng Chem Res*. 45: 2901–2913.

Lange, J.P. (2008). Lignocellulose Convertion: An Introduction to Chemistry, Porcess and Economics. In: Gabriel, C. and Rutger A. *Catalysis for Renewables. From Feedstock to Energy production*. Ed Wiley-VCH Verlag GmbH & Co. KGaA 1est Ed.

Li, Y., Horsman, M., Wu, N., Lan, C., Dubois-Calero, N. (2008a). Biofuels from microalgae. *Biotechnol. Prog.* 24(4): 815–820

Li, Y., Wang, B., Wu, N., Lan, C. (2008). Effects of nitrogen sources on cell growth and lipid production of Neochloris oleoabundans. *Appl. Microbiol. Biotechnol.* 81(4): 629–636.

Li, Z., Wakao, S., Fischer, B.B., Niyogi, K.K., (2009). Sensing and responding to excess light. *Annu. Rev. Plant Biol.* 60: 239–260.

Liu, Z-Y., Wang, G., Zhou, B. (2008). Effect of iron growth and lipid accumulation in *Chlorella vulgaris*. *Bioresource Technol*. 99(11): 4717-4722

Lunn, J. (2002). Evolution of Sucrose Synthesis. *Plant Physiol.* 128: 1490-1500.

Mata, T., Martins, A., Caetano, N. (2010). Microalgae for biodiesel production and other applications: A review. *Renew. Sust. Energy Rev*. 14: 217–232.

Meher, L., Vidya, S., Naik, S. (2006). Technical aspects of biodiesel production by transesterification-a review. *Renew. Sust. Energy Rev*.10: 248–68.

Melis, A. (2002). Green alga hydrogen production: progress, challenges and prospects. *Int J Hydrogen Energy*. 27:1217–28.

Molina, E., Belarbi, E-H., Acién Fernández, F., Robles Medina, A., Chisti, Y. (2003). Recovery of microalgal biomass and metabolites: process options and economics. *Biotechnol. Adv*. 20: 491–515.

Molina, E. (1999). Microalgae, mass culture methods. In: Flickinger MC, Drew SW, editors. *Encyclopedia of bioprocess technology: fermentation, biocatalysis and bioseparation,* vol. 3.Wiley; p. 1753–69.

Molina, E., Acién Fernández, F., García Camacho, F., Chisti, Y. (1999). Photobioreactors: light regime, mass transfer and scale up. *J. Biotechnol.* 70: 231–247.

Murata, N., Takahashi, S., Nishiyama, Y., Allakhverdiev, S.I (2007). Photoinhibition of photosystem II under environmental stress. *Biochim. Biophys. Acta.* 1767: 414– 421.

Müller, J., Staehelin, C., Mellor, R., Boller, T., Wiemken, A. (1994). Trehalose **and** trehalase in root **no**dules from various legumes. *Physiol. Plant*. 90: 86-92

Ogbonna, J.C., Soejima, T., Tanaka H. (1999). An integrated solar and artificial light system for internal illumination of photobioreactors. *J. Biotechnol*. 70(1-3): 289-297

Ogbonna, J.C., Yada, H., Masui, H., Tanaka, H., (1996). A novel internally illuminated stirred tank photobioreactor for large-scale cultivation of photosynthetic cells. *J. Ferment. Bioeng*. 82: 61–67.

Oren, A. (2007). Diversity of organic osmotic compounds and osmotic adaptation in cyanobacteria and algae. In: *Algae and cyanobacteria in extreme environments*. Springer, Seckback (ed). p. 639-655.

Page-Sharp, M., Behm, C., Smith, G. (1999). Involvement of compatible solutes trehalose and sucrose in the response to SALT stress of cyanobaterial *Scytonema* species isolated from desert solils," *Biochim. et Bioph. Acta*. 1472 : 519-528.

Porchia, A., Salerno, G. (1993). *Proc. Natl. Acad. Sci.* 93: 13600-13604.

Pruvost, J., Van Vooren, G., Cogne, G., Legrand, J. (2009). Investigation of biomass and lipids production with *Neochloris oleoabundans* in photobioreactor. *Bioresource Techn.* 100: 5988–5995.

Razif, H., Jason, W., Cherrington, T., Danquah, M. (2011). Microalgal biomass as a cellulosic fermentation feedstock for bioethanol production: a renewable and sustainable resource. *Energy Reviews* in press online.

Richmond, A. (1999). Physiological principles and modes of cultivation in mass production of photoautotrophic microalgae. In: *Chemical for Microalgae*. Cohen, Z., Taylor and Francis Phiiladelphia pp 353-386.

Richmond, A. (2004). *Handbook of microalgal culture: biotechnology and applied pshycology*. Blackwell Science Ltd.

Rippka, R., Deruelles, J., Waterbury, J., Herdman, M., Stanier, R. (1979). "Generic Assignments, Strain Histories and Properties of Pure Cultures of Cyanobacteria." *J. Gen. Microbiol.* 111: 1-61.

Rojan, P., John , G., Anisha, K., Madhavan Nampoothiri, A. (2011). Micro- and macroalgal biomass: A renewable source for bioethanol Pandey. *Bioresource Technol.* 102: 186–193.

Roessler, P., Brown, L., Dunahay, T., Heacox, D., Jarvis, E., Schneider, J., Talbot, S., Zeiler, K. (1994). Genetic-engineering approaches for enhanced production of biodiesel fuel from microalgae. *ACS Symp. Ser*. 566: 255–70.

Sánchez Mirón, A., Contreras Gómez, A., García Camacho, E., and Chisti, Y. (1999). Comparative evaluation of compact photobioteactors for large scale monoculture of microalgae. *J. Biotech*. 70: 249-270

Satoh, K., Murata. K. (1998). *Stress Response of Photosynthetic organism*. Elsevier Science, Amsterdam.

Sawayama, S., Inoue, S., Dote, Y., Yokoyama, S-Y. (1995). CO_2 fixation and oil production through microalgae. *Energy Convers Manag*. 36: 729–31.

Searchinger, T., Heimlich, R., Houghton, R., Dong, F., Elobeid, A., Fabiosa, J., Tokgoz, S., Hayes, D., Yu, T. (2008). Use of U.S. croplands for biofuels increases greenhouse gases through emissions from land-use change. *Science*. 319: 1238–1240

Shakeel, A., Rashmi, Z., Hussain, M.Z., Prasad, S., Banerjee, U.C. (2009). Prospects of biodiesel production from microalgae in India. Prospects of biodiesel production from microalgae in India. *Renew. Sust. Energy Rev*. 13: 2361–2372

Sheehan, J., Dunahay, T., Benemann, J., Roessler, P. (1998). A look back at the U.S. Department of Energy's Aquatic Species Program- biodiesel from algae. *National Renewable Energy Laboratory*, Golden, CO. Report NREL/TP-580–24190.

Simionato, D., Sforza, E., Corteggiani Carpinelli E., Bertucco, A., Giacometti, G., Morosinotto, T. (2011). Acclimation of Nannochloropsis gaditana to different illumination regimes: Effects on lipids accumulation. *Bioresource Technol.* 102: 6026–6032.

Singh, A., Singh Nigam, P., Murphy, J. (2011). Renewable fuels from algae: An answer to debatable land based fuels. *Bioresource Technol.* 102: 10–16.

Spolaore, P., Joannis-Cassan, C., Duran, E., Isambert, A. (2006). Commercial applications of microalgae. *J. Bioscien. Bioengin*. 101(2): 87-96

Tredici, M., Carlozzi. P., Chini Zittelli, G., Materassi, R. (1991). A vertical panel (VAP) for outdoor mass cultivation of microalgae and cyanobacteria. *Bioresource Technol*. 38: 153-196

Tredici, M., Materassi, R. (1992). From open ponds to vertical alveolar panels: the Italian experience in the development of reactors for the mass cultivation of phototrophic microorganisms. *J. Appl. Phyco*l. 4: 221–231.

Timilsina, G., Shrestha, A. (2011). How much hope should we have for biofuels?. *Energy*. 36(4): 2055-2069

Ugwu, C., Aoyagi, H., Uchiyama, H. (2008). Photobioreactors for mass cultivation of algae. *Bioresource Technol.* 99(10): 4021-4028.

Van Gerpen, J. (2005). Biodiesel processing and production. *Fuel Process Technol*. 86:1097-1107.

Visser, E., Filhi, D. Martins, M. Steward, B. (2011). Bioethanol production potential from Brazilian biodiesel co-products. *Biomass and Bioenergy*. 35: 489-494.

Wang, B., Li, Y., Wu, N., Lan, C. (2008). CO_2 bio-mitigation using microalgae. *Appl. Microbiol. Biotechnol*. 79(5): 707–718.

Watanabe, Y., Hall, D.O., 1996. Photosynthetic production of the filamentous cyanobacterium Spirulina platensis in a cone-shaped helical tubular photobioreactor. *Appl. Microbiol. Biotechnol.* 44, 693-698.

Watanabe, Y., Saiki, H. (1997). Development of a photobioreactor incorporating *Chlorella* sp. for removal of CO_2 in stack gas. *Energy Conversion and Management*. 38(1): S499-S503

Wei Chen, Chengwu Zhang, Lirong Song, Milton Sommerfeld, Qiang Hu. (2009). A high throughput Nile red method for quantitative measurement of neutral lipids in microalgae. *J. of Microbiological Methods*. 77: 41–47.

World Energy Outlook (2010). *International Energy Agency*. OECD, IEA, 2010

Yusoff, S. (2006). Renewable energy from palm oil innovation on effective utilization of waste. *J. Cleaner Prod*. 14(1): 87-93.

Zarrouk, C. (1966). Contribution a` l'etude d'une cyanophycée. Influence de divers facteurs physiques et chimiques sur la croissance et la photosynthese de *Spirulina maxima*. Doctoral Dissertation, *School of Sciences, University of Paris.*

In: Biotechnology
Editors: L. G. Torres Bustillos and I. Garcia Peña ISBN 978-1-62081-071-2

Chapter 7

METHANE AND BIO-HYDROGEN PRODUCTION FROM FRUIT AND VEGETABLE WASTE (FVW): FROM CH_4 AND H_2 TO ELECTRICITY

Ines García-Peña[1*], Edgar Salgado-Manjarrez[2], Juan Aranda-Barradas[2], Isaac Chairez[1], Prathap Parameswaran[3] and Rosa Krajmalnik-Brown[3]
[1]Bioprocesses Department, [2]Bioenginering Department, UPIBI-IPN, Mexico
[3]Swette Center for Environmental Biotechnology at Arizona State University, AZ, US

ABSTRACT

Bio-hydrogen and/or methane production from the organic fraction of solid waste, such as fruit and vegetable waste (FVW), constitutes an attractive and economical technology for energy production. The application of this process would allow for the simultaneous treatment of organic residues and energy production. In this chapter, data obtained to evaluate methane production from FVW are presented. Besides, the feasibility of bio-hydrogen production by using the anaerobic digester sludge as seed, and the culture conditions to enhance Bio-H_2 production in batch and a continuous system were established and are described. The effect of increasing initial concentration of FVW on H_2 production was evaluated in batch experiments and is also presented and discussed. Finally, the feasibility of producing H_2 from FVW and obtaining electricity by coupling proton exchange membrane fuel cells (PEMFC) was demonstrated.

INTRODUCTION

Mexico City is a mega city with a population of more than 20 million people in the whole metropolitan area. This population has many requirements in services, energy, food, and waste disposal. Around 30,000 tons of meal products must be commercialized daily to fulfill

* inesppu3@yahoo.com.mx, egarciap@ipn.mx

the food supplies. A tremendous amount of solid waste, more than 12,000 tons, is also produced daily (INEGI, 2009). Additionally, an increasing need in the production and use of alternative energy sources exist due to the limited supply of fossil fuels and their negative effects on the environment (Rittmann, 2008). The limited fossil fuel reserves, environmental degradation, and concerns about climatic changes are driving the search for a variety of biofuels and their production is currently the object of study by many researchers (Manish et al., 2007).

Food supply in Mexico City is possible due to the establishment of many markets around and within the city. However, the main market for food distribution in Mexico City, where the products are delivered directly from the producers, is the so named Central de Abasto (CEDA, for its initials in Spanish). CEDA is the second largest market in the world, receiving 30,000 tons of food products and producing 895 tons of organic solid waste each day; 84% of the total solid waste produced by CEDA is organic waste, and more than 50% of that is from the fruit and vegetable fraction, as shown in Table 1.

Table1. Waste distribution in CEDA: total, organic and inorganic fractions

	Total waste produced		Organic		Inorganic	
Areas (commercialized product)	Tons/day	(%)	Tons/day	(%)	Tons/day	(%)
Cereals	35	3.9	30	3.3	5	0.6
Fruits and vegetables	580	64.8	468	52.3	112	12.5
Flowers market	228	25.5	228	25.5	0	0
Processed food	17	1.9	15	1.7	2	0.2
Containers	11	1.2	0	0	11	1.2
Services	12	1.3	0	0	12	1.3
Jamaiquita Market	12	1.3	8	0.89	4	0.4
Total	895	100	754	83.7	141	16.2

Modified from the MS thesis by Elizabeth Silva Rodriguez.

Fruit and vegetable waste (FVW) is produced in large amounts in markets in many large cities (Mata-Alvarez et al., 1992; Misi and Forster, 2002; Bouallagui et al., 2003; Bouallagui et al., 2005) just as in Mexico City. Fruit and vegetable waste (FVW) could be a good candidate for energy production using biological processes because of its high carbohydrate content, relatively easy to degrade. Additionally, this feedstock is mechanically sorted at the source (due to the market's distribution in sections, where only a specific product is commercialized) or in central waste sorting plants (OFMSW) and large quantities are produced. This constitutes large amounts of cheap substrate for energy production. The application of an anaerobic digestion process for simultaneous waste treatment and renewable energy production from the organic fraction of these residues could be of interest (Bouallagui et al., 2005).

Early studies in the literature and during the first stage of the present project, the anaerobic digestion process was evaluated and extensively studied for its potential to produce methane (CH_4) from the FVW. However, CH_4 production has limitations, like the long time period to finally obtain CH_4, and the possible inhibition of the process due to the stricter conditions needed for methane producers to grow and exert a strong metabolic activity. The high biodegradability of FVW promotes the rapid production of volatile fatty acids (VFAs) resulting in a rapid decrease in pH, which in turn could inhibit methanogenic activity

(Bouallagui et al., 2003; Bouallagui et al., 2009). Some strategies have been reported to avoid acidification of the system, such as alkali addition, buffer solutions supplementation, and addition of co-substrates, these allow for a constant neutral pH that enhances the biogas and methane productivities. Data obtained in the present study showed that the co-digestion of FVW and meat residues results in an efficient digestion process, improving the methane yields obtained due to the positive synergistic effects of the mixed materials with complementary characteristics and the supply of missing nutrients by the co-substrate, as has been reported by other authors (Agdag and Sponza, 2005; Habiba et al., 2009; Bouallagui et al., 2009).

Recently, research activity has focused on the bio-hydrogen (Bio-H_2) production from renewable sources, biomass, such as FWV. Bio-H_2 is considered a sustainable energy supply because it could be used to reduce the current reliance on fossil fuels and become an environmentally friendly alternative to obtain energy (Rittmann, 2008). It has the highest energy content per weight unit of any known fuel (142 kJ/g) and its conversion to heat or power is simple and clean. The Bio-H_2 production process shows several technical, socio-economic, and environmental benefits. Bio-H_2 can be used either as the fuel for direct combustion in an internal combustion engine or as the fuel for a cell (Lin et al., 2007; Levin et al., 2004).

Fermentative hydrogen production research has been focused on determining appropriate inoculum sources, inoculum pretreatment methods, fermentable substrates, and reactor environmental conditions (Wang and Wan, 2009; Garcia-Peña et al., 2009).

Some research has reported the use of vegetable-based waste as a source for Bio-H_2 production. The interest in the use of this feedstock has increased due to the availability of high amounts of carbohydrates in the rich vegetable-based waste and the need for its treatment (Mohan et al., 2009). Some authors have reported the feasibility of H_2 production from the carbohydrate-rich wastes. Zhu et al., 2002, evaluated the Bio-H_2 production by anoxygenic phototrophic bacteria using tofu wastewater as substrate. Van Ginkel et al., 2005, and Yang et al., 2007, reported H_2 production by using different types of wastewater, such as food processing wastewater and cheese processing wastewater, respectively. Vijayaraghavan et al., 2007, studied the use of mixed fruit peel-waste as substrate for H_2 production. More recently, Mohan et al., 2009, reported the use of FVW as fermentative feedstock to obtain Bio-H_2 by a dark-fermentation process.

Bio-hydrogen production in batch and continuous glucose fermentation systems is obtained mainly by the activity of *Enterobacter* sp. and *Clostridium* sp species. It is well known that good yields in hydrogen production are obtained with *Clostridum* sp., theoretically, 4 mol H_2/mol glucose, but experimentally around 3 mol H_2/mol glucose are reported (Chen et al., 2005; Evyernie et al., 2007; Pan et al., 2008). Lower yields are obtained for some *Enterobacter* species (Kotay and Das, 2007) depending on the culture conditions except for *Enterobacter cloaecae* (Nath et al., 2006). These yields are low to make the hydrogen production a profitable commercial application, thus, some strategies are proposed to overcome this limitation.

One of the potential applications of hydrogen is electricity production by using proton exchange membrane fuel cells (PEMFC). To determine the feasibility of the PEMFC to produce electricity, it is necessary to establish a continuous Bio-H_2 production system with a steady supply of sufficient Bio-H_2, to determine the Bio-H_2 production rates in the biological systems, and to develop a hydrogen fuel cell device.

In this chapter, data obtained during the evaluation of the feasibility to obtain methane (CH_4) and Bio-H_2 from FVW is presented. In the first section, the enhancement of biogas and methane production by the addition of inocula, buffer solution and nitrogen is described and discussed; followed by results on the start-up and operation of a 30-L anaerobic digestion (AD) system with FVW and meat residues (MR), as co-substrate. Additionally, the characterization of the microbial population involved in the process is included. In a second section, results obtained during tests carried out to determine the conditions that favor Bio-H_2 production using glucose as substrate are presented. The effect of the two different pretreatments of the anaerobic digester seed sludge, the incubation temperatures (35 and 55°C), and the effect of increasing initial glucose concentrations (IGC) are integrated and analyzed. The environmental conditions determined for enhanced Bio-H_2 production and the kinetic parameters, measured for the mixed culture, were used to evaluate a system for continuous production of Bio-H_2, the data obtained are also described. Finally, the performance of a long-term operation process in a semi-continuous regimen to determine the feasibility to produce electricity by using a proton exchange fuel cell is described.

METHODOLOGY

Continuous Culture Set Up

The semi-continuos CH_4 production experiments were carried out in a 30-L AD system described elsewhere (Garcia-Peña et al 2011).

A continuous, completely mixed liquid reactor with a total volume of 3 L was used for Bio-H_2 production, mineral medium supplemented with glucose as "model" substrate was continuously fed to the reactor. The mineral medium composition has been reported elsewhere (Garcia-Peña et al., 2009). The anaerobic reactor (AR) was operated with a constant volume of 1.5 L, peristaltic pumps were used to feed the reactor and to draw off the effluent. The reactor was continuously stirred at 150 rpm by using a magnetic stirring bar, the pH of the reactor was set and maintained at 5.5 during all the operation time. The AR was inoculated with heat-pretreated sludge from a bench-scale anaerobic digestion treated FVW of the central market (CEDA, Mexico City), as previously reported by Garcia-Peña et al., 2009. The bioreactor was incubated in thermostatically regulated water bath at 35°C to maintain a constant temperature.

Batch Experiments Set Up

Different conditions to enhance the CH_4 production were evaluated in batch systems (Garcia-Peña et al 2011).

The effect of the different initial glucose concentration over the Bio-H_2 production was determined in closed batch systems (serological bottles) as previously described (Garcia-Peña et al., 2009). The effect of increasing substrate concentration (gCOD/L) of FVW on H_2 production was also evaluated in batch experiments. Increasing initial substrate concentrations, in a range of 5 to 50 gCOD/L, were obtained by using different dilutions of FVW, and prepared by mixing the FVW with a percentage by volume of water. The experiments were carried out in modified Erlenmeyer flasks of 500 mL of total volume, 300 mL of the FVW dilutions were inoculated with 60 mL (30 g SSV/L) of a heat pretreated microbial population (described in the previous section). Bottles were closed with modified rubber caps and flushed with a N_2 gas stream to guarantee anaerobic conditions. Biogas from the systems was collected by connecting 1-L Tedlar bags to the top of the Erlenmeyer flasks. All the experiments were performed in duplicates.

Analytical Methods

Gas samples from the AR and the Tedlar bags were taken periodically to analyze the biogas composition. H_2 and CH_4 production was analyzed using a gas chromatograph (Gow Mac 550 series) equipped with a thermal conductivity detector and a silica-gel 60/80 column, 18'×1/8'×0.085" (Alltech). The temperature conditions used were the same as previously reported (Garcia-Peña et al., 2009). Glucose consumption was determined in liquid samples and quantified by using the DNS technique. Acetic and butyric acid production as metabolic intermediates was measured with a high performance liquid chromatography (HPLC) series 200 of Perkin Elmer equipped with a UV detector and a Prevail Organic Acid (Alltech), 150 x 4.6 mm column. The analysis conditions were: aqueous solution of KH_2PO_4 with a concentration of 25 mMol (pH = 2.5) as mobile phase, wavelength of 210 nm, a flow rate of 1 mL/min and an injection volume of 30 μL of sample.

The following parameters were determined in the effluents of each system: pH, total chemical oxygen demand t-COD, total and volatile solids (TS, VS). All the analyses were carried out in accordance with the Standard Methods of the APHA (American Public Health Association) (APHA, 1989).

Molecular Analysis

DNA extraction was developed in samples of the AD system at steady state, as described previously in Garcia-Peña et al., 2011.

A Pyrosequencing (bTEFAP) analysis of the extracted DNA sample was performed as described by Garcia-Peña et al., 2011. The combined V2 and V3 regions of 16S rDNA were sequenced by the previously described procedure (Wolcott et al., 2009). 611 readouts were obtained for the bacterial community in the anaerobic digester, after removing the failed sequence reads, low quality sequence ends, and tags. These readouts were aligned and clustered the by using the MOTHUR software (the Silva alignment and Operating Taxonomic Unit (OTU)-based clustering). And then the sequences were classified by the Ribosomal Database Project (RDP) classifier software at an 80% confidence threshold (Cole et al, 2009). The sequences were analyzing with BLAST and SEQMATCH applications provided by National Center for Biotechnology Information (NCBI) and RDP, respectively.

Proton Exchange Membrane Fuel Cell (PEMFC) Performance

A proton exchange membrane fuel cell (PEMFC) was used to evaluate the feasibility of electricity generation by using the Bio-H_2 produced in the continuous system. The characteristics of the PEMFC were the same as previously described by Garcia-Peña et al, 2009. To evaluate the performance of the PEMFC, different feeding gas fluxes, in a range of 0.3 to 0.9 L/h, were used to determine the Bio-H_2 gas flux necessary to obtain the maximum cell voltage (V) and current.

Results and Discussion

Anaerobic Digestion for Methane Production from the FVW

It is well known that pH control is one of the most important parameters to achieve high biogas and methane production during the anaerobic digestion process (Verrier et al., 1987; Mata-Alvarez, 1992). Specifically, the anaerobic digestion of FVW could be limited, due to acidification, by the production of large amounts of volatile fatty acids (VFA), which, in turn, inhibit the biological activity of methanogens (Bouallagui et al., 2005). Some authors report the addition of sodium hydroxide and/or the use of buffer solutions as an option to control the pH of the FVW (Verrier et al., 1987). Another characteristic of the FVW is the absence of a nitrogen source, an adequate C/N relation is necessary to enhance the anaerobic digestion process. During the development of the study, a variety of conditions were tested to promote biogas and methane production using the FVW in batch test, including buffered and nitrogen-supplemented systems, in experimental systems with and without inoculation.

In general the results showed that the inoculated systems, with pH control, reached the highest biogas productivity as compared with those systems without inoculation (Figure 1). The final pH in these systems was around 6.9. The highest biogas productivity (0.42 m^3/kgVS) was determined in the inoculated system, with pH control and nitrogen addition (IpHN). This result corroborated the fact that the FVW has some limitation in nitrogen source.

The optimum C/N ratio for microbial activity involved in the bioconversion of vegetable biomasses to methane ranges from 25 to 32 (Hong-Wie and David, 2007). Mata-Alvarez et al., 1992, stated that the C/N ratio of FVW is balanced at approximately 25, and therefore no nitrogen needs to be added to the process. However, some authors have reported that the C/N ratio of FVW is high, ranging from 30 to 36 (Rao and Singh, 2004; Gunaselan, 2007; Bouallagui et al., 2009).

Higher removal percentages were measured in the inoculated systems, which was correlated with higher biogas productions. The highest VS removal (86%) was also determined in the IpHN system. Taking into account these results, the co-digestion of the FVW and meat residues (MR) was evaluated in batch test, the biogas production was enhanced twice as compared with the highest obtained in the salt- supplement systems (Figure 1).

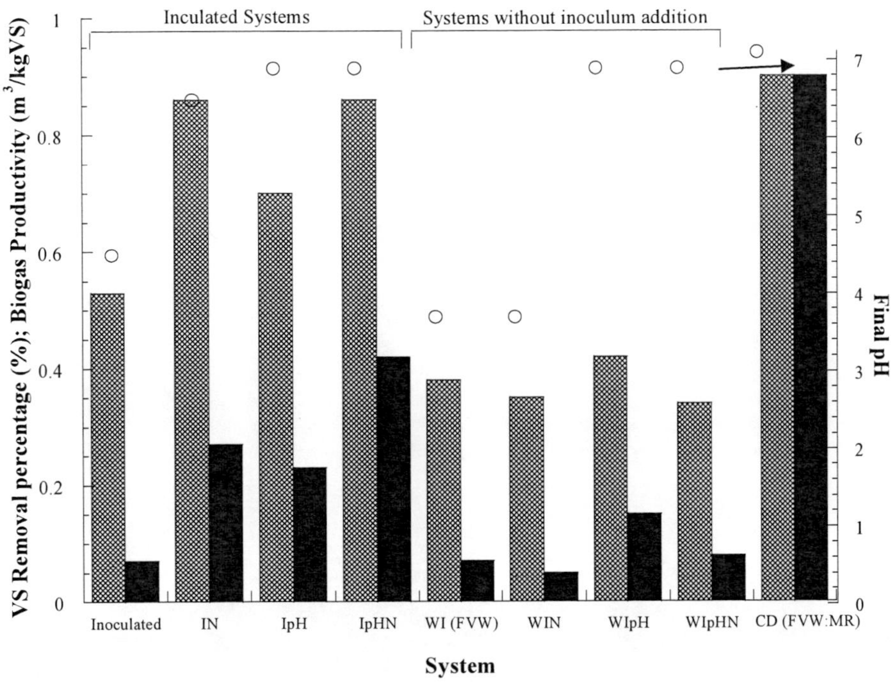

Figure 1. VS removal percentage (%/100); Biogas productivity m^3/kgVS; (○) final pH obtained for different conditions evaluated in batch systems. I, FVW inoculated with cow manure (10%); IN, FVW inoculated and supplemented with NH_4Cl as a nitrogen source; IpH FVW inoculated and salts added (buffer) to control pH; IpHN, FVW inoculated, buffering salts, and NH_4Cl added. WI, FVW without inoculation (WI) (Control); WIN, FVW and NH_4Cl; WIpH, FVW and buffering salts and WIpHN, FVW buffering salts and NH_4. Modified from Garcia-Peña et al., 2011.

Based on data obtained in batch experiments showing that the co-digestion of FVW and MR enhances biogas production, the feasibility of an anaerobic digestion process using FVW and MR was evaluated in a 30-L AD system. To determine the effect of the MR addition on biogas and methane production, experiments were carried out using different MR proportions. The anaerobic digestion process was performed in a semi-continuous regimen. The biogas productivity and methane percentages obtained under different conditions during 130 days of operation are shown in Figure 2. During the start-up of the system, the highest biogas production of 1 m^3/kgVs was obtained, however the analysis of the composition of the biogas showed that only CO_2 was produced due to the initial hydrolysis of the feedstock. The VS removal percentage was of 89%. After 20 days of operation, the AD system was inoculated with cow manure, and, in this period, a biogas yield of 0.64 m^3/kgVS was determined.

The CH_4 production started as a result of the methanogenic population introduced in the system with the inocula, a methane percentage of 16% (v/v) was measured in the biogas. The addition of 25% of MR (ratio of 3:1; FVW:MR) increased methane percentage to 28%, a new increment in the MR proportion to a ratio of 1:1 allowed to obtain a methane percentage of 30%. To evaluate the response of the system to the composition of the feedstock, the proportion of FVW and MR was again reduced and only FVW was added to the AD system. As expected, the CH_4 content in the resulting biogas was low, at only 14% (v/v).

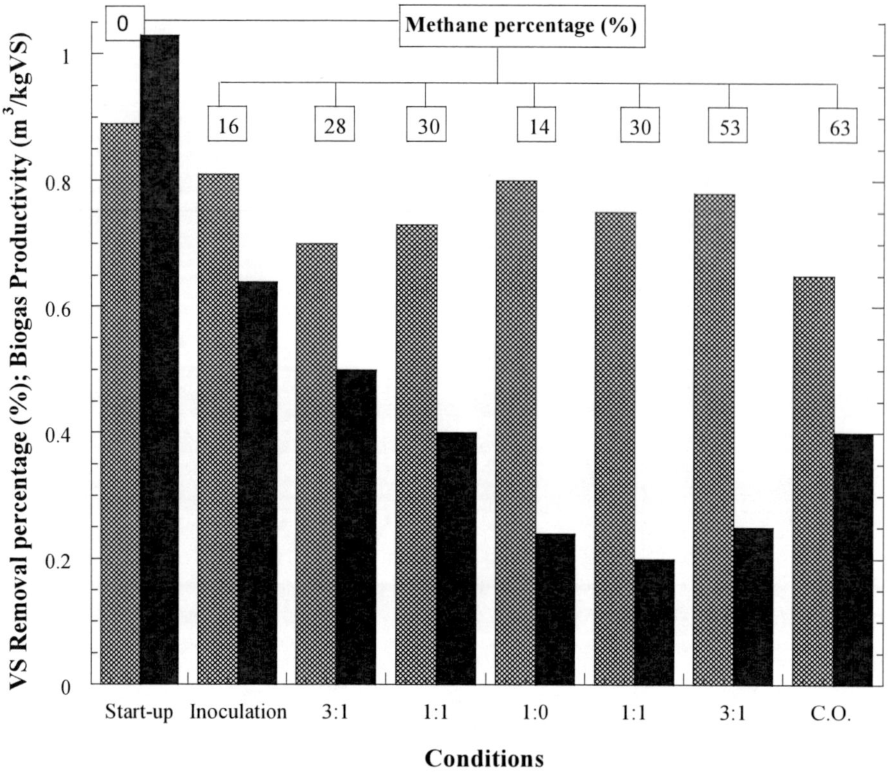

Figure 2. VS removal percentage (%/100) and Biogas productivity m^3/kgVS obtained in a DA system with different FVW and MR ratios. The reactor was operated in a fed-batch mode during 130 days.

Biogas production decreased as the MR proportion increased within the feedstock from 0.5 to 0.2 m^3/kgVS. These results suggest that the increase in the proportion of meat residues inhibited the initial hydrolysis due to the presence of more complex substrates, such as polypeptides instead of only carbohydrates (the main components in the FVW). However, the methane yield increased as a result of the better conditions established in the AD system that allowed an increment in the methanogenic population. Once stable operation was achieved after 83 days, the biogas production showed a constant value of approximately 0.25 m^3/kgVS and a methane percentage of 53%. During the co-digestion evaluation period, the pH in the AD system was stable at 6.9 and naturally regulated due to the combined hydrolysis of the carbohydrates from the FVW and the protein from the MR. The stable operation was maintained with Organic Loading Rates (OLRs) in the range of 2.4 and 2.7 g COD/L day (Hydraulic retention times, HRT, in the range of 15–20 days) (Garcia-Peña et al., 2011).

During the start-up of the AD system, the pH was continuously regulated by the addition of a NaOH solution (0.8 M); at this time, a total volatile fatty acids (VFAs) concentration of 4000 mg/L (acetate and butyrate 2:1) in the liquid phase was determined. When the co-digestion process was initiated, the ammonia release from MR showed an appropriate buffering capacity that allows for a highly stable operation of the system; under these last conditions, the average VFAs concentration decreased from approximately 2000 to 1500 mg/L. The addition of MR could resulted in the production of NH_4 during the protein hydrolysis, which increased the digester buffering capacity and, hence, the stability of the AD system. An average concentration of total VFAs of 1300 mg/L (mainly acetic and butyric acids) was maintained at the steady-state of the AD system. In a well balanced anaerobic digestion process, total VFA levels range from 55 to 1800 mg/L (Dinsdale, 2000; Misi and Foster, 2001; Buoallagui et al., 2009). The stable operation was maintained with Organic Loading Rates (OLRs) in the range of 2.4 and 2.7 g COD/L day (Hydraulic retention times, HRT, in the range of 15–20 days).

The increment of the methane yield after the stable conditions reached in the reactor could be explained by an increase in the methanogenic population and its adaptation to the operating conditions in the AD system. The high VS removal, the increase in the methane yield, and the natural pH control during the stable period of the AD system was due to an adequate ratio of nutrients and the availability of proteins for new cell synthesis.(Garcia-Peña et al., 2011). An optimum C/N ratio in the feedstock can be achieved by mixing low and high C/N ratio materials (Bouallagui et al., 2009; Habiba et al., 2009). Bouallagui et al. (2009) demonstrated that a C/N ratio between 22.5 and 25 enhanced the VS removal efficiency and biogas yield in an anaerobic co-digestion process of FVW and waste anaerobic sludge (WAS).

Some authors reported methane yields ranged from 0.16 to 0.762 m^3/kgVS for anaerobic digestion of FVW (Mata-Alvarez et al., 1992; Boallagui et al., 2009). The methane yields (0.05 to 0.1 m^3/kgVS) determined during the initial operation of the AD system were lower than those published. Since the 30-L reactor was operated at room temperature, the lower methane yields could be partially explained by the low operational temperature. This fact was corroborated by the results obtained in the batch experiment, where a controlled temperature of 30°C allowed reaching higher biogas and methane yields (0.9 and 0.45 m^3/kgVS, respectively) in the co-digestion of FVW and MR (50:50) system (Garcia-Peña et al., 2011). As it is well known, temperature has a strong effect on methane production. Many authors have reported that the highest methane yields are obtained at 35 and 55°C, for mesophilic and thermophilic conditions, respectively. However, under field conditions systems without temperature control (less energy consumption systems) could be more interesting. Low cost systems operated at temperatures below mesophilic conditions with a simple design and process control, robustness to harsh conditions and lower investment costs have been shown to be successful in on-farm manure treatment, although the degradation efficiency is lower and higher retention times are required (Bohn et al., 2007). These last authors also showed that the methane yield could be increased almost twice to 0.4 m^3/kgVS at 30°C compared with the yield obtained at 20°C (0.25 m^3/kgVS).

After a long period of operation, the gradual acclimation of the biomass and the enrichment of the methanogenic population could explain the increase in methane production in the AD system compared with the production observed during the start-up of the AD system. Under the current conditions, biogas and methane yields of 0.41 m^3/kgCOD and 0.258 m^3/kgCOD were determined in the AD system, respectively. These results are similar to the ones reported by Dinsdale et al. (2000) for the co-digestion of FVW and WAS in two-stage tubular digesters, but still lower as compared to the values reported (0.35-0.4 m^3/kgVS) by Callaghan et al. (2002) and Buoallagui et al. (2009) for the co-digestion of FVW and cattle slurry and chicken manure, as well as FVW and waste activated sludge (WAS). Ongoing work is being carried out to determine the performance of the system under mesophilic conditions to achieve the highest biogas production and methane yields mesasured in batch experiments.

Recently, the microbial ecology of the AD system, under stable operation conditions, was determined. The overall results showed that the structure of the sequenced bacterial community was correlated with the specific activity determined in the AD system and is typical of a bacterial community degrading plant- and animal-derived wastes (Garcia-Peña et al., 2011).

The bacterial and *Archea* populations were analyzed due to the methane production is the result of both activities. The bacterial community in the AD system was constituted by phylum: *Firmicutes* (89.5%), *Actinobacteria* (6.9%), *Bacteroidetes* (2.3%). Related to the predominant phyla, *Bacilli* (76.1%) and *Clostridia* (13.3%) were found as the major classes (Table 2). The activity of *Lactobacillus* species (72% of the total reads in *Bacilli*) could be responsible for the initial fermentation of carbohydrates, as it was previously reported (Jo et al., 2007). Additionally, *Firmicutes* are well known as syntrophic bacteria that can degrade volatile fatty acids. Some *Clostridium* species (in the class Clostridia), such as *Clostridium aminobutyricum* and *C. sticklandii*, have been reported as microorganisms capable of utilizing amino acids and producing acetate, butyrate, and ammonia (Shin et al., 2010). The syntrophic role of *Firmicutes* involves Bio-H_2 removal and has immediate implications for the composition of the methanogenic community.

The phylum *Actinobacteria* is mainly represented by *Bifidobacterium* species (62% of total reads) as it is shown in Table 2.

These microbial communities are mainly saccharolytic and cannot perform proteolytic activity; those constituted the major groups of the intestinal flora and have also been isolated from anaerobic digesters treating bean-curd wastes (Ling et al., 1996).

Table 2. Genus level distribution of the bacterial population in a 30-L anaerobic digester operated at steady state

Bacterial community[a]			
Phyla	% abundance	Class	% abundance of classes within each phylum
Firmicutes	89.5	*Bacilli* *Clostridia*	76.1 13.3
Actinobacteria	6.9	*Bifidobaterium*	62
Bacteroidetes	2.3	*Prevotella* *Anaerofilum*	63 75
Archaeal community[b]			
Genus	% abundance	Species	% abundance
Methanobacteriales	93%	*Methanobacterium curvum* *Methanobacterium congolense*	77 11.5

[a] Pyro-sequencing analysis. [b] Quantitative-Polymerase Chain Reaction (QPCR) analysis. Modified from Garcia-Peña et al., 2011.

The minor phylum found in the AD system *Bacteroidetes* was associated with degradation of meat residues (MR) used during the co-digestion studies and with the proteolityc degradation of plant residues. *Bacteroidetes* are proteolytic bacteria (Kindaichi et al, 2004) and also are associated to the attack the 1,4 α glycosidic bonds of plant polysaccharides (Charleston, PhD thesis). Therefore, those microbial populations were probably involved in the degradation of polypeptides in MR and the proteolytic degradation of plant residues (Debroas and Blanchart, 1993). The majority of proteolytic microorganisms are also able to metabolize carbohydrates to produce VFAs.

The *Archea* population was determined by a QPCR analysis. The genera archaea represent approximately 1.3 $x10^6$ gene copies/mL of the digested sludge, within the genus *Methanobacteriales* accounted for greater than 93% of the archaeal presence in the digester ($1.09x10^6$ gene copies/mL). Under this genus, *Methanobacterium curvum* was found as the major specie, 77% of the sequences targeting the *Archaea* 16S rRNA gene showed similarity with *Methanobacterium curvum.* While 11.5% of the sequences were similar to *Methanobacterium congolense*. Hydrogentrophic methanogens dominated the methanogenic community despite the fact that the digester was inoculated with cow manure, which usually contains acetoclastic methanogens. These results are in accordance with previous work characterizing FVW anaerobic digestion, which established that acetate-utilizing methanogens, *Methanosaeta*, were dominant in seed sludges of both types but decreased drastically during processing in the digestion tank. Consequently, *Methanosarcina* and *Methanobrevibacter/Methanobacterium* were the main contributors to methane production in this system (Ike et al., 2010).

In general, the data obtained during this first stage of the project with the FVW coming from the CEDA showed that an efficient methane production could be attained by adding salts and inocula to the feedstock. The results were enhanced by the addition of meat residues that allowed a natural pH control in a bench scale AD reactor. At the start up of the AD system, the biogas and methane yield were low as an effect of the low operation temperature. However after a long term operation the system is currently operated with biogas and methane yield within the range reported for other systems.

The results showed that the acidic conditions of the feedstock, the initial pH around 4.5, (closer to the optimum pH were needed for the Bio-H_2 production) favor Bio-H_2 production instead of CH_4. The biogas composition analysis showed that Bio-H_2 was produced, as a metabolic intermediate, after the hydrolysis phase in the anaerobic reactor (Garcia-Peña et al., 2011). The bacterial species determined by the molecular analysis in the AD system showed that the main phyla were *Firmicutes* (89.5%), constituted by *Bacilli* (76.1%), and *Clostridia* (13.3%) This last microbial group is well known as Bio-H_2 producer. Considering the aspects mentioned before, we evaluated the option to produce Bio-H_2 rather than methane from the FVW. The AD sludge was pretreated and its capacity to produce Bio-H_2 under different conditions was evaluated. Pretreatment of the inoculum has been reported as a strategy to increase Bio-H_2 producers and eliminate the methanogenic population, which consumes the Bio-H_2 to produce CH_4. Two different methods were used to obtain an inoculum for the H_2 production, a heat and an acid pretreatment. Additionally, different environmental conditions were tested to determine the conditions that allowed and enhanced the Bio-H_2 production.

Conditions for Enhanced Bio-Hydrogen Production

As the Bio-H_2 production from the FVW seems to be more attractive than CH_4 production, some conditions were evaluated to produce this compound by using the anaerobic digestion seed sludge (SS) as inoculum. To reach this goal, heat and acid pretreatments were applied on the anaerobic digester SS. The pretreatment method and incubation temperature effect was evaluated simultaneously, then the experimental systems accounted for: the heat pretreated inocula (heat pre-treated microbial population, HPMP) and acid pretreated inocula (acid pre-treated microbial population, APMP), both individually incubated at 35 and 55°C. Environmental conditions, such as type of inoculum, incubation temperature, and the initial glucose concentration (IGC), exert an effect on the maximum glucose utilization rate, Bio-H_2 production rate, and the metabolic pathway followed by the microorganisms for glucose transformation (Van Ginkel et al., 2001).

Data showed that both, heat and acid pretreatments, exerted a positive influence on Bio-H_2 production, reaching H_2 molar yields in the range of those reported in the literature. The HPMP systems showed higher Bio-H_2 production rates and glucose consumption kinetic constants at both 35 and 55°C, as compared to those obtained in the APMP systems (Table 3). The HPMP incubated at 35ºC (HPMP35) showed the highest Bio-H_2 volume (289 mL), Bio-H_2 molar yield (2.85 mol H_2/mol glucose) and Bio-H_2 production rate (0.014 L/h)

Table 3. Bio-H_2, CH_4 production and glucose consumption obtained under the different conditions evaluated

Experimental system	H_2 (mL)	CH_4 (mL)	Kinetic constants k $(h^{-1})^a$	H_2 yield[b] (molH$_2$/molglucose)	Possible microbial communities involved in process
HPMP 35	289	0	0.145	2.8	*Clostridium species*[c]
HPMP55	81	0	0.061	0.82	
APMP35	221	0	0.114	2.2	*Clostridium species*[c], *Enterobacteriaceae*, *Streptococcaceae species (Streptococcus bovis)*
APMP55	70	0	0.061	0.7	
NP35	168	6	0.084	1.7	Methane producers. *Methanobacteriales*, *Hydrogentrophic methanogens*[d]
NP55	56	30	0.064	0.6	Methane producers. *Methanobacteriales*, *Hydrogentrophic methanogens*[d]

[a] Kinetic constants for glucose consumption. [b] H_2 yield = amount of H_2 formed (mol) ÷ amount of glucose consumed (mol). [c] Determined as one of the bacterial population in the SS of the AD system (Garcia-Peña et al, 2011). [d] Identified as the major Archae population in the SS of the AD system. Modified from Garcia-Peña et al., 2009.

The data obtained were similar to those reported by Iyer et al., 2004; Mu et al., 2007. These authors reported Bio-H_2 molar yields for the HPMP of 1.8 and 2 mol H_2/ mol glucose, respectively, whereas, for an APMP, Fang et al., 2002, and Mu et al., 2007, reported H_2 molar yields of 1.4 and 1.3 mol H_2/mol glucose, respectively.

The pretreatment methods promote a difference in the initial microbial communities that might be directly responsible for different fermentation types and hydrogen yields (Garcia-Peña et al., 2009). The heat pretreatment method used during the experiments could favor the predominance of *Clostridium* species identified as part of the microbial population in the SS of the AD system used as inoculum (Garcia-Peña et al., 2011). *Clostridium* has been reported as the predominant microbial population for the heat-pretreated cultures (Iyer et al., 2004; Mu et al., 2007). Microorganisms from the *Clostridium* genus showed higher Bio-H_2 molar yields (Pan et al., 2008; Evyernie et al., 2007; Chen et al., 2005) compared to other Bio-H_2 producers as *Enterobacter* or *Bacillus* strains (Kotay and Das, 2007) depending on the culture conditions, except for *Enterobacter cloaecae* (Nath et al., 2006). Therefore, higher Bio-H_2 production for the system in the HPMP treatment incubated at 35°C could be partially explained by the fact that some species of *Clostridium* are predominant in the heat pretreated inocula

Iyer et al., 2004, reported that in a heat-treated anaerobic sludge, *Clostridium acetobutylicum* was identified as the major bacterium, which was responsible for the butyrate-type fermentation. Fang et al., 2002, based on a phylogenetic analysis of the rDNA sequences, demonstrated that an acid pretreated mixed culture consisted in 64.6% of *Clostridium* species affiliated with three different families, 18.8% corresponded to *Enterobacteriaceae* species and 3.1% belonged to *Streptococcus bovis* (*Streptococcaceae*). All the microorganisms present in the microbial population were closely related to species of the genera *Citrobacter*, *Clostridium*, or *Klebsiella*. The dominant bacteria identified in an alkaline pretreated sludge were *Eubacterium multiforme* and *Penibacillus polymyxa* (Cai et al., 2009).

CH_4 production was also measured in all the systems to corroborate the efficiency of the pretreatments on eliminating the methanogenic population. CH_4 was neither determined in the HPMP nor in the APMP systems. Methane volumes were obtained only in control systems (NP35 and NP55), as can be seen in Table 3. The results obtained in this work are in agreement with data reported on CH_4 production conditions; Mata-Alvarez et al., 2000, reported that CH_4 production yields seem more favorable at thermophilic temperatures.

The highest Bio-H_2 yield was obtained in the system inoculated with HPMP at 35°C, therefore, these conditions were established in the rest of the experiments. The effect of the IGC was then evaluated.

IGC determines the yield and production rate of Bio-H_2 (Fabiano and Perego, 2002), additionally it constitutes a factor to establish the economic and technical feasibility of the Bio-hydrogen production process (Hawkes et al., 2002, Levin et al., 2004, Nath et al, 2006).

Data obtained showed that low IGC, 3 and 5 g/L, result in low Bio-H_2 production rates (0.008 and .015 L/h, respectively); a similar effect was observed at high glucose concentrations, at 20 and 30 g/L, the Bio-H_2 production rates were 0.005 and 0.006 L/h, respectively (Garcia-Peña et al, 2009). These results agree with those reported by Nath et al., 2006, and Jo et al., 2008. Higher glucose concentrations induced lower Bio-H_2 production rates and conversion efficiency, which suggests that the carbon flux is directed towards production of reduced by-products such as ethanol or organic acids rather than hydrogen (Van Ginkel et al., 2001, Kim et al., 2006). Nevertheless, Pan et al., 2008, found the highest yield (4.16 L H_2/L medium) with an IGC of 22 g/L using sewage sludge as inoculum. It has been reported that substrate inhibition becomes predominant at higher glucose concentration because this modifies the metabolic pathways (Oh et al., 2003).

For IGC's of 7, 10, and 15 g/L, the H_2 production rates were very similar (0.021, 0.023, and 0.024 L/h, respectively) and around three times higher than those determined for low and high IGC. Wu and Lin, 2004, observed a similar trend between substrate concentration and H_2 production rate; in their study the highest production rate was found at 4% (m/v) of molasses using a mixed culture.

The metabolic pathway used to produce Bio-H_2 as final product is closely associated to the predominant microbial population in the inoculum after the corresponding enrichment treatment. In the systems inoculated with the HPMP, the main metabolic products were acetic and butyric acids. The acetic acid concentration for the HPMP incubated at 35°C was 6000 mg/L and the butyric acid concentration was of around 2500 mg/L. The heat treatment favors the acetic acid fermentation, whereas small concentrations of the other metabolic products were determined. The highest acetic acid concentration correlated with the corresponding higher hydrogen yields according to the theoretical hydrogen balance in the systems inoculated with HPMP and incubated at 35°C. The HPMP metabolic intermediates profile obtained during the process suggests that the microorganisms responsible for H_2 production were mainly from the *Clostridium* species.

The final acetic acid concentrations for APMP at both incubation temperatures, 35 and 55°C, were similar and around 2600 mg/L. Butyric acid was produced all over the process in the APMP incubated at 35°C, A35, which suggests butyric fermentation. The presence of other metabolites such as butyric acid, in general reduces the Bio-H_2 yield. In general, the HPMP showed acetic acid as the major metabolic product, whereas in the APMP system both, butyric and acetic, acids were produced. A mixture of metabolic intermediates determined in the APMP systems suggests that other microorganisms such as *Enterobacter* and *Klebsiella* besides *Clostridium* could have been responsible for the fermentation. Therefore, controlling the metabolism of the microflora towards acetate formation is a key factor to achieve a high Bio-H_2 production. Mu et al., 2007, evaluated three different methods (heat, acid, and alkaline) for the enrichment of a microbial culture used for H_2 production. The authors found that the distributions of VFA and ethanol were different in the three cases, resulting in various fermentative types. Typical butyrate-type fermentation was observed, since the major acidogenic products were butyrate and acetate for the heat- and acid-treated sludge. The percentage of butyrate was larger than that of acetate for the heat-treated sludge, whereas a similar level of butyrate and acetate was found for the alkaline-treated one. However, for the acid-treated sludge, the concentrations of acetate, propionate, butyrate, and valerate were almost at the same levels, suggesting a mixed-type fermentation. These authors attributed the differences in metabolic products among the various systems to the diversity of microbial population in the inoculum after the corresponding enrichment treatment. A high molar H_2 yield (2 mol H_2/mol glucose) was obtained for the heat-pretreated system.

Bio-H_2 Production under Continuous Culture Conditions

The kinetic parameters of the bio-hydrogen producers mixed microbial culture were determined using increasing glucose concentrations. The experiments were carried out in batch experiments inoculated with heat pre-treated inocula, systems were incubated at 35°C, and stirred at 150 rpm, the pH was set and maintained at 5.5 (Canul-Chan, 2010). The kinetic parameters were obtained by the linearization of the growth rates at different glucose concentrations, as shown in Figure 3.

The continuous culture conditions used in this work were determined by using the kinetic parameters. Taking $\mu max = 0.101\ h^{-1}$, $S = 10$ g/L, and $ks = 2.56$ g/L, experimentally obtained, the critical dilution rate was calculated as $D = 0.08\ h^{-1}$. Then the feed flow rate Q must be kept and maintained at 1.345 mL/min to avoid washout in the reactor. This condition is necessary to have a substrate concentration supporting steady-state active biomass at the maximum growth rate and producing H_2 as the main product.

The corresponding hydraulic retention time (HRT) for the calculated flow rate was of 12.38 h. The optimum dilution rate, Dopt, was then calculated as $0.05\ h^{-1}$ and the corresponding HRT was 17.9 h, considering the kinetic parameters experimentally measured for the H_2 producer mixed culture. Using a safety factor (SF) of 30% to avoid a washout of the reactor and to give stable biomass accumulation, an HTR of 24 hours was used for the performance of the reactor.

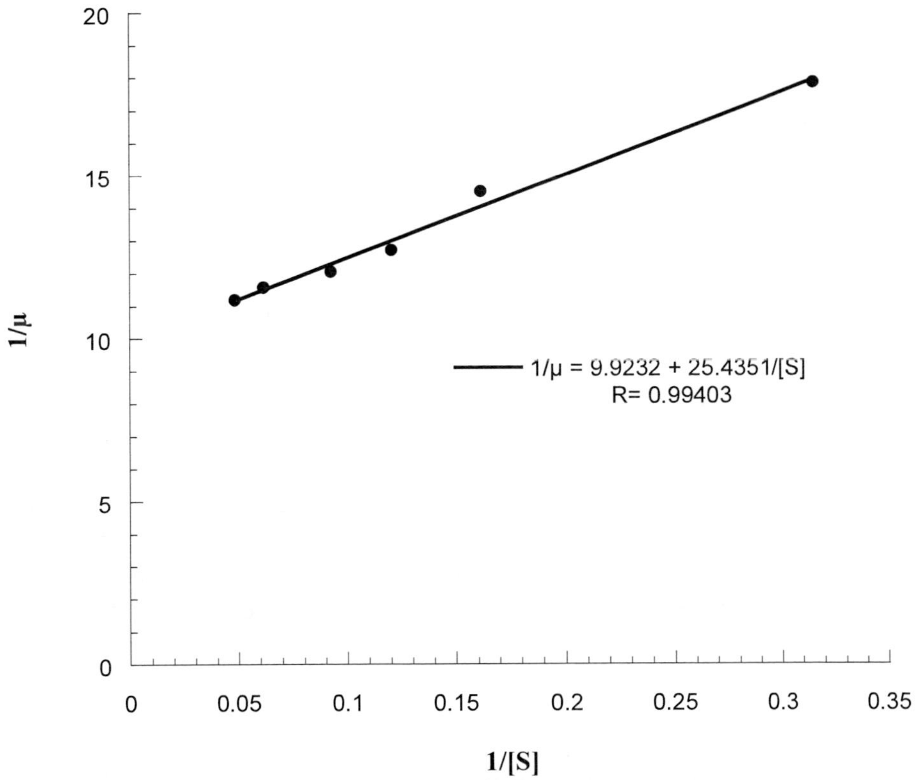

Figure 3. Linear relation between the 1/[S] and the growth rates (1/μ) measured at different initial glucose concentrations by using the Bio-H_2 producer microbial mixed culture.

Experiments were carried out in an anaerobic reactor (AR) of 3 L of total volume, continuously fed at a flow rate (Q) of 0.7 mL/min. The performance of the AR is summarized in Table 4. During the first 48 h of culture, the reactor was operated in a batch mode, during this period glucose was almost completely consumed (93%) and 162 mL of H_2 was produced, reaching a H_2 production rate of 0.0034 L/h. The Bio-H_2 content in the biogas was 35% (v/v).

When sustainable microbial activity was determined in the reactor, the continuous culture was initiated constantly feeding mineral medium with glucose concentration of 10 g/L at a flow rate Q of 0.9 mL/min. The HRT was established at 24 h.

Table 4 summarizes the glucose consumption percentage, the Bio-H_2 cumulative production, and the H_2 production rate for four changes of total volume in the reactor. Glucose removal amounted to between 88 and 92%. In the first period, only 85 mL of Bio-H_2 was obtained, the low production could be attributed to the change in environmental conditions from batch mode to continuous operation, however the Bio-H_2 production rate was similar to the one obtained in batch conditions (0.0035 L/h consumption). The total cumulative produced H_2 increased from 124 to 196 during the second and the third period, and the H_2 production rate also increased almost twice of the H_2 production rates measured at the beginning of the process. This could be explained as an increment of biomass in the system, measured at the final time of each period.

Table 4. Batch and continuous culture performance. Glucose consumption, cumulative Bio-H_2 production, Bio-H_2 production rate, and biomass production

Batch Mode operation				
Time of operation (hours)	Glucose consumption (%)	Cumulative Bio-H_2 production (mL)	Bio-H_2 production rate (L/h)	Biomass (g/L)
48	93	162	0.0034	0.37
Continuous mode operation				
24	91	85	0.0035	0.5
25	88	124	0.0049	0.17
25	88	196	0.0078	0.23
25	92.2	163	0.0065	0.26
Total 144		729		

After the complete operation of the system, 144 h, a total H_2 production of around 700 mL was obtained. The H_2 content during the continuous operation was maintained constant around 50% (v/v) and CH_4 was not detected. The highest H_2 molar yield was 2.3 mol H_2/mol glucose. The metabolic products were monitored throughout all the experiment, butyric acid was found as the main metabolic product. The butyric acid production could partially explain the lower molar H_2 yield obtained in this experiment compared with those determined in previous studies (Garcia-Peña et al., 2009). Ongoing work is being conducted to determine the reason why the metabolic pathway was modified in the continuous culture. Controls performed in batch systems showed that under this condition the main metabolic intermediate produced by the microbial population was acetic acid, which is in agreement with the previous published results (Garcia-Peña et al., 2009) (data not shown).

Steady-state conditions were reached (when the deviations of the measured parameters were around 5%) after 96 hours of operation, at this time the amount of biomass and glucose consumption was almost constant in the AR. The HRT used in this study was higher than the one reported by Lin et al., 2007, these authors carried out tests with a continuous stirred anaerobic bioreactor operated at an HRT (hydraulic retention time) of 6 h with a sucrose influent concentration of 30 g COD/l, obtaining an H_2 molar yield of 3.1 ± 0.18 mol H_2/mol sucrose and an H_2 content in biogas of 40.6 ± 1.4 % (v/v). The H_2 molar yield was also lower that the one reported by Lin et al., 2007, however the Bio-H_2 content was around 50% (v/v) higher.

Bio-H_2 Production using FVW as Feedstock

After determining the conditions that favor bio-hydrogen production in a model system using glucose as substrate, the feasibility to produce Bio-H_2 with the mixed culture coming from the anaerobic digestion system from the organic fraction of solid waste produced in CEDA, a source- sorted fraction containing mainly fruit and vegetable waste (FVW) was used. The characterization of the FVW showed a TS content of approximately 73-100 g/kg waste (around 7.3-10%), pH of 4, and a moisture content of 90%. These data are in agreement with those reported in the literature for similar organic solid wastes (Rao and Singh, 2004; Gunaselan, 2007; Bouallagui et al., 2009).

Increasing substrate concentrations (from 5 to 50 g COD/L) were used to determine H_2 production and organic matter removal (Table 5). In general, low initial substrate concentration (5, 18 g COD/L) showed lower Bio-H_2 production as compared with those obtained at higher substrate concentrations (28, 38, and 50 g COD/L). The system with an initial substrate concentration of 37.3 g COD/L reached the highest H_2 volume (310 mL) in approximately 240 h of culture. Bio-H_2 production rates increased as a function of the increment in the initial substrate concentration, the highest was obtained at an initial concentration of substrate of 37.3 g COD/L. At a concentration of 50 g COD/L, the Bio-H_2 production rate was 20% lower than the highest value. Data obtained using the FVW as substrate to produce H_2 are in agreement with the data reported by Mohan et al., 2008, and Kim et al., 2008.

Table 5. Performance of the batch systems used to evaluate the effect of increasing initial substrate concentration over Bio-H_2 production by using FVW

Initial substrate concentration (gCOD/L)	Final substrate concentration (gCOD/L)	Removal percentage (%)	Bio-H_2 production rate (mmol H_2/day)	H_2 production (mL)
5.0	4.44	11.7	0.12	9
11.3	10.07	11.3	0.11	9
17.8	15.02	15.6	0.32	37
19.9	17.52	12.3	0.39	67
22.5	20.65	8.2	0.38	58
28.0	23.84	14.9	1.24	218
37.3	27.94	25.1	1.72	310
49.9	39.09	21.7	1.46	199

Low COD removal percentages were obtained under the different substrate concentrations evaluated. The system with an initial substrate concentration of 37.3 g COD/L showed the highest COD removal percentage (25%), which was correlated with the highest H_2 production (310 mL) and H_2 production rate (1.7 mmol/day). In general, the removal percentages were lower compared to those reported by other authors, this fact could be partially explained by the lignocellulosic content of the feedstock. Additionally, more experiments have to be performed to determine the initial concentration of substrate used to produce volatile fatty acids as metabolic intermediates of the process.

Finally, an anaerobic reactor (AR), operated in feed batch mode, was used to evaluate the performance of a Proton Exchange Membrane Fuel Cell (PEMFC). After sustained operation was reached in the AR, the system was coupled to a PEMFC. The complete system is shown in Figure 4. The anaerobic reactor (AR) was connected to a closed vessel containing a NaOH solution (2 N) by using a valves system to eliminate CO_2 of the biogas stream. The effluent of the rector was bubbled into the NaOH solution, thereafter the biogas stream was mixed with humidified air and both were introduced at different flow rates into the fuel cell. To evaluate the PEMFC performance, different feeding gas fluxes were used to determine the H_2 gas flux necessary to obtain the maximum cell voltage (V) and current.

Figure 4. Coupled system of the AR system and the PEMFC.

The evaluation of the time required to reach the highest voltage is shown in Figure 5, the time of stable performance and the time in which the PEMFC is deactivated once the H_2 feeding is stopped was determined. At a flux of 0.3 L/h the PEMFC reached a maximum voltage of approximately 0.3 V, using twice the initial flux, a maximum voltage of 0.9 V was obtained in 10 min. The highest voltage of 1.0 V was determined at a flux of 0.9 L/h, the cell maintained the charge during 7 min. The maximum circuit potential (cell voltage) reached 1.0 V, which corresponds to 86 % of the maximum cell voltage (Emax) according to the analysis reported by Levin et al., 2004. The maximum current determined in the PEMFC was 68 mA; taking into account an area of 5 cm^2, a current density of 13.6 mA/cm^2 was determined. Based on the polarization curve, an internal resistance of 10 Ω was obtained. These results are in agreement with the data reported by Lin et al., 2007.

To confirm the feasibility of using the electrical energy generated by the PEMFC, a small motor (which consumed 9.32 mW and 23.3 mA) was operated during 20 min using a flux of 0.6 L/h. Based on the results, to generate a current of 68 mA during 1 hour it is necessary to feed 600 mL of H_2 at a flux of 0.6 L/h. Data obtained in the present work indicate that the lab scale AR-PEMFC system was stable enough and allowed generating electricity.

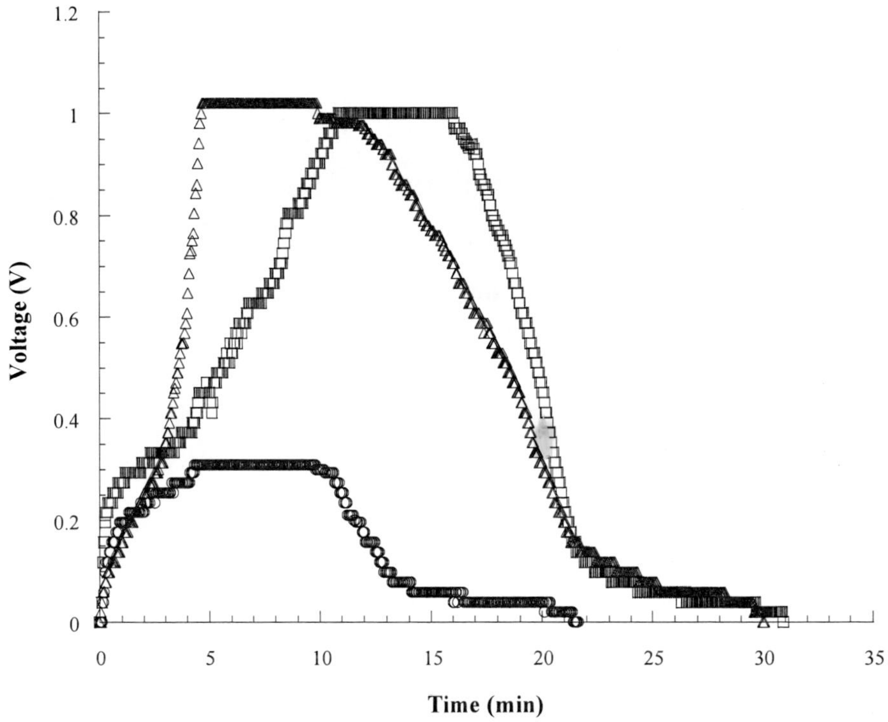

Figure 5. Voltage profiles obtained by using different H_2 gas fluxes; 0.3 L/h (○), 0.6 L/h (□), 0.9 L/h (△).

CONCLUSIONS

This chapter analyzes and discusses the feasibility of methane and Bio-H_2 production from FWV using an anaerobic digestion system (ADS). For methane production, the highest biogas production (0.42 m^3/kgVS) and VS removal (80%) were obtained in batch systems supplemented with buffering salts and nitrogen (IpHN). Co-digestion of the FVW with MR enhanced the biogas production and methane yield. This process was also evaluated in a 30-L AD system revealing a natural pH control and stable performance. The highest methane percentage (v/v) achieved in the AD system was 63% (v/v). The characterization of the microbial population in the AD system showed that the main phylum in the bacterial community was Firmicutes (89.5%), which was responsible for acidogenesis or syntrophic acid degradation. *Methanobacteriales* constituted the main methanogenic population (93%). Bio-H_2 production using the sludge of the AD system was enhanced using a heat-pretreated inoculum and an initial glucose concentration of 10 g/L under mesophilic conditions. A continuous system to produce H_2 was successfully operated, an HTR of 24 h was used to avoid washout of the reactor and to obtain stable biomass accumulation. Steady-state conditions were reached in the AR after 96 h of culture, a H_2 production of 700 mL was obtained during the process. Results obtained in batch experiments with FVW showed a strong effect of the initial substrate concentration on H_2 production and removal percentages. The system with an initial substrate concentration of 37.3 showed the highest H_2 volume (310 mL), H_2 production rate (1.7 mmol/day), and COD removal percentage (25%). At relatively lower substrate concentrations, the H_2 production rates were lower than the ones determined for higher substrate concentrations. PEMFC coupled to the H_2 production system showed a good performance, producing 1.0 V, which corresponded to 84% of the Emax.

Acknowledgements

This work was supported through funding provided by the Instituto Politécnico Nacional, grant SIP 20113067 and CONACYT grant 60976.

References

Agdag, O.N., Sponza, D.T. (2005). Co-digestion of industrial sludge with municipal solid wastes in anaerobic simulated landfilling reactors. *Process Biochem.* 40: 1871- 1879.

Bouallagui, H., BenCheikh, R., Marouani, L., Hamdi, M. (2003). Mesophilic biogas production from fruit and vegetable waste in tubular digester. *Bioresour. Technol.* 86: 85-90.

Bouallagui, H., Lahdheb, H., Ben Romdan, E., Rachdi, B., Hamdi, M. (2009). Improvement of fruit and vegetable waste anaerobic digestion performance and stability with co-substrates addition. *J. Environ. Manage*. 90: 1844-1849.

Bouallagui, H., Touhami, Y., Ben Cheikh, R., Hamdia, M. (2005). Bioreactor performance in anaerobic digestion of fruit and vegetable wastes: Review. *Process Biochem*. 40: 989-995.

Cai, M.L., Liu, J.X., Wei, Y.S. (2004). Enhanced biohydrogen production from sewage sludge with alkaline pretreatment. *Environ. Sci. Technol.*, 38: 3195-3202.

Callaghan, F.J., Wasea, D.A.J., Thayanithya, K., Forster, C.F. (2002). Continuous codigestion of cattle slurry with fruit and vegetable wastes and chicken manure. *Biomass Bioenergy* 27: 71–77.

Canul-Chan, M. (2010). *Estudio de los parámetros de operación de un reactor anaerobio para la producción de hidrógeno a partir de residuos orgánicos.* Master in Science Thesis, Instituto Politecnico Nacional, Mexico.

Charleston, L.O. (2008). *Hyperthermophilic anaerobic digestion of food waste.* PhD thesis submitted to McGill University.

Chen, W.M., Tseng, Z.J., Lee, K.S., Chang, J.S. (2005). Fermentative hydrogen production with Clostridium butyricum CGS5 isolated from anaerobic sewage sludge. *Int J Hydrogen Energy*, 30: 1063-70.

Debroas, D., and Blanchart, G. (1993). Interactions between proteolytic and cellulolytic rumen bacteria during hydrolysis of plant cell wall protein. *Reprod Nutr Dev* 33: 283-288.

Dinsdale, R.M., Premie, G.C., Hawkes, F.R., Hawkes, D.L. (2000). Two-stage anaerobic co-digestion of waste activated sludge and fruit/vegetable waste using inclined tubular digesters. *Bioresour. Technol.* 72: 159–168.

Evyernie, D., Yamazaki, S., Morimoto, K., Karita, S., Kimura, T., Sakka, K., Ohmiya, K. (2000). Identification and characterization of Clostridium paraputrificum M-21, a chitinolytic, mesophilic and hydrogenproducing bacterium. *J Biosci Bioeng.* 89: 596–601.

Fabiano, B., Perego, P. (2002). Thermodynamic study and optimization of hydrogen production by Enterobacter aerogenes. *Int. J Hydrogen Energy.* 27: 149-156.

Fang, H.H., Zhang, T., Liu, H. (2002). Microbial diversity of a mesophilic H2- producing sludge. *Appl Microbiol Biotechnol.* 58: 112-118.

Garcia-Peña, E.I. , Parameswaran, P., Miceli, J., Canul Chan, M. Krajmalnik, R. (2011). Anaerobic digestion process from vegetable and fruit residues; Process and microbial ecology studies. *Biores. Technol.* 102: 9447-9455.

García-Peña, E.I., Ramirez, D., Guerrero-Barajas, C., Arriaga-Hurtado, L.G. (2009). Semi-continuos biohydrogen production as an approach to generate electricity. *Bioresour. Technol.*, 100: 6369-6377.

Gunaseelan, V.N. (2004). Biochemical methane potential of fruits and vegetable solid waste feedstocks. *Biomass Bioenergy* 26: 389-399.

Gunaseelan, V.N. (2007). Regression models of ultimate methane yields of fruits and vegetable solid wastes, sorghum and napiergrass on chemical composition. *Bioresour. Technol.* 98: 1270-1277

Habiba, L., Bouallagui, H., Hamdi, M. (2009). Improvement of activated sludge stabilization and filterability during anaerobic digestion by fruit and vegetable waste addition. *Bioresour. Technol.* 100: 1555-1560.

Hawkes, F.R., Dinsdale, R., Hawkes, D.L., Hussy, I. (2002). Sustainable fermentative hydrogen production: challenges for process optimization. *Int J Hydrogen Energy.* 27: 1339-1347.

Hong-Wie, Y., David, E.B. (2007). Anaerobic co-digestion of algal sludge and waste paper to produce methane. *Bioresour. Technol.* 98: 30-134.

Ike, M., Inoue, D., Miyano, T., Liu, T.T., Sei, K., Soda, S. (2010). *Microbial population dynamics during startup of a full-scale anaerobic digester treating industrial INEG*I, 2009

Iyer, P., Bruns, M.A., Zhang, H., van Ginkel, S., Logan, B.E. (2004). H2-producing bacterial communities from a heat-treated soil inoculum. *Appl Microbiol Biotechnol.* 66: 166-173.

Jo, J.H., Lee, D.S., Park, D., Choe, W.S., Park, J.M. (2008). Optimization of key process variables for enhanced hydrogen production by Enterobacter aerogenes using statistical methods. *Bioresour Technol.* 99: 2061–2066

Kim, S.H., Han, S.K., Shin, H.S. (2006). Effect of substrate concentration on hydrogen production and 16S rDNA-based analysis of the microbial community in a continuous fermenter. *Process Biochem.* 41: 199–207.

Kindaichi, T., Ito, T., and Okabe, S. (2004). Ecophysiological interaction between nitrifying bacteria and heterotrophic bacteria in autotrophic nitrifying biofilms as determined by microautoradiography fluorescent in situ hybridization. *Appl. Environ. Microbiol.* 70: 1641-1650.

Kotay, S.M., Das, D. (2007). Microbial hydrogen production with Bacillus coagulans IIT-BT S1 isolated from anaerobic sewage sludge. *Bioresour. Technol.* 98: 1183–1190.

Levin, D.B., Pitt, L., Love, M. (2004). Biohydrogen production: prospects and limitations to practical applications. *Int J Hydrogen Energy*. 29: 173-185.

Lin C.N., Wu S.Y., Lee K.S., Lin P.J., Lin C.H., and Chang J.S. (2007) Integration of fermentative hydrogen process and fuel cell for on-line electricity generation. *Int J Hydrogen Energy*. 32: 802–808.

Ling, D., Zhou, Y., and Yuan, Q. (1996). Numerical analysis and determinative tests for bifidobacteria of human and animal origin. *In Germfree Life and its Ramifications,* vol. XII 123-128.

Manish S., Venkatesh K.V., Banerjee, R. (2007). Metabolic flux analysis of biological hydrogen production by Escherichia coli. *Int J Hydrogen Energy* 32: 3820-3830.

Mata-Alvarez, J, Cecchi, F., Llabrés, P., Pavan, P. (1992). Anaerobic digestion of the Barcelona central food market organic wastes: plant design and feasibility study. *Bioresour. Technol.* 42: 33-42.

Mata-Alvarez, J., Mace, S., Llabrés, P. (2000). Anaerobic digestion of organic solid wastes. An overview of research achievements and perspectives. *Bioresour. Technol.* 74: 3-16

Misi, S.N., Forster, C.F. (2002) Semi-continuous anaerobic co-digestion of agro-waste. *Environ Technol* ,23: 445–51.35.

Mohan, V.S., Mohanakrishna, G, Goud, R.K., and Sarma, P.N. (2009). Acidogenic fermentation of vegetable based market to harness biohydrogen with simultaneous stabilization. *Bioresour. Technol.*, 100: 3061-3068

Mu, Y., Yu, H.Q., Wang, G. (2007). Evaluation of three methods for enriching H2-producing cultures from anaerobic sludge. *Enz Microbiol Technol.* 40: 947-953.

Nath, K., Kumar, A., Das, D. (2006). Effect of some environmental parameters on fermentative hydrogen production by Enterobacter cloacae DM11. *Can J Microbiol.* 52: 525-532.

Oh, S.E., van Ginkel, S., Logan, B.E. (2003). The relative effectiveness of pH control and heat treatment for enhancing biohydrogen gas production. *Environ Sci Technol.* 37: 5186-5190.

Pan, C.M., Fan, Y.T., Xing, Y., Hou, H.W., Zhang, M.L. (2008). Statistical optimization of process parameters on biohydrogen production from glucose by *Clostridium* sp. Fanp2. *Biores Technol.* 99: 3146–3154

Rao, M.S., and Singh, S.P. (2004). Bioenergy conversion studies of organic fraction of MSW: kinetic studies and gas yield-organic loading relationships for process optimization. *Bioresour. Technol.* 95: 173-185.

Rittmann, B.E., Krajmalnik-Brown, R., Halden, R. U. (2008). Pregenomic, genomic and post-genomic study of microbial communities involved in bioenergy. *Nature Microbial Reviews*, 6: 604-612.

Silva-Rodriguez, E.M. (2007) *Reducción de masa de residuos sólidos orgánicos por biosecado en invernadero.* Master degree thesis, UPIBI-IPN, van Ginkel, S.W., Oh, S.E., and Logan, B.E. (2005). Biohydrogen gas production from food processing and domestic wastewaters. *Int. J. Hydrogen Energy* 30: 1535-1542.

van Ginkel, S., Sung, S., Lay, J.J. (2001). Biohydrogen production as a function of pH and substrate concentration. *Environ Sci Technol.* 35: 4726–30.

Verrier D., Ray, F., Albagnac, G., 1987. Two-phase methanization of solid vegetable wastes. *Biol. Wastes* 22: 163–77.

Vijayaraghavan, K., Ahmad, D., and Soning, C. (2007). Bio-hydrogen generation from mixed fruit peel waste using anaerobic contact filter. *Int. J. Hydrogen Energy* 32: 4754-4760.

Wang, J., and Wan, W. (2009). Factors influencing fermentative hydrogen production: A review. *Int J Hydrogen Energy*, 32: 799-811.

Wu, J.H., Lin, C.Y. (2004). Biohydrogen production by mesophilic fermentation of food wastewater. *Water Sci Technol.* 49: 223-228.

Yang, P., Zhang, R., McGarveyc, J.A., and Benemann, J.R. (2007). Biohydrogen production from cheese processing wastewater by anaerobic fermentation using mixed microbial communities. *Int. J. Hydrogen Energy* 32: 4761-4771.

Zhu, H., Ueda, S., Asada, Y., and Miyake, J. (2002). Hydrogen production as a novel process of wastewater treatment-studies on tofu wastewater with entrapped *R. sphaeroides* and mutagenesis. *Int. J. Hydrogen Energy* 27: 1349-1357.

In: Biotechnology
Editors: L. G. Torres Bustillos and I. Garcia Peña
ISBN 978-1-62081-071-2

Chapter 8

BIOLOGICAL DEGRADATION OF AROMATIC ACIDS

C. Garibay-Orijel*[*] *and A. Ordaz-Cortés
Departamento de Bioingeniería, Unidad Profesional Interdisciplinaria de Biotecnología, Instituto Politécnico Nacional, México City, Mexico

ABSTRACT

Aromatic acids are commodity chemicals, principally in the food and plastic industries, they are used in some cases in the food industry as food preservatives, and in other cases, as the raw materials for plastic envelopes and films. The production of these compounds is widely distributed around the globe. The net production of all of the aromatic acids is more than 50 million tons, therefore a great amount of waste acids are released into the environment. This chapter deals with the biodegradation of these kind of compounds in wastewater streams either by aerobic or anaerobic processes.

CHEMISTRY OF AROMATIC ACIDS

Aromatic acids are commodity chemicals, principally in the food and plastic industries. The most well-known aromatic acids are benzoic acid (CAS: 65-85-0), phthalic acid (CAS:88-99-3), isophthalic acid (CAS: 121-91-5) and terephthalic acid (CAS: 100-21-0), they are formed by an aromatic ring and one or two carboxylic groups (Figure 1). Other aromatic acids with less industrial importance are trimesic acid (CAS: 554-95-0) and hemimellitic acid (CAS:569-51-7) (Park and Sheehan 2000).

Although phthalic, isophthalic and terephthalic acids are isomers, their physical properties such as melting point, enthalpy of formation and solubility are different. Table 1 shows the most important physical properties of benzoic, phthalic, isophthalic and terephthalic acids.

* claudiogaribay@yahoo.com

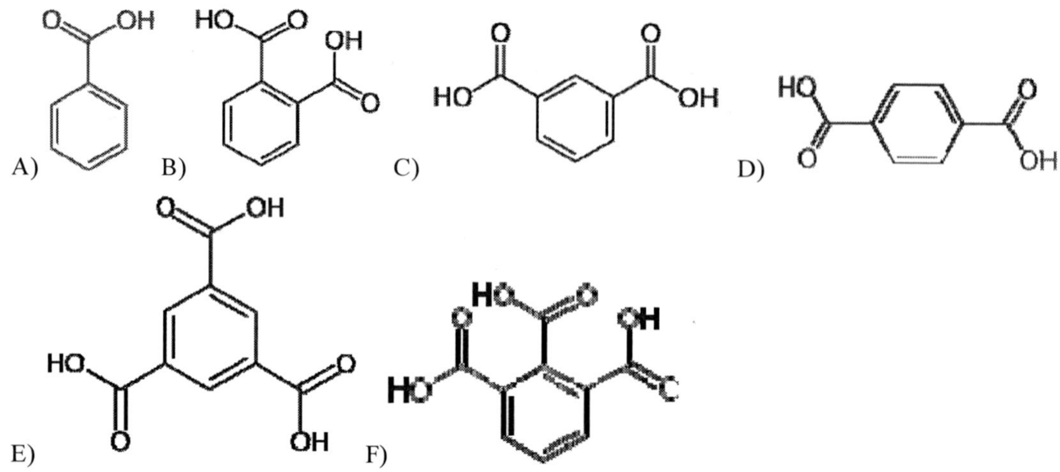

Figure 1. A) Benzoic Acid, B) Phthalic Acid, C) Isophthalic Acid, D) Terephthalic Acid, E) Trimesic Acid, F) HemimelliticAcid.

Table 1. Physical properties of aromatic compounds (Bergström et al. 2003; Lorz et al. 2000; Park and Sheehan 2000; Sheehan 2000)

Compound	Terephthalic Acid	Benzoic Acid	Isophthalic Acid	Phthalic Acid
Molecular formula	$C_6H_4(COOH)_2$	C_6H_5COOH	$C_6H_4(COOH)_2$	$C_6H_4(COOH)_2$
Molar mass (g/mol)	166.13	122.12	166.13	166.13
Density (g/cm3)	1.52	1.32	1.52	1.59
Melting point (°C)	300	122.35	384	191 - 230
ΔH_f° at 25°C (kJ/mol)	-816	-382	-803	-782
Solubility (g/100 g water)	0.0017	0.34	0.012	0.7

Important chemical compounds derived from the phthalate acids are the phthalate ester isomers, which are dimethyl phthalate (DMP), dimethyl isophthalate (DMI) and dimethyl terephthalate (DMT). Phthalic acid esters (PAEs) are a class of refractory organic compounds widely used as additives or plasticizers in the manufacturing of plastics (Amir et al. 2005). DMP is typically used in polyvinyl chloride (PVC) and cellulose ester-based plastics, including cellulose acetate and cellulose butyrate as a plasticizer. DMI is used in the production of high molecular weight polyester elastomers, and it is also used as a polyester intermediate to improve the intensity and transparency of PET soda pop bottles. In the field of engineering, plastic DMI is used to strengthen the intensity of fibers and films. DMT is used mainly to produce polyester resins, fibers and films as well as engineering polymers (Li et al. 2005).

INDUSTRIAL IMPORTANCE OF AROMATIC ACIDS

Benzoic Acid

Approximately 638,000 tons of benzoic acid is produced globally per year (Qi et al. 2009). It is produced commercially by the oxidation of toluene (Xu et al. 2009). Benzoic acid and its salts have been widely used in the food industry for many years as an important food preservative for fruit juices, pickles, wine and pharmaceutical preparations in order to inhibit various bacteria, yeasts and fungi growth in acidic media (at concentrations of 5-10mM) (Krebs et al. 1983; Warth 1991). It is either added directly to or created from reactions with its sodium, potassium, or calcium salt. The growth inhibition mechanism begins with the absorption of Benzoic Acid into the cell, when the intracellular pH changes to 5 or lower, the fermentation of glucose through phosphofructokinase is decreased by 95%, thus, the efficacy of benzoic acid and benzoate is higher when the pH of the food is low (Pastorova et al. 1997).

Phthalic Acid

Phthalic acid has remained unimportant industrially, it is formed as a byproduct in the manufacturing of phthalic anhydride. Nonetheless, a wide variety of dialkyl phthalates are produced and marketed, for example: diethyl phthalate, dibutyl phthalate, di-2-ethylhexyl phthalate, diisononyl phthalate, diisodecyl phthalate, among others (Lorz et al. 2000).

Isophthalic Acid

Isophthalic acid and terephthalic are produced by the oxidation of xylenes, such as m-xylene and p-xylene, with the Co-Mn-Br homogeneous catalyst system (Lv et al. 2011)

Isophthalic acid is used as a feedstock for unsaturated polyesters as well as a co-monomer in some saturated products. Also, isophthalic acid is used as a co-monomer with terephthalic acid in bottles and specialty resins. It is also a feedstock for coatings and for high-performance unsaturated polyesters, being first reacted with maleic anhydride and then this resin is cross-linked with styrene (Sheehan 2000). Worldwide capacity available for production of isophthalic acid was about 270,000 tons in 1994 and has risen (Park and Sheehan 2000).

Terephthalic acid

Terephthalic acid is used almost exclusively to produce saturated polyesters. Polyethylene terephthalate, the alternating copolymer of terephthalic acid and 1,2-ethanediol, accounts for >90% of this use, with a worldwide demand of 12.6×10^6 tons in 1992. Textile and industrial fiber accounted for 75% of this demand. Polyester is the largest volume synthetic fiber. Food and beverage containers accounted for 13%. Other uses are for poly butylene terephthalate, a high-performance molding resin made by reaction with 1,4-

butanediol (Sheehan 2000). Annually, more than 50 million tons of pure terephthalic acid is produced in the world (Zhang et al. 2010b). The production of terephthalic acid is based on the well-established process developed by the American Amoco. The process comprises of two steps; in the first step, crude terephthalic acid (CTA) is produced by wet oxidation of p-xylene into acetic acid, and in the second stage, CTA conversion into terephthalic acid through hydrogenation using palladium as a catalyst (Karthik et al. 2008).

The manufacturing process of pure terephthalic acid (PTA) comprises three major stages: (i) Manufacturing of (CTA); (ii) Purification of CTA to PTA; (iii) Recovery and regeneration of CTA oxidation catalyst. In the CTA stage, para-xylene is oxidized by air in the presence of cobalt and manganese as the catalysts, and acetic acid as the solvent, at 200 °C and 15 kg/cm^2 pressure in the reactor. The off-gases are led into an off gas turbine from where it is vented. The CTA thus produced is crystallized and filtered to separate CTA from the solvent and catalysts. Thereafter, CTA is dried and sent for purification. This stage involves an acetic acid distillation process, where the byproduct water is formed and removed. The CTA is purified to obtain PTA by dissolving it in pure water and converting the intermediate products (4-carboxyl benzaldehyde) to para-toluic acid, at around 85 kg/cm^2 and at a temperature of over 290 °C, in the presence of a lead catalyst in a hydrogen atmosphere. The para-toluic acid formed is recycled into the process for further oxidation to produce PTA. The PTA is subsequently crystallized, dried and packed in container bags. The catalyst is recovered and regenerated (Pophali et al. 2007).

Due to the global utilization of aromatic acids in large quantities, these have been detected in every environment in which they have been sought (Fan et al. 2004), hence, they produce several health and environmental problems. The environmental occurrence of terephthalic and isophthalic acids is less abundant than phthalic acid esters (Kleerebezem et al. 1999b). However a number of point sources releasing large amounts of terephthalic and isophthalic acids into the environment exist (Tsuno and Kawamura, 2009; Pophali et al., 2007). The main source consists of the concentrated waste generated during the production of terephthalic acid and isophthalic acid. Also, accidental spills may occur around chemical production plants. Specific characteristics of the waste stream generated during terephthalic acid production are the high temperature (40–50 °C) and the high concentration of organic pollutants (5-20 $kg\text{-}COD{\cdot}m^{-3}$). Main compounds in this wastestream are, in decreasing order of concentration, terephthalic acid, acetic acid, benzoic acid and para-toluic acid (Kleerebezem et al. 1999b).

Aromatic Acids and Health Diseases

Since benzoic acid is an FDA-approved compound, it is safety for human health, however, in some cases, adverse effects may appear including asthma, urticaria, metabolic acidosis and convulsions (Chang and Zylstra 1998; Tfouni and Toledo 2002). An extensive review of the toxicity of benzoate showed that at high doses, it interferes with an intermediary metabolism, including the urea cycle, gluconeogenesis, fatty acid metabolism and the tricarboxylic acid cycle (Qi et al. 2009). Also, it has been found that benzoic acid produces benzene by reacting with ascorbic acid in the presence of transition-metal catalyst (Gardner and Lawrence 1993).

The major source of human exposure to phthalate isomers and their esters is through ingestion, inhalation, and dermal exposure for their whole lifetime, since the intrauterine life. The main focus has moved away from the hepatotoxic effects to the endocrine disrupting potency of these chemicals (Latini 2005). Humans exposed to phthalate isomers showed teratogenic, estrogenic and reproductive problems (Dillingham and Autian 1973; Koch et al. 2003; Matsumoto et al. 2008). Recent studies suggest that several phthalates or their metabolites, respectively, can cross the placenta barrier and reach the human fetus (Wittassek et al. 2009).

Small chain alkyl esters of phthalates are shown to be more toxic because of their high solubility compared to longer chain homologs. Exposure during the development results in irreversible damage (Colborn et al. 1993). In mice these compounds induced malformations of the reproductive tract, causing hypospadias and cryptorchidism (Chang and Zylstra 1998). They are detected in body fluids and also are known to induce bladder stone formation, irritation of skin, eyes, respiratory tract and blurred vision (Dai et al. 2005).

Known as endocrine-disrupting chemicals, some phthalates may also interfere with the reproductive system and development of animals and humans (Andreoni and Gianfreda 2007; Jobling et al. 1995).

Isophthalic acid is reported to be a competitive inhibitor of glutamate dehydrogenase (GDH) of eukaryotic organisms (Cunliffe et al. 1983; Rogers et al. 1972). GDH is an extremely important enzyme that catalyses the conversion of α-ketoglutarate to glutamate and vice versa (Hudson and Daniel 1993).

Terephthalic acid has been studied for toxicity effects in animals and humans, and it was found that generates bladder stones and bladder tumor in rats (Chin et al. 1981), damages of spermatogenic cells, and motility of spermatozoa in rats and mice (Cui et al. 2004; Dai et al. 2005; Zhang et al. 2010a; Zhang et al. 2010b) and slight effects on lung functions in humans (Dai et al. 2005). Also it has been indicated that exposure to terephthalic acid can cause impairment of liver (Lamb et al. 1987), lung and kidney (Meyer et al. 1984), as well as DNA damage (Zhang et al. 2010b).

Biodegradation of Aromatic Acids

Since these compounds are of environmental concern, several efforts have been made to diminish their presence in the environment. Biological wastewater treatment has proven to be a good alternative for degradation of aromatic acids (Andreoni and Gianfreda 2007; Macarie et al. 1992; Pophali et al. 2007). The abiotic removal of phthalic acids proceeds very slowly, with half-life for chemical hydrolysis ranging from 4 months to over 100 years (Wolfe et al. 1980), the principal method for removal is microbial activity. However advanced oxidation processes (AOP), supercritical water oxidation, UV-assisted ozonation (UV/O_3), ozone-assisted photochemical oxidation ($UV/O_3/H_2O_2$), photo-fenton oxidation ($UV/H_2O_2/FeSO_4$), ozone-assisted photo-fenton oxidation ($UV/O_3/H_2O_2/FeSO_4$) and radiation treatment using gamma-ray have been used for the treatment of wastewater containing aromatic acids (Karthik et al. 2008). Since physicochemical methods are expensive, and the generation of toxic intermediates and sludge leads to secondary pollution, the biological wastewater treatment has proven to be a techno-economic alternative for degradation of aromatic acids (Andreoni and Gianfreda 2007; Pophali et al. 2007). The physicochemical treatments are useful as an option for the pretreatment of PTA wastewater to enhance the biodegradation rate (Karthik et al. 2008). Biological wastewater treatment includes anaerobic or aerobic treatment with activated sludge, however, it has been proven that a combination of both processes is more efficient (Pophali et al. 2007). Phthalate isomers and their esters are widely spread in the environment, thus, microbes have evolved and adapted new pathways to degrade these compounds. Bacterial degradation is efficient as these compounds are either biotransformed or mineralized completely.

Benzoic acid is anaerobically reduced to carboxycyclohexene level, followed by ring cleavage (Schink et al. 1992). On the other hand, several phthalic acids can be metabolized in both aerobic and anaerobic environments. The most common pathway for aerobic degradation of phthalate is through the protocatechuate pathway, followed by ring cleavage and complete mineralization to carbon dioxide and water (Vamsee-Krishna and Phale 2008a). Mineralization of phthalate under anaerobic conditions has only been described for denitrifying cultures and methanogenic cultures (Aftring et al. 1981; Nozawa and Maruyama 1988; Taylor and Ribbons 1983).

In methanogenic consortia, bacteria degrade all three phthalate isomers and their esters, producing acetate and methane as the end products. The first and the foremost step in the degradation of phthalate isomers is the production of esters of CoA by an inducible acyl CoA synthetase. Decarboxylation of these esterified phthalates by decarboxylase results in the formation of benzoyl CoA. In a few cases, phthalates are first decarboxylated in the presence of specific decarboxylases followed by esterification with CoA. Reduction of benzoyl CoA produces cyclohex-1-ene-carboxylate ester of CoA. This product undergoes ring cleavage and forms pimelyl CoA via 2-oxocyclohexanecarboxylate and enters into the β-oxidation pathway. External addition of acetate, pyruvate or lactate repressed the utilization of phthalates. Shorter chain compounds are metabolized at higher rates as compared to their longer chain counterparts (Vamsee-Krishna and Phale 2008a).

Benzoic acid as the sole carbon source can be degraded both aerobically (Chidambara Raj et al. 1997) and anerobically (Elder and Kelly 1994; Van der Woude et al. 1994) in presence of microorganism principally from the Pseudomonas genus and also some fungi such as *Aspergillus niger* and *Rhodotorula graminis* (Wright 1993). Studies have demonstrated that Benzoic acid can be mineralized using co-cultures of *Methanospirillum hungatei* and *Pelotomaculum terephthalicum* or *Pelotomaculum isophthalicum* (Qiu et al. 2006).The biodegradation of benzoic acid has been shown to occur faster than the biodegradation of terephthalic acid when a mixture of these acids is used as the carbon source (Chidambara Raj et al. 1997).

Degradation of the three phthalic isomers under anaerobic conditions using methanogenic consortia (Kleerebezem et al. 1999a; Kleerebezem et al. 1999d; Qiu et al. 2004) and co-cultures of *Methanospirillum hungatei* and *Pelotomaculum terephthalicum* or *Pelotomaculum isophthalicum* (Qiu et al. 2006) have been demonstrated using low concentrations of phthalic isomers. However, there are few reports of phthalic isomers biodegradation using pure cultures (Table 2).

Under aerobic conditions, degradation of phthalic isomers was possible mainly by utilizing pure cultures (Table 2), while under anaerobic conditions, the use of pure cultures have not shown to be a good alternative. There is a lack of information regarding the biodegradation of phthalic isomers using aerobic consortia (Zeng et al. 2007).

Unlike phthalic acid and terephthalic acid, isophthalic acid was found to be resistant to aerobic degradation since very few microbial species have been reported to utilize this isomer as the sole source of carbon and energy (Vamsee-Krishna and Phale 2008b).

Table 2. Microorganisms suitable for phthalic isomers degradation

Microorganism	Isomer	Environment	Ref.
Pseudomonas sp.	Phthalic Acid	Anaerobic	1
Burkholderia cepacia	Phthalic Acid	Aerobic	2
Comamonas acidovorans	Phthalic Acid	Aerobic	3
Rhodococcus sp.	Phthalic Acid	Aerobic	4
Arthrobacter keyseri	Phthalic Acid	Aerobic	5, 6
Moraxella sp.	Phthalic Acid	Aerobic	7
Comamonas testosteroni	Phthalic Acid	Aerobic	8
Mycobacterium vanbaalenii	Phthalic Acid	Aerobic	9
Comamonas testosteroni	Terephthalic Acid	Aerobic	8
Comamonas testosteroni	Isophthalic Acid	Aerobic	8

1: (Nozawa and Maruyama 1988); 2: (Chang and Zylstra 1998); 3: (Wang et al. 2003); 4: (Choi et al. 2005); 5: (Eaton 2001); 6: (Eaton and Ribbons 1982); 7: (Rani et al. 1996); 8: (Wang et al. 1995); 9: (Stingley et al. 2004)

Terephthalic acid is an important chemical usually produced via a wet oxidation process, where p-xylene is oxidized by air in the presence of acetic acid (Frank and Stadelhofer 1988). During the terephthalic acid manufacturing procedures, a high-strength organic wastewater is generated in the ratio of 3–10 m^3 wastewater per ton of purified terephthalic acid (Kleerebezem et al. 2005). The major organic compounds in the effluent of terephthalic acid wastewater are terephthalic acid, acetic acid, benzoic acid and p-toluic acid (Zhang et al., 2006; Kleerebezem et al. 1999c; Macarie et al. 1992). Other minor organic pollutants such as trimellitic acid, o-phthalic acid and 4-carboxybenzaldehyde are also present in the wastewater (Lee et al. 2005).

The terephthalic acid wastewater is also characterized by high chemical oxygen demand (COD), alkaline pH as well as a strong brown color, posing serious environmental problems.

Traditionally, the wastewater containing terephthalic acid has been treated by using microorganisms since this is a techno-economic alternative (Pophali et al. 2007).

The treatment of aromatic acids effluent from the industry using various anaerobic and aerobic processes came to practice within the last two decades (Deshmukh et al. 2005). It is desirable to undertake a systematic investigation of aromatic acid plant waste for which practically no information is available in the published literature. Macarie et al. (1992), studied the anaerobic treatment of a terephthalic acid wastewater treatment plant using two Upflow Anaerobic Sludge Blanket (UASB) reactors, however, the process was not successful; the failure was explained by sludge inhibition caused by toxic characteristics of the wastewater. For instance, they concluded that for a successful application of this technology to the terephthalic acid removal, disposal of primary settled solids and reduction in chemical consumption for effluent neutralization is necessary. Karthik et al. (2008) proposes the use of a coagulation–flocculation process as pretreatment for the wastewater containing terephthalic acid, in order to eliminate the problems found by Macarie et al. (1992). Also, Zhang et al., (2006) studied the biodegradation of benzoic acid, phthalic acid and terephthalic acid in a pilot wastewater treatment system with a novel strain denominated strain Fhhh.

Experiences on the Degradation of Terephthalic Acid at UPIBI

It is important to note that there are no previously reports regarding the estimation of kinetic and stoichiometric parameters during aromatic acids degradation using a mixed culture, neither with pure cultures. In our research group (Laboratory of Bioengineering, UPIBI-IPN, Mexico), a kinetic and stoichiometric characterization of the biodegradation of terephthalic acid was carried out at laboratory scale, using two bacteria that were isolated from a soil contaminated with terephthalic acid. Degradation studies were done at different pH, agitation rates and concentrations of terephthalic acid. Figure 2 shows the results obtained for *Rhodococcus sp.* growing at different pH. From Figure 2 it was observed that the optimal growth occurred at pH of 7 while at pH of 9 there is no growth, for instance, it was selected a pH value of 7 as the optimal for both *Rhodococcus sp.* (Campos-Tejeda, 2011).

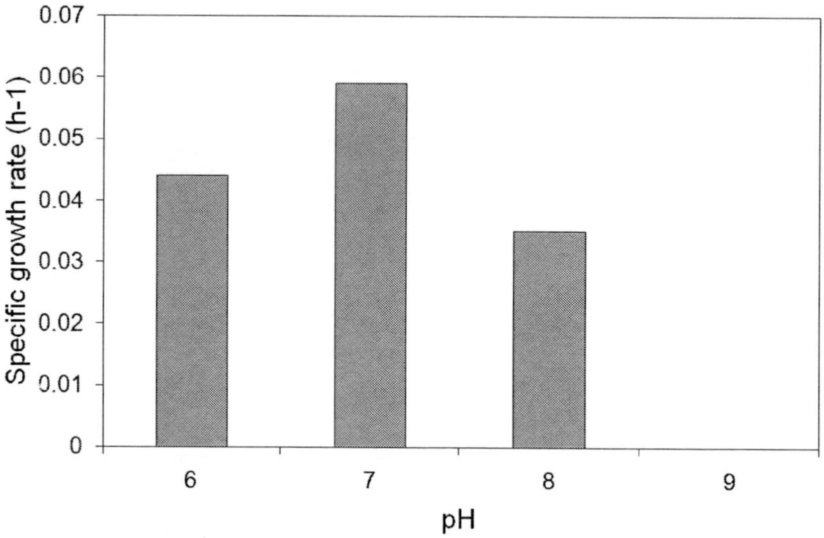

Figure 2. Growth of *Rhodococcus sp.* at different pH values.

Buchs et al., (2001) pointed out problems related to oxygen limitation in biotechnological processes and degrading cultures, such as metabolic slowing down and production of undesired toxic intermediary products. It is therefore of major importance to gather kinetic and stoichiometric data of the terephthalic acid aerobic degradation process, including potential inhibitory effect and oxygen requirements. To examine the influence of the oxygen transfer rate in the shaking flask, two experiments at different agitation speeds were carried out. Figure 3 shows that the specific growth rate can be increased almost 100% when the agitation speed is change form 150 rpm to 200 rpm (Campos-Tejeda, 2006).

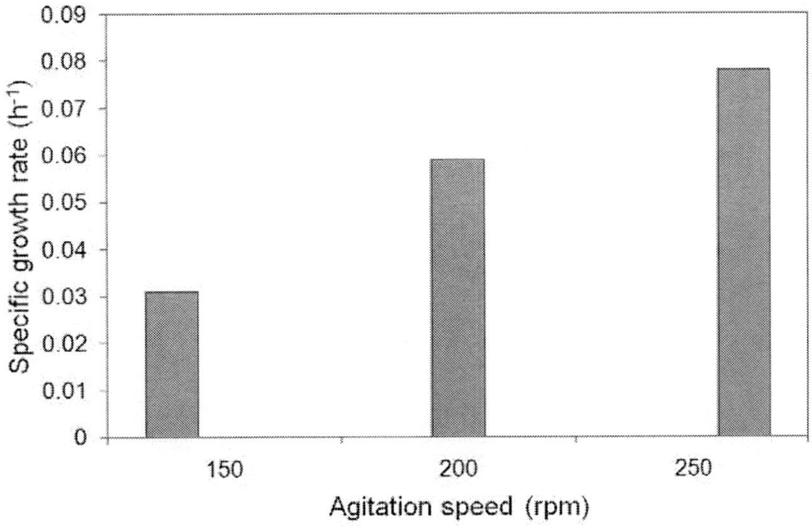

Figure 3. Growth of *Rhodococcus sp* at different agitation speeds.

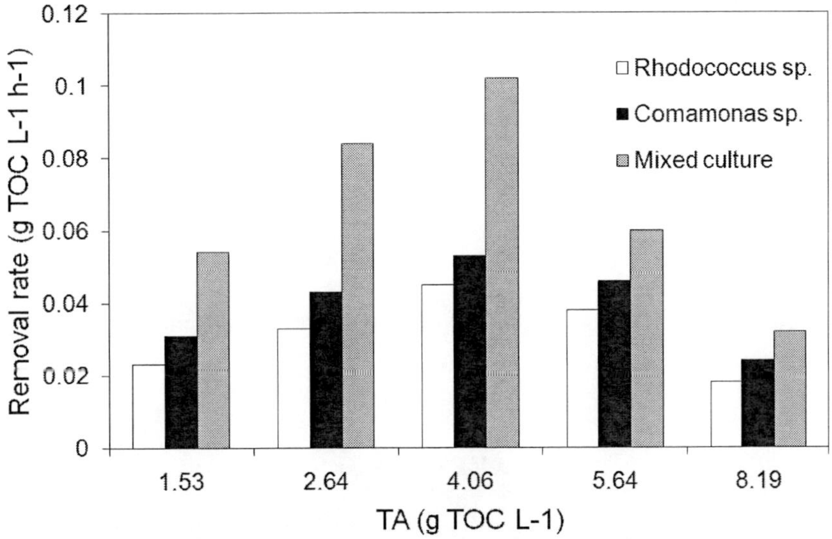

Figure 4. Terephthalic acid biodegradation at different concentration of terephthalic acid.

Figure 4 shows the results of terephthalic acid degradation with the two bacteria (*Rhodococcus sp.* and *Comamonas sp.*) and the mixed culture of them. The terephthalic acid was measured as total organic carbon (TOC units). The results in Figure 4 show that both microorganisms are able to fully degrade the added TA as complete mineralization was observed with the TOC analysis. However, a greater degradation capacity with the *Comamonas sp.* culture was observed (21 ± 5 %). The degradation rate was increased as the initial concentration increased until 4.058 g TOC L^{-1}. At 5.642 and 8.191 g TOC L^{-1}, the degradation rates were slow for both cultures. Inhibition by TA in anerobic process has been previously reported by Tsuno and Kawamura (2009) from 0.9 to 3.934 g TOC L^{-1}. To our knowledge, there are no previously reports on inhibition by terephthalic acid in the aerobic process, however, the data presented in this work suggested an inhibition by TA after 4.058 g TOC L^{-1}.

During the mixed culture, both substrate and biomass concentration were measured and used to estimate kinetic and stoichiometric parameters, according to a double Haldane model. Four important parameters were estimated: terephthalic acid affinity constant, 1.67 ± 0.04 g TOC L^{-1}; terephthalic acid inhibition constant, 2.57 ± 0.31 g TOC L^{-1}; terephthalic acid maximum removal rate, 0.29 ± 0.02 g TOC L^{-1} h^{-1} and growth yield 0.32 ± 0.08 g TOC g^{-1} TOC.

Conclusions

Aromatic acids are important commodity chemicals used in several industries and their production and applications are constantly increasing, this trend has lead to a pollution increment. People that work in aromatic acids factories are in continuous contact with these kind of chemicals. Even though there are studies of its toxicity in rats and different animals, there are only few works of its effect in human health. For instance, it is desirable to start studies on the toxic effect of aromatic acids in human health.

For further development and optimization of bioprocesses for wastewater treatment containing aromatic acids, it is necessary to determine the kinetic and stoichiometric parameters of the microorganisms used in these processes. It is necessary to develop new technologies that can deal with the waste generation of these compounds and to understand in a better way the ones that are currently used.

Acknowledgements

The authors gratefully acknowledge the Instituto de Ciencia y Tecnología del Distrito Federal for the financial support (PICS08-56).

References

Aftring, R.P., Chalker, B.E,. Taylor, B.F. (1981). Degradation of phthalic acids by denitrifying, mixed cultures of bacteria. *Appl Environ Microbiol* 41: 1177-83.

Amir, S., Hafidi, M., Merlina, G., Hamdi, H., Jouraiphy, A., El Gharous, M., Revel, J.C. (2005). Fate of phthalic acid esters during composting of both lagooning and activated sludges. *Process Biochemistry* 40: 2183-2190.

Andreoni, V., Gianfreda, L. (2007). Bioremediation and monitoring of aromatic-polluted habitats. *Applied Microbiology and Biotechnology* 76: 287-308.

Bergström, C.A.S., Norinder, U., Luthman, K., Artursson, P. (2003). Molecular Descriptors Influencing Melting Point and Their Role in Classification of Solid Drugs. *Journal of Chemical Information and Computer Sciences* 43: 1177-1185.

Buchs, J. (2001). Introduction to advantages and problems of shaken cultures. *Biochemical Engineering Journal* 7: 91-98.

Campos-Tejeda, P. (2011). Biodegradación de ácido tereftálico es sistemas modelos aerobios. *Master degree Thesis*. UPIBI-IPN.

Colborn, T., vom Saal, F.S., Soto, A.M. (1993). Developmental effects of endocrine-disrupting chemicals in wildlife and humans. *Environ Health Perspect* 101: 378-84.

Cui, L., Dai, G., Xu, L., Wang, S., Song, L., Zhao, R., Xiao, H., Zhou, J., Wang, X. (2004). Effect of oral administration of terephthalic acid on testicular functions of rats. *Toxicology* 201: 59-66.

Cunliffe, D., Leason, M., Parkin, D., Lea, P.J. (1983). The inhibition of glutamate dehydrogenase by derivatives of isophthalic acid. *Phytochemistry* 22: 1357-1360.

Chang, H.K., Zylstra, G.J. (1998). Novel organization of the genes for phthalate degradation from Burkholderia cepacia DBO1. *J Bacteriol* 180: 6529-37.

Chidambara Raj, C.B., Ramkumar, N., Haja jahabar siraj, A., Chidambaram, S. (1997). Biodegradation of Acetic, Benzoic, Isophthalic, Toluic and Terephthalic acids Using a Mixed Culture: Effluents of PTA Production. *Process Safety and Environmental Protection* 75: 245-256.

Chin, T.Y., Tyl, R.W., Popp, J.A., Heck, H.D.A. (1981). Chemical urolithiasis: 1. Characteristics of bladder stone induction by terephthalic acid and dimethyl terephthalate in weanling Fischer-344 rats. *Toxicology and Applied Pharmacology* 58: 307-321.

Choi, K.Y., Kim, D., Sul, W.J., Chae, J.C., Zylstra, G.J., Kim, Y.M., Kim, E. (2005). Molecular and biochemical analysis of phthalate and terephthalate degradation by Rhodococcus sp. strain DK17. *FEMS Microbiol Lett* 252: 207-13.

Dai, G., Cui, L., Song, L., Cheng, J., Zhong, Y., Zhao, R., Wang, X. (2005). Bladder epithelial cell proliferation of rats induced by terephthalic acid-calculi. *Food and Chemical Toxicology* 43: 217-224.

Deshmukh, N.A., Goel, V.S., Joshi, J.B., Mathew, T. (2005). Kinetics of Aerobic Biological Oxidation of Purified Terephthalic acid Plant Waste. *Process Safety and Environmental Protection* 83: 224-230.

Dillingham, E.O., Autian, J. (1973). Teratogenicity, mutagenicity, and cellular toxicity of phthalate esters. *Environ Health Perspect* 3:81-9.

Eaton, R.W. (2001). Plasmid-Encoded Phthalate Catabolic Pathway in Arthrobacter keyseri 12B. *J. Bacteriol.* 183: 3689-3703.

Eaton, R.W., Ribbons, D.W. (1982). Metabolism of dimethylphthalate by Micrococcus sp. strain 12B. *J. Bacteriol.* 151: 465-467.

Elder, D.J.E,. Kelly, D.J. (1994). The bacterial degradation of benzoic acid and benzenoid compounds under anaerobic conditions: Unifying trends and new perspectives. *FEMS Microbiology Reviews* 13: 441-468.

Fan, Y., Wang, Y., Qian, P-Y., Gu, J-D. (2004). Optimization of phthalic acid batch biodegradation and the use of modified Richards model for modelling degradation. *International Biodeterioration & Biodegradation* 53: 57-63.

Frank, H.G., Stadelhofer, J.W. (1988). p-Xylene and its Derivatives: Terephthalic acid. In: Franck, H.G., editor. *Industrial Aromatic Chemistry*. Berlin, Germany.: Springer-Verlag.

Gardner, L.K., Lawrence, G.D. (1993). Benzene production from decarboxylation of benzoic acid in the presence of ascorbic acid and a transition-metal catalyst. *Journal of Agricultural and Food Chemistry* 41: 693-695.

Hudson, R.C., Daniel, R.M. (1993). L-glutamate dehydrogenases: distribution, properties and mechanism. *Comp Biochem Physiol B* 106: 767-92.

Jobling, S., Reynolds, T., White, R., Parker, M.G., Sumpter, J.P. (1995). A variety of environmentally persistent chemicals, including some phthalate plasticizers, are weakly estrogenic. *Environ Health Perspect* 103.

Karthik, M., Dafale, N., Pathe, P., Nandy, T. (2008). Biodegradability enhancement of purified terephthalic acid wastewater by coagulation-flocculation process as pretreatment. *Journal of Hazardous Materials* 154: 721-730.

Kleerebezem, R., Beckers, J., Hulshoff Pol, L.W., Lettinga, G. (2005). High rate treatment of terephthalic acid production wastewater in a two-stage anaerobic bioreactor. *Biotechnol Bioeng* 91: 169-79.

Kleerebezem, R., Hulshoff Pol, L.W., Lettinga, G. (1999a). Anaerobic degradation of phthalate isomers by methanogenic consortia. *Appl Environ Microbiol* 65: 1152-1160.

Kleerebezem, R., Hulshoff Pol, L.W., Lettinga, G. (1999b). Energetics of product formation during anaerobic degradation of phthalate isomers and benzoate. *FEMS Microbiology Ecology* 29: 273-282.

Kleerebezem, R., Ivalo, M., Hulshoff Pol, L.W., Lettinga, G. (1999c). High-rate treatment of terephthalate in anaerobic hybrid reactors. *Biotechnol Prog* 15: 347-57.

Kleerebezem, R., Pol, L.W., Lettinga, G. (1999d). Anaerobic biodegradability of phthalic acid isomers and related compounds. *Biodegradation* 10: 63-73.

Koch, H.M., Drexler, H., Angerer, J. (2003). An estimation of the daily intake of di(2-ethylhexyl)phthalate (DEHP) and other phthalates in the general population. *Int J Hyg Environ Health* 206: 77-83.

Krebs, H.A., Wiggins, D., Stubbs, M., Sols, A., Bedoya, F. (1983). Studies on the mechanism of the antifungal action of benzoate. *Biochem J* 214: 657-663.

Lamb, J.C., Chapin, R.E., Teague, J., Davis Lawton, A., Reel, J.R. (1987). Reproductive effects of four phthalic acid esters in the mouse. *Toxicology and Applied Pharmacology* 88: 255-269.

Latini, G. (2005). Monitoring phthalate exposure in humans. *Clin Chim Acta* 361: 20-9.

Lee, M.W., Joung, J.Y., Lee, D.S., Park, J.M., Woo, S.H. (2005). Application of a Moving-Window-Adaptive Neural Network to the Modeling of a Full-Scale Anaerobic Filter Process. *Industrial & Engineering Chemistry Research* 44: 3973-3982.

Li, J., Gu, J-D., Pan, L. (2005). Transformation of dimethyl phthalate, dimethyl isophthalate and dimethyl terephthalate by Rhodococcus rubber Sa and modeling the processes using the modified Gompertz model. *International Biodeterioration & Biodegradation* 55: 223-232.

Lorz, P.M., Towae, F.K., Enke, W., Jäckh, R., Bhargava, N., Hillesheim, W. (2000). Phthalic Acid and Derivatives. *Ullmann's Encyclopedia of Industrial Chemistry*: Wiley-VCH Verlag GmbH & Co. KGaA.

Lv, H-F., Wu, S-Q., Liu, N., Long, X-L., Yuan, W-K. (2011). A Study on the m-xylene Oxidation to Isophthalic Acid under the Catalysis of Bromine-free Homogeneous Catalytic System. *Chemical Engineering Journal In Press,* Accepted Manuscript.

Macarie, H., Noyola, A., Guyot, J.P. (1992). Anaerobic treatment of a petrochemical waste-water from a terephthalic acid plant. *Water Science and Technology* 25: 223-235.

Matsumoto, M., Hirata-Koizumi, M., Ema, M. (2008). Potential adverse effects of phthalic acid esters on human health: A review of recent studies on reproduction. *Regulatory Toxicology and Pharmacology* 50: 37-49.

Meyer, B.R., Fischbein, A., Rosenman, K., Lerman, Y., Drayer, D.E., Reidenberg, M.M. (1984). Increased urinary enzyme excretion in workers exposed to nephrotoxic chemicals. *The American Journal of Medicine* 76: 989-998.

Nozawa, T., Maruyama, Y. (1988). Anaerobic metabolism of phthalate and other aromatic compounds by a denitrifying bacterium. *J Bacteriol* 170: 5778-5784.

Park, C-M., Sheehan, R.J. (2000). Phthalic Acids and Other Benzenepolycarboxylic Acids. *Kirk-Othmer Encyclopedia of Chemical Technology*: John Wiley & Sons, Inc.

Pastorova, I., de Koster, C.G., Boon, J.J. (1997). Analytical Study of Free and Ester Bound Benzoic and Cinnamic Acids of Gum Benzoin Resins by GC–MS and HPLC–frit FAB–MS. *Phytochemical Analysis* 8: 63-73.

Pophali, G.R., Khan, R., Dhodapkar, R.S., Nandy, T., Devotta, S. (2007). Anaerobic-aerobic treatment of purified terephthalic acid (PTA) effluent; a techno-economic alternative to two-stage aerobic process. *J Environ Manage* 85: 1024-1033.

Qi, P., Hong, H., Liang, X., Liu, D. (2009). Assessment of benzoic acid levels in milk in China. *Food Control* 20: 414-418.

Qiu, Y.L., Sekiguchi, Y., Hanada, S., Imachi, H., Tseng, I.C., Cheng, S.S., Ohashi, A., Harada, H., Kamagata, Y. (2006). Pelotomaculum terephthalicum sp. nov. and Pelotomaculum isophthalicum sp. nov.: two anaerobic bacteria that degrade phthalate isomers in syntrophic association with hydrogenotrophic methanogens. *Arch Microbiol* 185: 172-82.

Qiu, Y.L., Sekiguchi, Y., Imachi, H., Kamagata, Y., Tseng, I.C., Cheng, S.S., Ohashi, A., Harada, H. (2004). Identification and isolation of anaerobic, syntrophic phthalate isomer-degrading microbes from methanogenic sludges treating wastewater from terephthalate manufacturing. *Appl Environ Microbiol* 70: 1617-26.

Rani, M., Prakash, D., Sobti, R.C., Jain, R.K. (1996). Plasmid-mediated degradation of o-phthalate and salicylate by a Moraxella sp. *Biochem Biophys Res Commun* 220: 377-81.

Rogers, K.S., Boots, M.R., Boots, S.G. (1972). Molecular interactions of six aromatic competitive inhibitors with bovine liver glutamate dehydrogenase. *Biochim Biophys Acta* 258: 343-50.

Schink, B., Brune, A., Schnell, S. (1992). Anaerobic degradation of aromatic compounds. In: Winkelmann, G., editor. *Microbial degradation of natural products*. p 219-242.

Sheehan, J.R. (2000). Terephthalic Acid, Dimethyl Terephthalate, and Isophthalic Acid. *Ullmann's Encyclopedia of Industrial Chemistry*: Wiley-VCH Verlag GmbH & Co. KGaA.

Stingley, R.L., Brezna, B., Khan, A.A., Cerniglia, C.E. (2004). Novel organization of genes in a phthalate degradation operon of Mycobacterium vanbaalenii PYR-1. *Microbiology* 150: 3749-3761.

Taylor, B.F., Ribbons, D.W. (1983). Bacterial Decarboxylation of o-Phthalic Acids. *Appl Environ Microbiol* 46: 1276-81.

Tsuno, H., Kawamura, M. (2009). Development of an expanded-bed GAC reactor for anaerobic treatment of terephthalate-containing wastewater. *Wat Res* 43: 417-422.

Tfouni, S.A.V., Toledo, M.C.F. (2002). Determination of benzoic and sorbic acids in Brazilian food. *Food Control* 13: 117-123.

Vamsee-Krishna, C., Phale, P. (2008a). Bacterial degradation of phthalate isomers and their esters. *Indian Journal of Microbiology* 48: 19-34.

Vamsee-Krishna, C., Phale, P.S. (2008b). Carbon source-dependent modulation of NADP-glutamate dehydrogenases in isophthalate-degrading Pseudomonas aeruginosa strain PP4, Pseudomonas strain PPD and Acinetobacter lwoffii strain ISP4. *Microbiology* 154: 3329-3337.

Van der Woude, B.J., de Boer, M., Van der Put, N.M.J., Van der Geld F.M., Prins R.A., Gottschal, J.C. (1994). Anaerobic degradation of halogenated benzoic acids by photoheterotrophic bacteria. *FEMS Microbiology Letters* 119: 199-207.

Wang, Y., Fan, Y., Gu, J.D. (2003). Microbial Degradation of the Endocrine-Disrupting Chemicals Phthalic Acid and Dimethyl Phthalate Ester Under Aerobic Conditions. *Bulletin of Environmental Contamination and Toxicology* 71:0810-0818.

Wang, Y.Z., Zhou, Y., Zylstra, G.J. (1995). Molecular analysis of isophthalate and terephthalate degradation by Comamonas testosteroni YZW-D. *Environ Health Perspect* 103: 9-12.

Warth, A.D. (1991). Mechanism of action of benzoic acid on Zygosaccharomyces bailii: effects on glycolytic metabolite levels, energy production, and intracellular pH. *Appl Environ Microbiol* 57: 3410-3414.

Wittassek, M., Angerer, J., Kolossa-Gehring, M., Schafer, S.D., Klockenbusch, W., Dobler, L., Gunsel, A.K., Muller, A., Wiesmuller, G.A. (2009). Fetal exposure to phthalates--a pilot study. *Int J Hyg Environ Health* 212: 492-498.

Wolfe, N.L., Steen, W.C., Burns, L.A. (1980). Phthalate ester hydrolysis: Linear free energy relationships. *Chemosphere* 9: 403-408.

Wright, J.D. (1993). Fungal degradation of benzoic acid and related compounds. *World Journal of Microbiology and Biotechnology* 9: 9-16.

Xu, J., He, J., Zhang, W., Yang, T., Jiao, S., Hu, X. (2009). Development on the Technique of Total Recovery of Benzoic Acid Residue. *Chinese Journal of Chemical Engineering* 17: 608-612.

Zeng, P., Zhuang, W-Q., Tay, S.T-L., Tay, J-H. (2007). The influence of storage on the morphology and physiology of phthalic acid-degrading aerobic granules. *Chemosphere* 69: 1751-1757.

Zhang, X.X., Cheng, S.P., Wan, Y.Q., Sun, S.L., Zhu, C.J., Zhao, D.Y., Pan, W.Y., (2006). Degradability of five aromatic compounds in a pilot wastewater treatment system. *International Biodeterioration & Biodegradation* 58: 94-98.

Zhang, X.X., Sun, S.L., Zhang, Y., Wu, B., Zhang, Z.Y., Liu, B., Yang, L.Y., Cheng, S.P. (2010a). Toxicity of purified terephthalic acid manufacturing wastewater on reproductive system of male mice (Mus musculus). *J Hazard Mater* 176: 300-305.

Zhang, Z., Ma, L., Zhang, X.X., Li, W., Zhang, Y., Wu, B., Yang, L., Cheng, S. (2010b). Genomic expression profiles in liver of mice exposed to purified terephthalic acid manufacturing wastewater. *J Hazard Mater* 18: 1121-1126.

In: Biotechnology
Editors: L. G. Torres Bustillos and I. Garcia Peña
ISBN 978-1-62081-071-2

Chapter 9

FLAVONOIDS WITH PHARMACOLOGICAL ACTIVITY OF MEDICINAL PLANTS

Yolanda Gómez**, *Maria Esther Bautista Ramírez and Abraham Balderas López
Pharmacology Laboratory. UPIBI-IPN, Mexico

INTRODUCTION

Flavonoids are a group of polyphenolic compounds diverse in chemical structure and characteristics. They occur naturally in fruit, vegetables, nuts, seeds, flowers, and bark and are an integral part of the human diet. (Hackett 1986). They have been reported to exhibit a wide range of biological effects, including antibacterial, antiviral (Hanasaki et al. 1994), anti-inflammatory, antiallergic, (Hanasaki et al. 1994; Hope el at.1985) and vasodilatory actions, In addition, flavonoids inhibit lipid peroxidation (LPO) (Duarte et al. 1993; Salvayre et al . 1998) capillary permeability, and fragility, and the activity of enzyme systems including cyclo-oxygenase and lipoxygenase (Torel et al. 1985; Robak et al. 1988). Flavonoids exert these effects as antioxidants, free radical scavengers, and chelators of divalent cations. (CooK and Summan, 1996).

Less is known about the absorption and metabolism of flavonoids, at the usual levels of dietary intake. They are believed to be nontoxic and if absorbed and biologically active in vivo may prevent free radical mediated cytotoxicity and LPO, which is associated with cell aging and chronic diseases such as atherosclerosis. Much evidence suggests that peroxidation of low density lipoproteins (LDL) is positively associated with atherogenesis (Rankin et al. 1993). Frankel et al. 1993, reported that phenolic compounds (including flavonoids and nonflavonoid polyphenols) isolated from red wine inhibit copper catalyzed oxidation of LDL in vitro. It is postulated that the antioxidant and free radical scavenging properties of phenolic compounds, present in red wine, may partly explain the anomaly observed in the coronary heart disease (CHD) rate between the French population who consume wine regularly and have rates of CHD lower than other populations despite similar fat intakes.

---

* ygomezipn@hotmail.com

CHEMISTRY OF FLAVONOIDS

Flavonoids are low molecular weight polyphenolic substances based on the flavan nucleus. Figure 1 shows the generic structure of flavonoids and the numbering system used to distinguish the carbon positions around the molecule. The three phenolic rings are referred to as the A, B, and C (or pyrane) rings. The biochemical activities of flavonoids and their metabolites depend on their chemical structure and the relative orientation of various moieties on the molecule. Flavonoids are classified according to their chemical structure. The major flavonoid classes include flavonols, flavones, flavanones, catechins (or flavanols), anthocyanidins, isoflavones, dihydroflavonols, and chalcones.

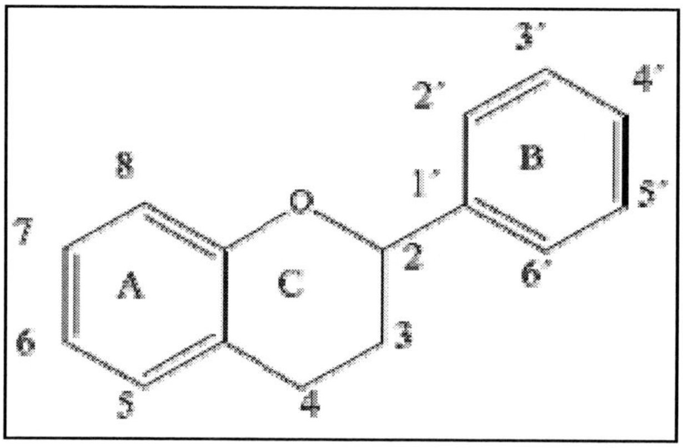

Figure 1. The generic structure of flavonoids.

Tables 1 and 2 show the major flavonoid classes and some structural variations that have been identified. The structure of flavonoids varies widely within the major classifications, and substitutions include hydrogenation, hydroxylation, methylation, malonylation, sulphation, and glycosylation. Many flavonoids occur naturally as flavonoid glycosides, and carbohydrate substitutions include D-glucose, L-rhamnose, glucorhamnose, galactose, lignin, and arabinose. Quercitrin, rutin, and robinin are the most common flavonoid glycosides in the diet. They are hydrolyzed by intestinal flora to produce the biologically active aglycone (sugar-free flavonoid). Quercetin is the subject of many studies investigating the biological effects of flavonoids, because it is the predominant flavonoid found in foods (Hertog et al.1993).

Flavonoids may be monomeric, dimeric, or oligomeric. Monomers vary greatly in size; for example flavone has a molecular weight of 222 whereas blue anthocyanin has a molecular weight of 1,759. Polymeric compounds, called tanins, are divided into two groups based on their structure: condensed and hydrolyzable. Condensed tannins are polymers of flavonoids and hydrolizable tannins contain gallic acid, or similar compounds, esterified condensed to a carbohydrate.

Galloyl groups have iron chelating properties *in vitro* and are believed to interfere with iron absorption in vivo.

Table 1. Structure of flavonoids

Flavonoid	Total No. of OH groups	Position of the OH groups	Substitutions of on the generic structure	Position of the substitutions
Myricetin	6	3,5,7,3',4',5'.		
Gossypetin	6	3,5,7,8,3',4'.		
Quercetagen	6	3,5,6,7,3',4".		
Quercetin	5	3,5,7,3',4'.		
Morin	5	3,5,7,3',4'.		
Robinetin	5	3,5,7,2',4".		
Myricetrin	5	3,7,3',4",5'.	O-Rh	3
Rutin	5	5,7,3',4'.	O-Ru	3
Kaempferol	4	3,5,7,4'.		
Quercetrin	4	5,7,3',4'.	O-Rh	3
Fisetin	4	3,7,3',4'.		
Datiscetin	4	3,5,7,2'.		
Rhamnetin	4	3,5,3',4'.	O-Me	7
Tamarixetin	4	3,5,7,3'	O-Me	4'
Silybin	3	3,5,7.	O-Lig-O	4'
Galangin	3	3,5,7.		
Kaempferide	3	3,5,7.	O-Me	4'
Diosrnin	2	3,3'.	O-Ru, O-Me	5,4'
Robinin	2	5,4'.	O-Gal-Rh, Rh	3,7'
Troxerutin	1	5.	O-Ru, O-He	3,7,3',4'

Rh = rhamnose = 6-deoxy-L-mannose (C6H1205); Lig = lignin; Ru = rutinose = 6-O-D-glucose ($C_{12}H_{12}0_{10}$); He = hydroxyethyl (CH_2CH_2OH);Me = methyl (CH_3); Gal = galactose ($C_6H_{12}0_6$).

Tea tannins consist of four main catechin components: epicatechin, epigallocatechin, epicatechin gallate, and epigallocatechin gallate. In tea epigallocatechin gallate is the predominant catechin accounting for more than half of the total catechin content structures. Enzymatic oxidation of tea catechins during fermentation of macerated tea leaves produces the dimeric thea flavins and the polymeric thearubigins of black, which produce the brightness and astringency, respectively. Thearubigins range widely in size between oligomers of four or five flavonoid units to molecules of up to 100 flavonoid units. In contrast to black tea, the flavonoids in green "Chinese" tea occur mostly as monomers because green tea is not fermented during processing. In red wine, tannins are formed by the polymerization of anthocyanins and other flavonoids producing the wine's characteristic colors, flavors, and astringency (Stalikas 2007).

Distribution and Function in Plants

Over 4,000 types of flavonoid compounds have been identified in vascular plants and these vary in type and quantity due to variations in plant growth, conditions, and maturity. Only a small number of plant species have been examined systematically for their flavonoid content and therefore the identification and quantification of all the types of flavonoids consumed by humans is incomplete.

Table 2. The structures of the major classes of flavonoids

Group	Structure	Group	Structure
Anthocyanidins	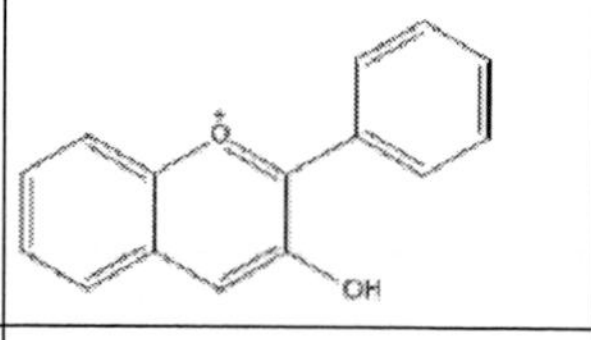	Dihydroflavonols	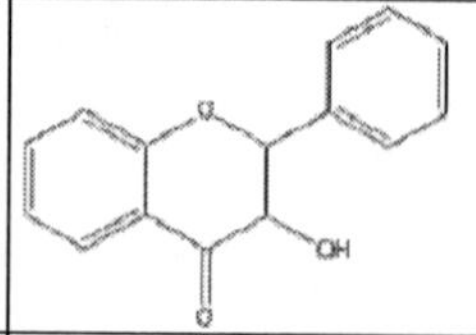
Auronas	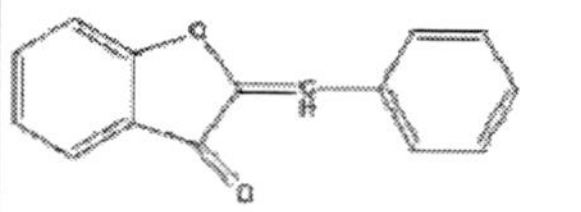	Flavones	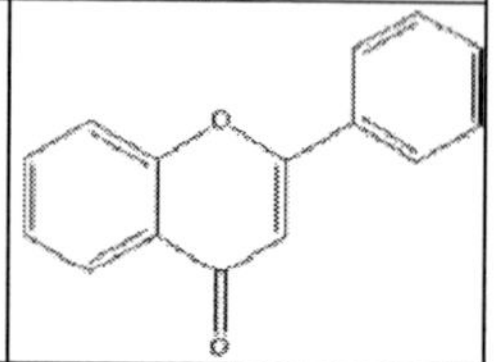
Biflavones	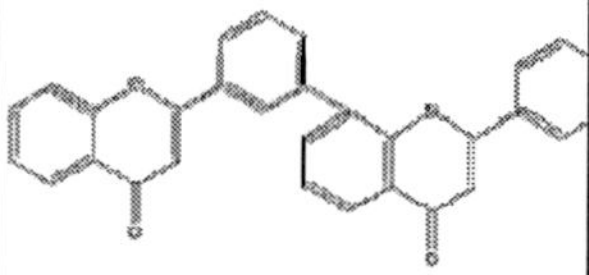	Flavonols	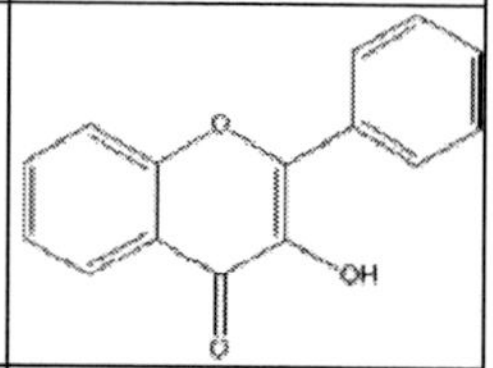
Chalcones	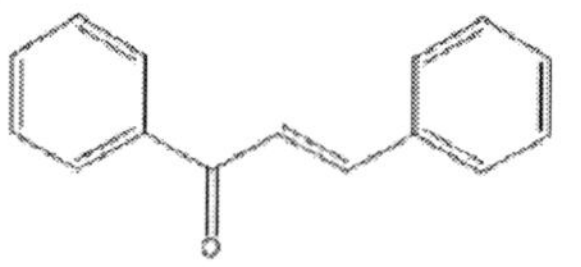	Flavondiols	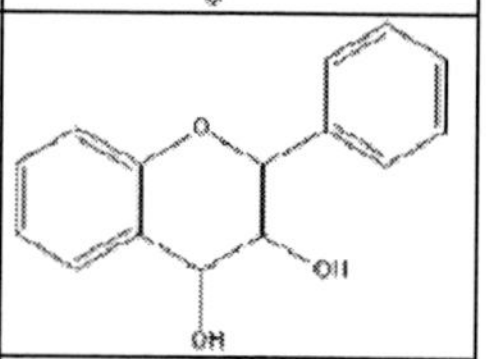
Dihidrochalcones	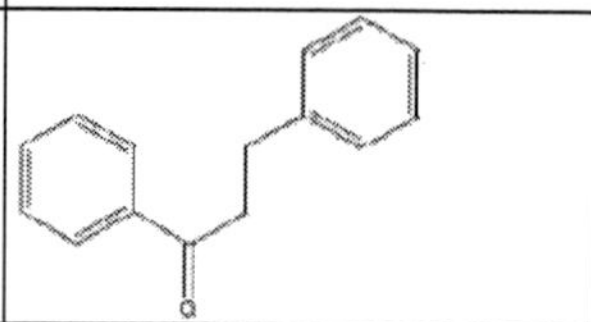	Isoflavones	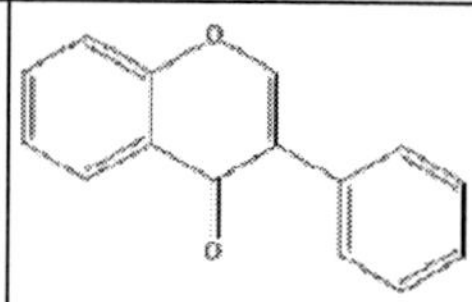
Flavanones	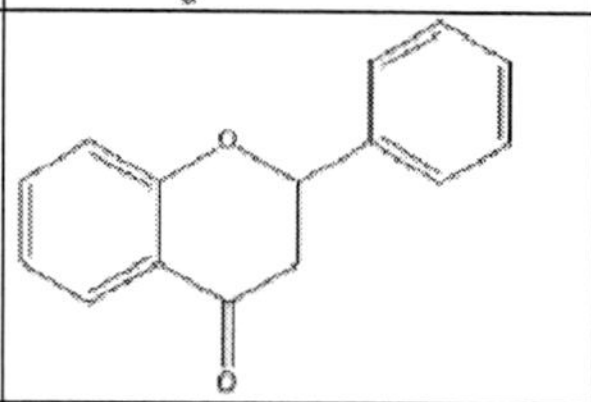	Prothocyanidins	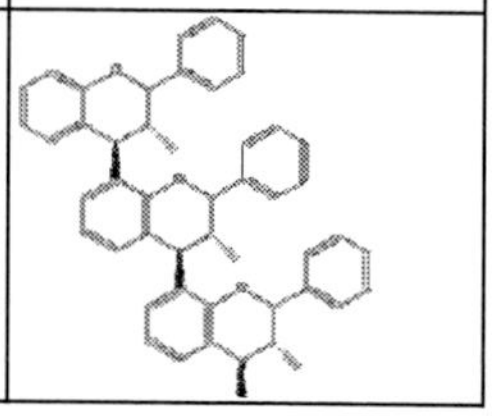

Plants have evolved to produce flavonoids to protect against fungal parasites, herbivores, pathogens, and oxidative cell injury. Conversely, flavonoids produce stimuli to assist in pollination and guide insects to their food source. For example, anthocyanins produce the pink, red, mauve, violet, and blue colors of flowers, fruits, and vegetables.

Studies in Humans

Despite the potentially significant effects of flavonoids on coronary heart disease, information about the absorption, metabolism, and excretion of individual flavonoids in humans is scarce. Some studies report that flavonoids are absorbed after oral administration, although others conclude that they are poorly absorbed and do not reach the general circulation unchanged at measurable concentrations. However, most studies of flavonoid metabolism in humans have examined the metabolism of individual flavonoids taken at pharmacological doses rather than at estimated levels of dietary intake of approximately 25 to 170 mg/day. Therefore, extrapolation of the results of these studies may be inappropriate to explain the absorption and metabolism of dietary fiavonoids.

LIPID PEROXIDATION

Polyunsaturated fatty acids (PUFA) present in cell membranes are oxidized by both enzymatic and auto-oxidative peroxidation and by free radical chain reactions. An overabundance of free radicals can lead to uncontrolled chain reactions and LPO resulting in pathological conditions that may include atherosclerosis and cancer. LPO proceeds in three stages: initiation, propagation, and termination.

In the initiation stage of LPO, free radicals abstract hydrogen from PUFA to form the lipid radical. In the propagation stage, the lipid radical reacts with molecular oxygen to form the lipid peroxy radical which breaks down to generate more free radicals thus maintaining the chain of reactions. In the termination stage, the free radical species react together or with antioxidants to form inert products. LPO can be suppressed by enzymatic inactivation of free radicals and antioxidants that inhibit the initiation stage and/or accelerate the termination stage. Thus, LPO can be prevented at the initiation stage by free radical scavengers and singlet oxygen quenchers, and the propagation chain reaction can be broken by peroxy-radical scavengers.

ANTIMICROBIAL ACTIVITY OF FLAVONOIDS

One of the funtions of flavonoids and polyphenols is their role in protecting plants against microbial invasion, the ability to inhibit spore germination of plant phatogens, and also for use against fungal and microbial pathogens of man. There is an ever increasing interest in plant flavonoids for treating microbial human diseases. Many compounds of plant flavonoids have been isolated with antifungal, antibacterial or antiviral activity (table 3)(Erasto P., et al. 2004, harborne J.B. et al 2000, Jensen P.R. et al 1998, Rice-Evans C., et al 1996, Süzgec S. et al 2005).

ANTIOXIDANTS

Peroxidative processes caused by free radicals are directly or indirectly involved in various human disease states (inflammation, carcinogenesis, ischemia/perfusión injury) and ageing. Oxygen, which is vital to aerobic organism, may become toxic through its metabolites such as superoxide anion, hydroperoxyl and hydrogen peroxide, which are reactive oxygen species (ROS) all involved in the called "oxidative stress". A peroxidative process takes place every time carbon-centred radicals are formed. Fig 2 (A) shows that this process occurs in a cycling fashion through the formation of carbon-centered radicals ($R^{\cdot}$) and peroxyl radicals ($ROO^{\cdot}$) while fig. 2(B) show how this process may be inhibited by using preventive antioxidants which inhibit the initiation process. The organic systems need to be protected by antioxidant in order to maintain their physical, chemical and functional properties. An antioxidant may therefore be defined as a compound which inhibits oxidation by either reacting with ROS to yield harmless products and/or by disrupting free radical chain reactions (Damiani E. et al 2008).

Table 3. Antimicrobial flavonoids

Flavonoid	Activity
Maackiain (3-hydroxy-8-9-methylenedioxioxypterocarpan)	Antifungal
Macronulatol (7,3′dihydroxy-2′,4′-dimehoxypterocarpan)	Antifungal
Luteolin (7-(2”-sulphatoglucoside)	Antifungal
3,5-dihydroxy-6,7,8-trimethoxyflavone	Antifungal, active against Candida albicans
2′4′dihydroxy-6′-methoxy-3′,5′-dimethyl 2′4′dihydroxy-6′-methoxy-5′-methylchalcone (2S)-4¨hydroxy-5-7,3′-trimethoxy (+_)-5,4′dihidroxy-7,3′-dimethoxyflavan	Antifungal, active against C. cucumerinum
Retrochalcone licochalcone C (4,4′-dihydroxy-2′-methoxy-3-prenyl)	Antibacterial , active against Sthapylococcus aureus
5,7 dihydroxy-3,8 dimethoxyflavone	Antibacterial , active against Sthapylococcus epidermidis
5,7,2′,6′-tetrahydroxy-6-prenyl-8-lavandolyl)-4′-methoxy-flavanone	Antibacterial , active against Sthapylococcus aureus
Isogacaonin C (5,7,4′-trihydroxy-6-[1-hydroxy-2-methylbuten-2-yl]isoflavone) Bolusanthin III (7,2′-dihydroxy-4′-methoxyisoflav-3-ene) Bolusanthin IV (6′,6′-dihydroxy-4′-methoxy-2-arylbenzofuran)	Antibacterial, active against E. Coli, B. subtilis, S. aureus, C. mycoderma
Licochalcone A,B,C,D and echinatin	Antibacterial and antifungal, active against B subtilis, S. aureus,M. luteus, E.coli, P. aeruginosa, S. cerevisiae, C. albicans and A. niger
Baicalin(5,6,7-trihydroxy—flavones 7-glucuronide	Antiviral
Robustaflavone and hinokiflavone	Antiviral
Quercetin 3-(2”galloylar-abinopyranoside)	Antiviral

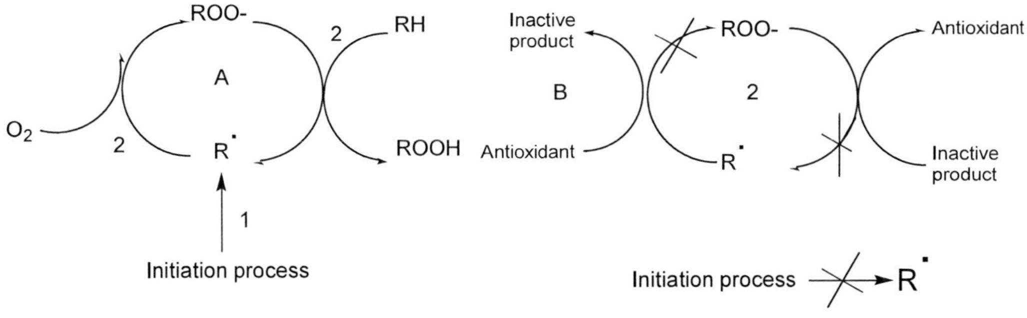

Figure 2. The peroxidative process (A), and inhibition (B) by preventive antioxidants (1), and chain-breaking antioxidants (2).

ANTIOXIDANT ACTIVITY OF FLAVONOIDS

Flavonoids have been show to act as scavengers of various oxidizing species, superoxide anion, hydroxyl radical or peroxy radicals. They may also act as quenchers of singlet oxygen. Diets rich in fruit and vegetables are protective against cardiovascular disease and certain forms of cancer. These protective effects have been atribuited, in large part, to the antioxidants including vitamin C, β-carotene and plant phenolics such as the flavonoids and phenylpropanoids. The flavonoids constitute a large class of compounds, ubiquitous in plants, containing a number of phenolic hydroxyl groups attached to ring structures, conferring the antioxidant activiy. Plants polyphenols are multifunctional and can act as reducing agents, hydrogen donating antioxidants, and singlet oxygen quenchers. Also in some cases metal chelation properties have been proposed (Montoro P. et al 2005, Rice-Evans C. et al 1996).

Antioxidant Activity Test

TEAC Test

The assay for the total antioxidant activity (TAA), or the Trolox equivalent antioxidant activity (TEAC), measures the concentration of Trolox solution with an equivalent antioxidant potential to a standard concentration of the compound under investigation. The TEAC reflects the ability of hydrogen-donating antioxidants to scavenge the ABTS $^{\cdot+}$ radical cation (fig. 3) , absorbing in the near-ir region at 734, 645, and 815 nm , compared with that of Trolox, the water soluble vitamin E analog. Antioxidants suppress the absorbance at 734 nm to an extent and on a time scale dependent on the antioxidant activity. The TEAC is defined as the concentration of Trolox solution with equivalent antioxidant potential to a 1 mM concentration of the compound under investigation.

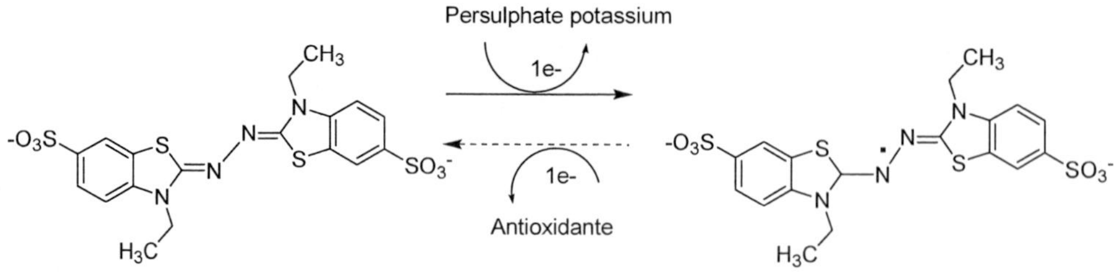

Figure 3. Radical scavenging action of antioxidants (ABTS).

DPPH Test

DPPH assay is based on the measurement of the scavenging ability of antioxidants towards the stable radical 2,2-diphenyl-1-picrylhydrazyl (DPPH). The free radical DPPH, which shows absorption at 517 nm, is reduced to the corresponding hydrazine when it reacts with hydrogen donors (fig 4). DPPH assay is considered a valid and easy assay to evaluate the RSA of antioxidants (Sanchez-Moreno, 2002). In agreement with the literature reports (Yang et al., 2001)

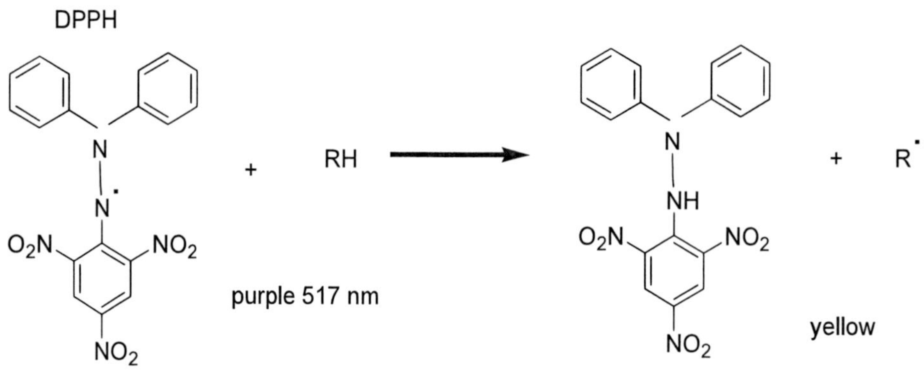

Figure 4. Radical scavenging action of antioxidants (DPPH).

Capillary Electrophoresis

Capillary electrophoresis (CE) (or high-performance CE, HPCE) is an instrumental evolution of traditional slab gel electrophoretic techniques. Since its introduction, CE has shown great potential not only in biopolymer analysis, in which electrophoresis has long since been applied, but also in areas (e.g. inorganic ion and drug analysis) where electrophoretic techniques have never been used before.

Intrumentation: the basic set-up of CE equipment is extremely simple (fig. 5). A CE instrument consists of an injection system, a separation capillary, a high voltage source, electrodes and electrode jars and detectIn CE. Two injection techniques are used, hydrodynamic and electrokinetic mode, the capillary is the compartment where separation occurs and where, often, detection takes place as well, fused-silica capillaries meet almost all these requirements are externally protected by a layer of polyimide. (Tagliaro F. et al 1998).

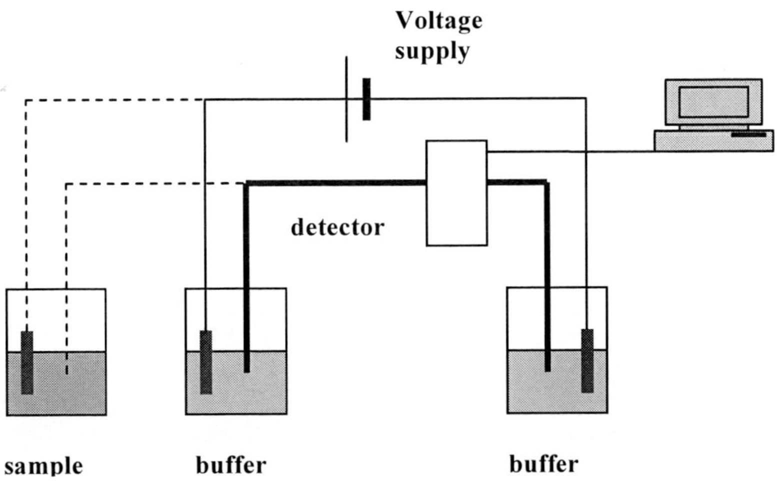

Figure 5. Scheme of a capillary electropherograph: solid lines indicate separation conditions; dotted lines indicate electrokinetic injection conditions,

SEPARATION TECHNIQUES

Using the same CE instrumental harware, several separation modes can be carried out, based on different physical-chemical pinciples, by simply changing the buffer and/or the capillary. Capillary zone electrophoresis(CZE), including separations based on inclusion complex formation, micellar electrokinetic chromatography (MEKC or MECC), capillary electrochomatography (CEC), capillary isoelectric focusing (CIEF), capillary gel electrophoresis (CGE) and capillary isotachophoresis (CITP) can be accomplished with a capillary electropherograp. (Muijselaar, 1996).

MICELLAR ELECTROKINETIC CHROMATOGRAPHY

Neutral solutions that cannot be resolved by CE using simple buffer, Therefore micellar surfactant solutions under the are used in MEKC to separate neutral components cromatographically. The surfactants are used at sufficiently high concentrations that the surfactant molecules group together to form micelles. Neutral solutes partition with this micelles in a chromatographic fashion and are separated based on differences in their partitioning between the aqueos phase and the micellar phase the mechanism of MEKC is schematically presented for an anionic surfactant (Fig. 6). The separation system consist of two moving phases, an electroosmotically moving aqueous phase and an electrophoretically moving pseudo-stationary micellar phase, the micelles in this example have a negatively charged surface and possess an effective mobility in the direction of the anode. Using fused silica capillaries, the EOF will be directed to the cathode. If the electroosmotic mobility of the micelles *occurs*, both the aqueous phase and the micellar phase will move in the direction of the cathode. Totally solubilized compounds migrate with the velocity of the micelles. All other compounds migrate with a linear velocity between these two limiting values, determined by the partitioning with the pseudo-stationary micellar phase. Different surfactants are used include anionic such as sodium dodecylsulphate (SDS), bile salts, cationic surfactants such as ceyltrimethylammonium bromide and neutral surfactants such as Tween. Alternatively a charged additive such as an anionic cyclodextrin (Altria K.D. 1999).

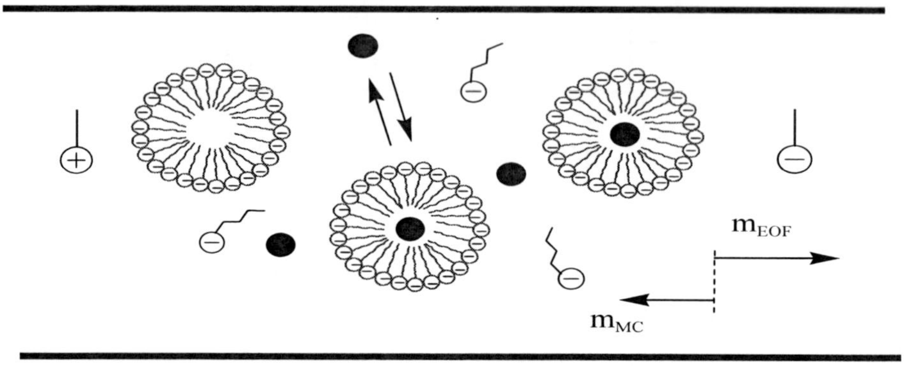

Figure 6. Schematic representation of the separation mechanism for neutral compounds in Micellar Electrokinetic Cromatography with an anionic surfactant system.

METHODOLOGY

Sample Preparation

Extraction procedure plant material (1g per sample) was reduced to a fine powder and extracted with 70% aqueous acetone (50 ml), under 20 min sonication in an ultrasonic bath at ambient temperature. The extracts were rapidly vacuum-filtered through a sintered glass funnel and kept refrigerated before assay. (Maksimovic Z. et al 2005)

Determination of Antioxidant Activity

The DPPH radical scavenging activity of plants was estimated according the method explained by Sánchez-Moreno 2002, with some modifications. Aliquots of 2 ml of 6×10^{-5} M DPPH methanol were mixed with 50 μl of the extracts. The mixtures were vigorously shaken and left to stand for 10 min under subdued light. The absorbance at 540 nm was measured against methanol as a blank. Decrease in colorization was measured spectrophotometrically at 517nm using a spectrophotometer. The radical scavenging activity (RSA) was calculated using the equation

%RSA= 100 x(1-A□/ AD)

where AE is the absorbance of the solution containing antioxidant extract whereas AD is the absorbance of the DPPH* solution

And expressed like mg equivalents of trolox/g dry weight (TEAC)

ABTS Test

The scavenging activity of extract was estimated according to the method of (Re et al. 1999) with some modifications. ABTS radical cation was generated by adding 7 mM ABTS to 2.45 mM potassium persulphate solution and the mixture was left to stand for overnight in the dark at room temperature. The ABTS radical cation solution was diluted with ethanol obtain an absorbance of 0.7 ($\pm$ 0.01) to 754 nm. Diluted ABTS radical cation solution (980 μl) was added to 50 μl of extracts or ascorbic acid standard solution. The absorbance was measured at 714 nm after 5 sec. The ABTS radical cation scavenging activity was expressed as ascorbic acid equivalent antioxidant activity (AEAC) and defined as the mg ascorbic acid equivalents per 100 g of sample (Zhao et al 2010).

Antimicrobial Activity Test

Disk diffusion assay. Extracts were tested for antibiotic activity against *Escherichia coli, Salmonella thypi, Staphylococcus aureus, Bacillus subtilis and Candida albicans*, all the microorganism were obtained from UPIBI collection 50 μl of extract was solubilized in ETOH and placed on the surface of the inoculated agar and incubated at 30°C, using antibiotic no. 1 medium, and the antibiotic activity was recoded as the diameter of clear zones of inhibited microbial growth around the paper disk expressed in μg of cephalosporine C/ml (Jensen P.R. 1998).

Instrumentation and Electrophoretic Procedure

The method electrophoretic was accorded according the method explained by Sun Y. 2008, with some modifications, all CE experiments were performed using a beckman P/ACE MDQ capillary electrophoresis instrument equipped with a photodiode array detector. Uncoated fused-silica capillaries with inner diameters of 0.75 μm and total lengths of the 60 cm were used. Sample injections were done by hydrodynamic pressure at 0.5 psi. an injection time of 5 s was used for all analyses. UV absorption was monitored at 214 nm. The separation voltage was 15 kV at a constant temperature of 25° C. Flavonoids standards mixture (catechin, naringenin, quercetin, apigenin, kaemferol, myricetin, rutin, luteolin, epigallocatechin) were prepared by dissolving the apropiate mass of the flavonoid in ethanol and diluted with doubly deionized water.

Results And Discussion

The antimicrobial activity of extract of plants were examined, their result are listed in Table 4, all plants inhibited the growth of Gram-positive and Gram-negative bacterias, *B. orellana, C. officinalis, O. basilicum and C. edulis* showed effect on fungal growth, *B. orellana* showed the most potent antibacterial activity, *E. coli* was insensitive to plant extract.

Table 4. Antimicrobial activity of medicinals plants extracts (μg equivalent of cefalosporine C/ml)

Microorganism Sample	C. albicans	S. thyphi	S. aureus	E. coli	B. subtilis
B. orellana (shell)	245.87	134.01	-	-	-
B. orellana (seed)	347.71	107.93	-	-	463.87
Caléndula officinalis (sheet)	173.85	86.94	86.94	-	70.02
Caléndula officinalis (steam)	-	-	70.02	-	138.3
Ocinum basilicum (flower)	292.38	70.02	-	-	-
Ocinum basilicum (sheet)	122.93	118.20	-	-	-
Ocinum basilicum (stem)	146.19	206.56	-	-	-
Casimiroa edulis (sheet)	122.93	70.02	-	-	-
Brassica oleracea	-	-	70.02	-	138.30
Satureja macrostema	-	27.9	-	-	-
Eringium heterophyllum (sheet)	-	30.6	31.2	-	28.8
Eringium heterophyllum (Flower)	-	30.6	31.4	-	27.9

The ABTS method is widely employed for measuring the relative radical scavenging activity of hydrogen donating and chain breaking antioxidants in many plant extracts. The ABTS radical cation scavenging activity of cetonic extract, expressed as mg ascorbic acid equivalents per 100 g of sample (AEAC) is presented in the table 5 and showed scavenging activity on the ABTS radical of all the extracts. These plants rich in flavoniods, could be a good source of compounds that whould help to increase the overall antioxidant capacity of an organism and protect it against oxidative stress.

Table 5. Radical scavening of ABTS of differents medicinal plants

Sample	TEAC
Brassica olerácea	2340.35
Satureja macrostema	2258.20
Eryngium heterophylum (sheet)	2197.72
Eryngium heterophylum (flower)	1913.12
Bougeinvillea spectabilis (flower)	785.9
Cymbopogon citratos (sheet)	389.1
Persea americana (sheet)	1511.5
Citrus sinensis (sheet)	329.9
B. orellana (sheet)	298,56
B.orellana (seed)	482,82
Calendula officinalis (sheet)	282,11
Ocinum basilicum (flower)	305,59
Ocinum basilicum (sheet)	386,50
Ocinum basilicum (stem)	217,65
Casimiroa edulis (sheet)	628,74
Calendula officinalis (flower)	472,89

Table 6. Antioxidant scavening of DPPH of differents medicinal plants

Sample	TEAC
Brassica olerácea	1241.83
Satureja macrostema	1365.82
Eryngium heterophylum (sheet)	1375.30
Eryngium heterophylum (flower)	1344.76
Bougainvillea spectabilis (flower)	1133.4
Cymbopogon citratus	601.24
Persea americana (sheet)	1341.66
Citrus sinensis (sheet)	583.01

The stable DPPH radical is widely used to evaluate the free radical scavenging activity of hidrogen donating antioxidants in many plant extracts. The DPPH radical scavenging activity of the extracts is presented in table 6. The TEAC values ranged from 583.01 mg equivalents of trolox/g dry weight.

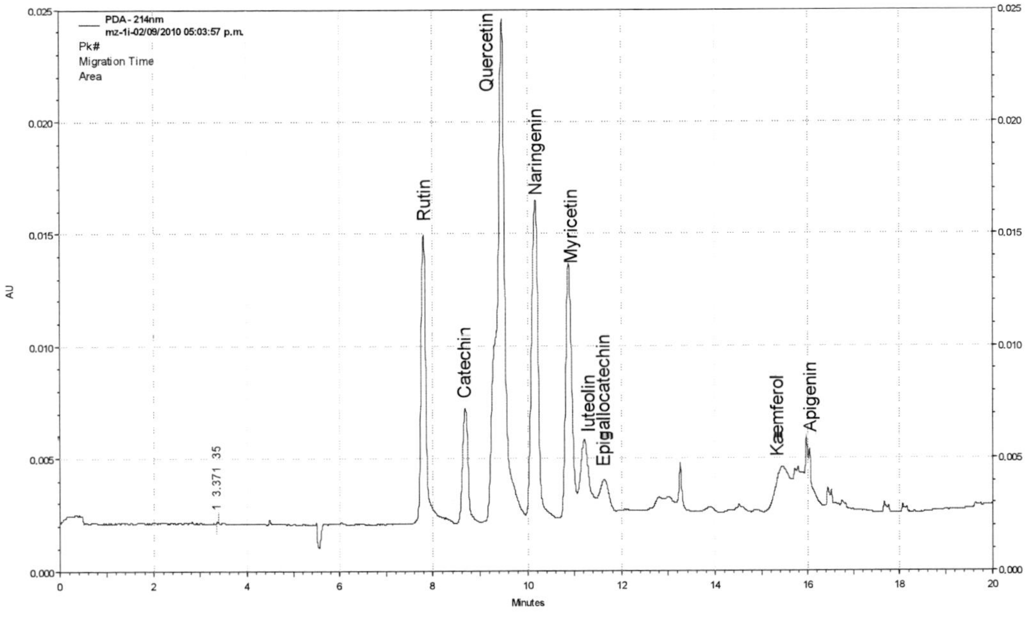

Figure 7. Electropherogram obtained from a standard mixture of nine flavonoids compounds rutin, catechin, quercetin, naringenin, myricetin, luteolin, epigallocatechin, kaemferol and apigenin.

MEKC Separation

Flavonoids have attracted increasing interest in recent years due to their antiviral, antibacterial, anti-inflamatory, antiallergic and antitumor properties. Several analytical methods, such as HPLC, HPLC–MS, GC–MS and multidimensional LC, have been described. Capillary electrophoresis (CE) is a very economical technique and it has many advantages such as small injection sample volume, high efficiency, and short analysis time, which can be useful in the rapid and efficient determination of flavonoids in complex systems (Sun Y. 2008).

The electromigration modes primarily used are capillary electrophoresis (CE), capillary zone electrophoresis (CZE) and micellar electrokinetic chromatography (MEKC) with, typically, phosphate or borate buffers, capillaries of 50–100 lmid, voltages of 10–30 kV, and injection volumes of 10–50 nL. Detection is usually performed with UV, but electrochemical and MS detectors are also used.

Most studies that use capillary electrophoretic methods for the analysis of phenolics fall in the field of natural product research, including the analysis of plants, vegetables, herbs, and other plant or fruit-derived products

A typical electropherogram for the 9 flavonoids separeted using 40 mM borate solution containing 40 mM sodium dodecyl sulfate (SDS) at pH 9.0 is presented at fig. 7

Table 7. Flavonoids identified by capillary electrophoresis

Medicinal Plant	Flavonoids detected
Bougainvillea spectabilis	Rutine Catechin Quercetin
Cymbopogon citratus (sheet)	Rutine Epigallocatechin
Persea americana (Sheet)	Quercetin Naringenin Myricetin
Citrus sinensis (Sheet)	Catechin Quercetin Naringenin Myricetin Luteolin Epigallocatechin
Brassica olerácea	Catechin Quercetin Naringenin Epigallocatechin Kaempferol
Satureja macrostema leaf	Rutine Quercetin Naringenin Myricetin Epigallocatechin
Eryngium heterophylum leaf	Kaempferol

References

Altria K D (1999) Overview of capillary electrophoresis and capillary electrochromatography. *Journal of Chromatography A*, 856: 443-463.

Damiani E, Astolfi P, Carloni P, Stipa P, Lucedio G (2008) *Oxidants in Biology, Springer Science-Business Media B.*V.251-266

Duarte J, Vizcaino F P, Utrilla P, Jimenez J, Tamargo J, Zarzuelo A (1993) Vasodilatory effects of flavonoids in rat aortic smooth muscle. Structure activity relationships. *Biochem. Pharmacol.* 24: 857-862

Erasto P, Bojase-Moleta G, Majinda RT (2004) Antimicrobial and antioxidant flavonoids from the root Wood of Bolusanthus speciosus. *Phytochemistry* 65: 875-880.

Hanasaki Y, Ogawa S, Fnkui S (1994) The correlation between active oxygens scavenging and antioxidative effects of flavonoids. *Free Radical Biol. Med.* 16: 845-850

Haraguchi H, Tanimoto K, Tamura Y, Mizutani K, Kinoshita T(1998) *Mode of antibacterial action of retrochalcones from Glycyrrhiza inflata Phytochemistry*, 48:125-129.

Hackett AM (1986) The metabolism of flavonoid compounds in mammals. In *Plant Flavonoids in Biology and Medicine: Biochemical Pharmacological, and Structural-Activity Relationships*, p.177-194, Alan R. Liss, New York, NY USA.

Harborne JB, Williams CA (2000) Advances in flavonoid research since 1992, *Phytochemistry* 55: 481-504.

Hertog MG Hollman PC, Katan MB, Kromhout D (1993) Intake of potentially anticarcinogenic flavonoids and their determinants in adults in The Netherlands. *Nutr. Cancer* 20: 21-29

Hope WC, Welton AF, Fielder-Nagy C, Batula-Bernardo C, Coffey J (1983) In vitro inhibition of the biosynthesis of slow reacting substances of anaphylaxis (SRS-A) and lipoxygenase activity of quercetin. *Biochem. Pharmacol.* 32: 367-371

Jensen PR, Jenkins KM, Porter D, Fenical W (1998) Evidence that a New antibiotic flavone glycoside chemically defends the sea grass Thalassia testudinum against zoosporic fungi. *Applied and environmental microbiology*, 64:1490-1496.

Maksimovic Z, Malencic D, Kovavevic N (2005) Polyphenol contents and antioxidant activity of Maydis stigma extracts ,*Bioresource Technology*, 96: 873-877.

Montoro P, Braca A, Pizza C, De Tommasi N (2005) Structure-antioxidant activity relationships of flavonoids isolated from different species. *Food chemistry*, 92: 349-355.

Muijselaar WGHM (1996) Micellar electrokinetic chromatography. *Fundamentals and applications.* NWO Iolanda 7-19.

Stalikas DC (2007) Extraction, separation, and detection methods for phenolic acids and flavonoids *J. Sep. Sci.* 30: 3268 – 3295.

Sun Y, Fang N, Chen DDY, Donkor KK, (2008) Determination of potentially anti-carcinogenic flavonoids in wines by micellar electrokinetic chromatography. *Food Chemistry*, 106: 415-420.

Re R, Pellegrini N, Proteggerte A, Yang M, Rice-Evans C (1999) Antioxidant activity appling and improved ABTS radical cation decolorization assay. *Free Radical Biology and Medicine,* 26: 1231-1237.

Rice-Evans C, Miller NJ, Paganga G (1996) Structure-antioxidant activity relationships of flavonoids and phenolic acids. *Free radical Biology & Medicine*, 20: 933-956.

Robak J, Korbut R, Shridi F, Swies J, and Rzadkowska-Bodalska H (1988) On the mechanism of antiaggregatory effect of myricetin. *Pol. J. Pharmacol. Pharm.* 40: 337-340.

Sánchez-Moreno C (2000) Review: methods used to evaluate the free radical scavenging activity in foods and biological systems. *Food Sci Tech Int* 8:121-137.

Süzgec S, Mericli AH, Hougnton PJ, Cubukcu B (2005) flavonoids of Helichrysum compactum and their antioxidant and antibacterial activity. *Fitoterapia* 76: 269-272

Tagliaro F, Manetto G, Crivellente F, Smith FP (1998) A brief introduction to capillary electrophoresis. *Forensic Science International*, 92: 75-78.

Torel J, Cillard J, Cillard P (1986). Antioxidant activity of flavonoids and reactivity with peroxy radical. *Phytochemistry* 25: 383-385

Zhao H, Chen W, Zhao M (2010) Phenolic profiles and antioxidant activities of commercial beers. *Food chemistry*,119: 1150-1158.

In: Biotechnology
Editors: L. G. Torres Bustillos and I. Garcia Peña
ISBN 978-1-62081-071-2

Chapter 10

MICROBIAL SULFATE REDUCTION COMBINED WITH REDUCTIVE DECHLORINATION OF TRICHLOROETHYLENE IN HYDROTHERMAL VENTS SEDIMENTS

***Claudia Guerrero-Barajas*[*1], *Jessica K. Suastes-Rivas*[1], *Lisiana E. Rosas-Rocha*[1], *Victor G. Rodríguez-García*[1], *Claudio Garibay-Orijel*[2], *Erika T. Quintana-Cano*[3] *and Luis A. Maldonado-Manjarrez*[4]**

[1] Bioprocess Department.UPIBI-IPN, Mexico
[2] Bioengineering Department.UPIBI-IPN, Mexico
[3] Microbiology department, Escuela Nacional de Ciencias Biológicas-IPN, Mexico
[4] Unidad Académica de Ecología y Biodiversidad Acuática, Instituto de Ciencias del Mar y Limnología, Universidad Nacional Autónoma de México, Mexico

ABSTRACT

Microbial sulfate reduction (MSR) is a natural process widely used in environmental biotechnology. The applications of MSR include biodegradation of hazardous organic compounds, biotransformation and recuperation of metals and bioremediation of sites damaged by the effect of acid mine drainage (AMD). Sulfate reducing bacteria (SRB) are microorganisms inhabiting a wide variety of environments including places such as deep sea hydrothermal vents and cold, harsh places in the Antarctic. Since these environments are a great source of microbial diversity, it was decided to take samples of shallow hydrothermal vents located in Mexico to carry out research on sulfate reduction and biodegradation of organic pollutants. In this chapter are presented some of the results on MSR and biodegradation of trichloroethylene (TCE) by enrichments of SRB from sediments collected at shallow hydrothermal vents located in Mexico (Punta Mita,

* claudiaguerrero29@yahoo.com.mx

Nayarit). The results show that the sediment samples incubated under sulfate reducing conditions can reduce TCE and sulfate simultaneously at 37 and 70°C. Results on 16S rRNA analysis of the TCE degrading sediments showed bacteria closely related to genera *Clostridium*, *Bacillus* and *Desulfuromonas*.

Also, it was demonstrated that samples of the hydrothermal vents sediments incubated with sulfate and never exposed to TCE encode α and β subunits of dissimilatory sulfite reductase and that reduction of sulfate to sulfite has been carried out by the SRB in the sediments. Experiments on MSR with some different samples taken at the same hydrothermal vents area suggest that the sulfate reduction and H_2S formation might be the result of syntrophies among several microorganisms besides SRB and fermentative bacteria.

INTRODUCTION

Microbial Sulfate Reduction Related to Biodegradation of Organic Pollutants and Biotransformation of Metals

Microbial sulfate reduction (MSR) is a natural process occurring within the Sulfur cycle, of upmost importance for both, terrestrial and marine environments including those of low and high temperatures (Purdy et al., 2003; Kallmeyer and Boetius, 2004; Shen and Buick, 2004; Finke and Jørgensen, 2008). Anthropogenic use of the Sulfur cycle has brought a great contribution to environmental biotechnology. Particularly, MSR has been greatly studied and widely used for the biodegradation of organic pollutants such as hydrocarbons (BTEX) and biphenyls (Rabus et al., 1993; Harms et al., 1999; Kleikemper et al., 2002; Knöller et al., 2003; Selesi and Meckenstock, 2009), chlorinated aliphatic compounds (particularly chlorinated solvents) (Gerritse et al., 1999; Drzyzga et al., 2001; Holliger et al., 1999), aromatic chlorinated compounds (polychlorinated biphenyls PCB´s), nitrogenated compounds (Azo-dyes and explosives in the form of cyclic nitramines) (Boopathy et al., 1993) and aromatic nitrogenated compounds (explosives as trinitrotoluene (TNT), Boopathy et al., 1993) in both, natural conditions and engineered systems.

Also, MSR is currently used in biotransformation of metals such as Cu, Cd, Cr, Hg, Pb, Fe, Mn, metalloids such as As and Se and radionuclides such as Uranium and Plutonium (Simonton et al., 2000; Chang et al., 2001; Thomson et al., 2001; Meriah-Arias and Tebo, 2003; Boonchayaanant et al., 2009).

The process has also been used in bioremediation of acid mine drainage of various polluted sites in the world (Benner et al., 2002; Doshi, 2006; Hockin and Gadd, 2007; Martins et al., 2009).

One of the most studied genus of sulfate reducing bacteria (SRB) is *Desulfovibrio*. It is known that *Desulfovibrio* and some other SRB use a great variety of electron donors (Rabus et al., 1993; Rueter et al., 1994; Beller et al., 1996; Zhang et al.,1997; Phelps et al.,1998; Galushko et al., 1999; So and Young, 1999). Although the preference of electron donor may vary among different genera of SRB, it is known that in general they can use electron donors such as hydrogen, ethanol, methanol, pyruvate, lactate and several other short chain fatty acids and compounds falling into the category of contaminants such as some of the mentioned above. Only few SRB species can use several organic acids as electron donors, including

Desulforhabdus amnigenus that degrades lactate, acetate, propionate and butyrate (Kleikemper et al., 2002).

Some SRB-consortia have been reported to degrade benzene (Lovely et al., 1995; Kazumi et al., 1997; Phelps et al. 1998; Kleinsteuber et al., 2008; Musat and Widdel, 2008), and some species such as *Desulfobacula toluolica* (Knöller et al. 2003), *Desulfobacula phenolica* (Rabus et al., 1993), *Desulfobacula cetonica* (Harms et al. 1999) and *Desulfobacterium phenolicum* (Rabus et al., 1993) have been associated to toluene degradation. For the biodegradation of some Xylene isomers the following SRB strains have been reported: SRB mXyS1 and SRB oXyS1 (Harms et al., 1999).

Sulfate Reducing Bacteria and Biodegradation of Chlorinated Compounds

A very interesting feature of some SRB is the capability of using some chlorinated compounds as electron acceptors obtaining energy for growth. *Desulfomonile tiedjei* was the first sulfate reducer from which a membrane bound enzyme (dehalogenase) expressed for the reduction of 3- chlorobenzoate was characterized when the microorganism used pyruvate as electron donor.

Desulfomonile tiedjei can also grow using hydrogen as electron donor (Mohn and Tiedje, 1992; Ni et al., 1995; Holliger et al., 1999). The dehalogenase isolated from this bacterium is constituted by two subunits and a heme –cofactor, which seems to play an important role on energy transfer.

Others SRB involved in dehalorespiration of chlorinated aliphatics such as tetrachloroethylene (PCE), trichloroethylene (TCE) and dichloroethenes (DCEs) are: *Desulfitobacterium frappieri* TCE1 (Gerritse et al., 1999; Drzyzga et al., 2001), *Desulfitobacterium frappieri*, PCP1 (Dennie et al., 1998), *Desulfitobacterium hafniense* Y51 (Nonaka et al., 2006), *Desulfitobacterium* sp. KBC1 (Tsukagoshi et. al., 2006), *Desulfitobacterium* sp.PCE-1 (Gerritse et al., 1996), *Desulfitobacterium* sp. viet1 (Loeffler et al., 1999), *Desulfitobacterium* sp. Y51 (Suyama et al., 2001), cited by Barton and Fauque (2009).

Sulfate Reducing Bacteria in Marine Environments

In marine environments SRB can use short chain fatty acids and hydrocarbons of low molecular weight such as ethane, propane and butane as electron donors, these hydrocarbons, are naturally formed as a result of the thermal decomposition of organic matter and are frequently found in marine hydrocarbon seep sediments besides the hydrocarbons that may be present as a result of anthropogenic activities (Kleikemper et al., 2002; Abildgaard et al., 2004; Rabus et al., 2006; Miralles et al., 2007; Barton and Fauque, 2009; Quistad and Valentine, 2011). The role of MSR carried out by SRB in marine sediments is actually important, it has been estimated that the process itself contributes to the oxidation of approximately 50% of the carbon in marine sediments where sulfate is not limiting (Purdy et al., 2003; Rabus et al. 2006).

Moreover, inocula obtained using sediments and soils coming from geothermal or hydrothermal areas have been demonstrated to be effective in the precipitation of metals (Takahashi et al., 2010), which make these environments a source of SRB consortia with direct application in biotechnology.

Just recently, it has been proved that some SRB can survive and even grow under low oxygen concentrations (Barton and Fauque 2009), a fact that may allow not considering them strict anaerobes and that has allowed to identify them under microaerophilic conditions (Dilling and Cypionka, 1990).

A comprehensive report on SRB issued by Barton and Fauque (2009) includes the biochemistry, physiology and biotechnology of SRB and specifically presents the current knowledge on soluble electron transport proteins and trans-membrane redox complexes involved in the dissimilatory sulfate reduction pathway for the genus *Desulfovibrio*. Further studies on the electron transport chain of these microorganisms may give an insight on the possible enhancement of biodegradation of certain pollutants by these bacteria using additional redox mediators for example, with some *Clostridium* sp. (*Clostridium* sp. EDB2) during the biodegradation of cyclic nitramines (Bhushan et al., 2006).

At this point, it should be mentioned, that either in biodegradation of organic compounds or biotransformation of metals, the SRB can yield very interesting results in pure cultures or in consortia associated to the fermentative activity of bacteria like *Clostridium* or *Bacillus* genera, as can be proved by the available information.

Sulfate Reducing Bacteria in Consortia with *Clostridium* Sp

The ubiquitous members of the genus *Clostridium* participate with SRB in biodegradation of pollutants and biotransformation of metals when contaminated soils or sediments are the source of microorganisms used for experimental or engineered work. In biotransformation of Cr (VI) for example, *Clostridium* sp. are present as members of the microbial communities in sulfidogenic and non-sulfidogenic consortia used for the reduction of Cr (VI) (Meriah Arias and Tebo, 2003; Stewart et al., 2010).

In some cases such as in Uranium (VI) reduction, the presence of *Clostridium* sp. in the consortia along with SRB like *Desulfovibrio* sp., actually promotes the stability of the reduced metal avoiding re-oxidation, an important fact to be considered in bioremediation of sites including metals (Boonchayaanant et al., 2009).

As for chlorinated compounds, *Clostridium bifermentans* DPH-1 is known to be a dehalorespirer bacterium capable of dehalogenating PCE (at a rate of 0.43 μmol/h mg protein) to TCE and finally to *cis* 1,2-dichoroethene (cDCE) besides being capable of reducing other halogenated aliphatic compounds (Chang et al., 2000; Okeke et al., 2001).

Interestingly, *Clostridium* relatives isolated from MSR cultures, are able to reduce polysulfides (S_n^{-2}) and possible intermediates in the Sulfur cycle that are used directly by bacteria in Sulfur respiration, a characteristic previously attributed only to *Wolinella succinogenes* (Takahashi et al., 2010).

SPECIFIC STUDIES ON MICROBIAL SULFATE REDUCTION AND BIODEGRADATION OF TRICHLOROETHYLENE WITH HYDROTHERMAL VENTS SEDIMENTS

Due to the role that SRB play in the environment and the applications on biotechnology derived from it, it is significant to consider hydrothermal vents sediments from shallow locations as source of SRB, in the next sections of this chapter, some results from our experimental work with sediments from Punta Mita, Nayarit, Mexico will be shown.

On the other hand, trichloroethylene (TCE), which is considered a worldwide priority pollutant (WHO, 2005), is present in groundwater streams in Mexico City, a fact that has become a major concern because it has reached over 19 tons/year according to a survey carried out by SEMARNAT, Mexico (2008). Figure 1 shows the zones of Mexico City in which TCE contamination has been detected in groundwater streams.

Blue color indicates TCE ton/year
Bright blue up to 10-12 tons/y

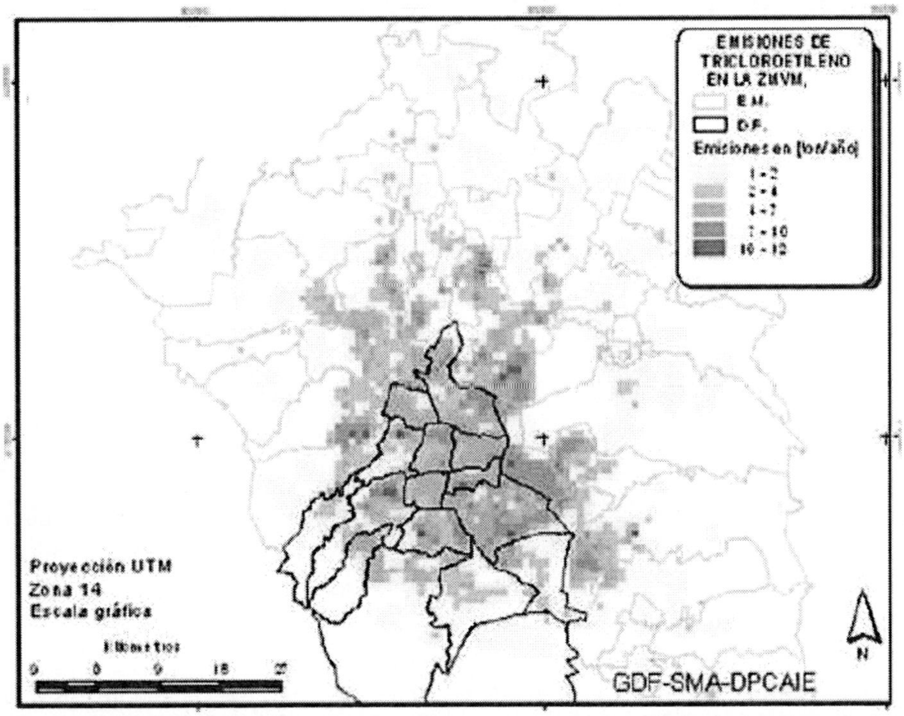

Figure 1. Groundwater contamination with TCE in Mexico City, source: SEMARNAT, Mexico. Annual Inventory of Toxic Emissions (2008).

Therefore, the objectives of this chapter are: 1) to present some of the relevant results obtained from experimental work on sulfate reduction and biodegradation of (TCE) including the identification of some bacteria, 2) to present some preliminary results on identification of SRB obtained from the 16S rRNA analysis of these sediment samples enriched under sulfate reducing conditions with no TCE addition and, 3) to share some interesting findings on sulfate reduction and conversion to sulfide (H_2S) from different samples of sediments collected from hydrothermal vents located in Punta Mita, Nayarit, Mexico.

Microbial Sulfate Reduction (MSR) Coupled to Biodegradation of TCE

Samples of sediments collected from hydrothermal vents located in Punta Mita, Nayarit, were cultivated at (37°and 70°C) in basal medium amended with trace metals and vitamins. Sulfate (as sodium sulfate, Na_2SO_4) was added at a final concentration of 4 g/L of SO_4^{-2} ion along with a mixture of volatile fatty acids (VFAs) as substrate (acetate, propionate and butyrate) and lactate in stoichiometric ratios in separated set of treatments. Some cultures were prepared without the addition of VFAs or lactate in order to have a background of MSR for the sediments. Once MSR was established in the cultures amended with substrates (over 80% of sulfate reduction), these were supplemented again with fresh medium and sulfate. The TCE was added at a final concentration of 300 μM. A detailed description of the location of the hydrothermal vents, the collection of the samples, the preparation of the cultures and the analytical procedures is presented in Guerrero-Barajas et al. (2011).

For this part of the experimental work, two samples of sediments taken from two different places in the vents were used. *Sediment 1* was obtained from 8 m depth and *Sediment 2* was obtained from 10 m depth in the vents.

MSR with VFAs as substrate was higher than with lactate for both sediments at both, 37 and 70°C, whereas biodegradation of TCE occurred at the two temperatures in both sediments to a different extension (Guerrero-Barajas et al., 2011). In the following sections some results on SR and biodegradation of TCE will be presented for VFAs at 37°C. For 70°C the data obtained for TCE biodegradation with lactate will be shown.

Microbial Sulfate Reduction and Biodegradation of TCE with VFAs at 37°C

Results showed that *Sediment 1* reduced sulfate up to 70% and *Sediment 2* reduced 87% in 120 days of incubation at 37°C. Figure 2 shows the accumulative sulfide concentration (H_2S in mg/L) for the cultures in the time course of the experiment. When TCE was added to the cultures, TCE biodegradation was of 50% in *Sediment 1* and 75% in *Sediment 2* yielding ethane as the main biodegradation product in a time span of 30 days.

The same cultures without VFAs yielded 30 and 50% of TCE reduction for *Sediment 1* and *Sediment 2* respectively, suggesting that the sediments oxidized the naturally occurring organic matter in them to reduce sulfate and TCE. Intermediate products of the TCE biodegradation were *cis* and *trans* DCEs, vinyl chloride (VC), ethene and ethane. The 16S rDNA analysis of the samples for which *the highest TCE removal was obtained*, showed the presence of bacteria closely related to the following genera: *Bacillus*, *Clostridium* and *Desulfuromonas* (Guerrero-Barajas et al. 2011). *Bacillus* and *Clostridium* are known fermentative bacteria, on the other hand, members of *Desulfuromonas* genus are known to participate in the Sulfur and Iron cycles in terrestrial and marine environments. From this work it was possible to conclude that TCE biodegradation and perhaps SR occurred as a result of syntrophies among the several microorganisms in the culture including members of the aforementioned genera.

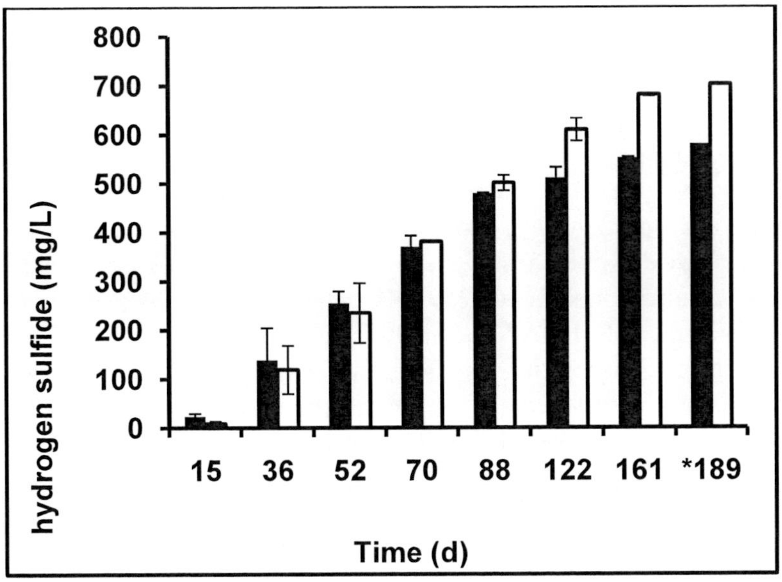

Figure 2. H_2S (hydrogen sulfide) concentration over time for sediments incubated with VFAs at 37°C. ■ *Sediment 1*; □, *Sediment 2*. *on day 189 TCE was added to the cultures when SR was above 80%.

Biodegradation of TCE with Lactate at 70°C

The results obtained with the sediments cultivated with lactate as a substrate at 70°C for biodegradation of TCE were 51% of TCE removed in cultures prepared with *Sediment 1* and 61% of TCE removed for the cultures containing the *Sediment 2*. A carbon balance on day 30 of incubation of the sediments with TCE is shown in Table 1 in which the intermediates identified were *cis* and *trans* DCEs, VC, ethene and ethane. From the results it can be seen that the undesirable VC tends to accumulate in all cases and this accumulation is higher in absence of lactate in *Sediment 1*. In the case of *Sediment 2*, *trans* DCE is the intermediate occurring at a higher concentration. Carbon differences may have occurred during sampling of the treatments when carbon recovery is approximately 70-75%.

Table 1. Carbon balance for TCE biodegradation with lactate at 70°C.

Carbon Balance (µM) on day 30 of incubation with lactate at 70°C							
Inoculum	TCE	*Cis*-DCE	*trans*-DCE	VC	ethene	ethane	% C recovery
Sediment 1[a]	132.9	0.142	4.0	120.5	33.5	33.4	100
Sediment 1[b]	145.3	0.0	2.8	14.9	52.8	33.8	83
Sediment 2[a]	132.8	0.0	4.4	34.9	33.4	33.4	79
Sediment 2[b]	101.1	1.8	64.3	24.5	33.5	0.7	75.3

[a] incubation without lactate, [b] incubation with lactate. Initial TCE concentration 300 µM.

Preliminary Molecular Evaluation of Enrichments of Sulfate Reducing Bacteria from Hydrothermal Vents Sediments of Punta Mita, Nayarit

SRB are very difficult to isolate and identify by traditional methods due to the culture conditions, they are strictly anaerobe and slow growing bacteria. After the revolution of the molecular methods, the culture-independent approaches have shown to be the more promising techniques for studying environmental samples. Sequences of the 16S rRNA gene have been used to characterize bacterial communities (Amann et al., 1995; Pace, 1997; Zhang and Young, 1997; So and Young, 1999;) and it is well known that analyses of PCR using the 16S rRNA gene has provided the most general framework for studies of natural microbial diversity and abundance (Dojka et al., 1998; Hugenholtz et al., 1998; Pace, 1997; Risatti et al., 1994). However, the 16S rRNA-bases analysis does not provide any information about the physiology or metabolic capabilities of a bacterium or a group of them. Hence, the functional gene approach has been developed and used to identify bacteria responsible for biochemical processes (Cottrell and Cary, 1999; Minz et al., 1999; Jimenez-Perez et al., 2001; Dhillon et al., 2003).

In order to study this, model organisms are the SRB because they are able to catalyze the reduction of sulfite to sulfide, by means of dissimilatory sulfate reductase machinery. As far as is known, *all* SRB encode a key gene for this essential enzyme system (Wagner et al., 1998; Klein et al., 2001; Dhillon et al., 2003).

In the present study, SRB were evaluated through the capability of the enrichment of reducing sulfate and converting it to H_2S at mesophilic conditions (37°C) and by using a specific set of primers (Wagner et al., 1998) for the amplification of a specific PCR product of 1.9-kb. It has been recognized that encoding α and β subunits of dissimilatory sulfite reductase can be amplified by PCR from all the lineages of Sulfate-reducing bacteria (Wagner et al., 1998; Klein et al., 2001; Dhillon et al., 2003).

Sample and Characteristics

A sediment sample from an anaerobic bioreactor acclimated to sulfate reducing conditions was taken after ten months of operation of the bioreactor. The culture medium was supplemented periodically with 4 g/L of SO_4^{-2} and a mixture of VFAs as substrate (acetate, propionate and butyrate) and incubated at 37°C. This VFAs mixture was used as substrate since short chain fatty acids are a common source of carbon found in marine sediments (Kleikemper et al., 2002; Abildgaard et al., 2004; Rabus et al., 2006; Miralles et al., 2007; Barton and Fauque, 2009; Quistad and Valentine, 2011). This bioreactor was never exposed to TCE. Sulfate reducing activity of the sediments at the time of sampling was 155 ± 23 mg COD-H_2S/ g VSS d.

Nucleic Acid Extraction, 16S rRNA and Dsr Gene Amplification

Total genomic DNA was extracted from 25 mg of sediment collected from the anaerobic bioreactor under sulfate reducing conditions by using a MOBIO DNA extraction Kit (MO BIO Labs Inc., Solana Beach, California; Figure 3. Bacterial oligonucleotides 27f [5'-AGAGTTTGATCMTTGGCTCAG-3'] y 1525r [5'-AAGGAGGTGWTCCARCC-3'] (Lane, 1991) were used to amplify the PCR of the 16S rRNA gene and later the set of primers *dsr*1F [5'-AC[C/G]CACTGGAAGCACG-3'] and *dsr*4R [5'-GTGTAGCAGTTACCGCA-3'] (Wagner et al., 1998) to amplify the *dsr* genes from the environmental sample. The mixture of the 16S rRNA (final volume of 50 μl) contained 0.5 μl of each primer solution (10 pmol/μl), 0.7 μl of DNA, 5 μl of 10X PCR buffer, 2.0 μl of $MgCl_2$, 1.25 μl of a 10X deoxynucleoside triphosphate mixture (2 mM dATP, 2 mM dCTP, 2 mM dGTP, 2 mM dTTP), and 0.2 μl of Taq DNA (2 U/μl). Amplification was carried out with a LabTech III thermocycler machine. After initial denaturation for 5 minutes at 95°C, 30 cycles were performed, with each cycle consisting of 1 min at 94°C, 1 min at 55°C, 1 min at 72°C and 10 min at 72 °C.

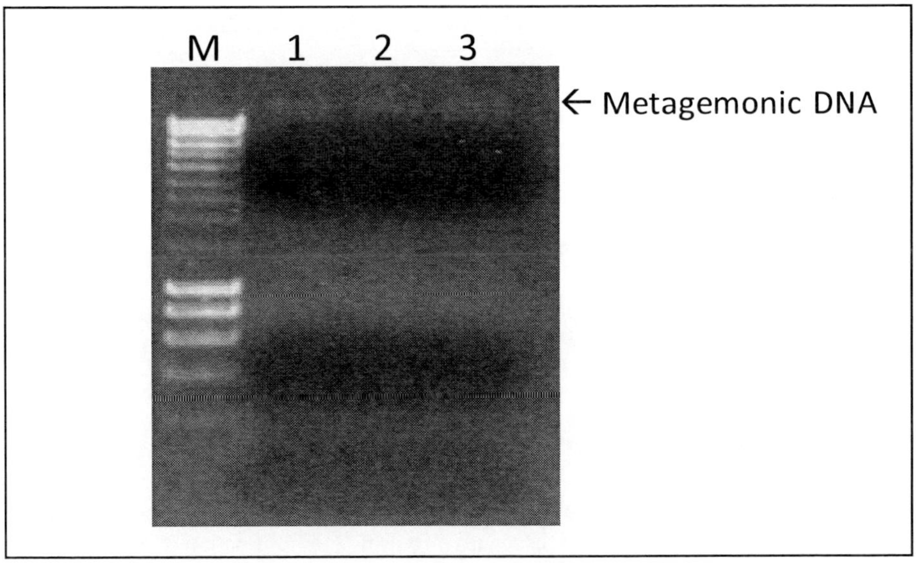

Figure 3. Genomic material extracted with the MOBIO DNA extraction Kit from the sample collected in the bioreactor under sulfate reducing conditions. Note: M, molecular ladder (100 kb, Bioline, USA), 1- 3, Genomic DNA of the environmental sample (per triplicate).

Similarly, the mixture of the *dsr* gene (final mixture 50 μl) contained 1.0 μl of each primer solution (10 pmol/μl), 1.0 μl of DNA, 5.0 μl of 10X PCR buffer, 2.0 μl of $MgCl_2$, 5.0 μl of a 10X deoxynucleoside triphosphate mixture (2 mM dATP, 2 mM dCTP, 2 mM dGTP, 2 mM dTTP), and 0.2 μl of Taq DNA (2 U/μl). Amplification was carried out with a LabTech III thermocycler machine. After initial denaturation for 1 minute at 94°C, 30 cycles were performed, with each cycle consisting of 1 min at 94°C, 1 min at 54°C, 3 min at 72°C and 5 min at 72 °C.

The specific PCR products (Figure 4) of the 16S rRNA (1.5 kb) and the *dsr* gene (1.9 kb) were visualized in an agarose gel (1% v/w, stained with Ethidium Bromide) under UV light in a transiluminator and the image recorded by a fotodocumentator (LabTech, USA).

The results showed an amplification of bacterial DNA using universal primers. More importantly, a specific PCR product of 1.9-kb was amplified using the specific set of primers *dsr*1 and *dsr*4 (Wagner et al., 1998). At the moment, this indicates that: **1)** bacterial DNA was present and extracted from the environmental sample, **2)** encoding α and β subunits of dissimilatory sulfite reductase are present in the bacteria evaluated and **3)** the reduction of sulfate to sulfite has been carry out by the Sulfate-reducing bacteria.

All our conclusions are in agreement with previous reports when studied SRB in marine sediments searching for the *dsr* genes. However, this is the first report of its kind from Punta Mita, Nayarit, Mexico. Our sediment was collected from 10 meters depth and has been previously reported as a potential inoculum for reducing sulfate and trichloroethylene simultaneously (Guerrero-Barajas and García-Peña, 2010; Guerrero-Barajas et al., 2011). The later, strongly supports our findings in the molecular approach. However, further research will focus on cloning the specific PCR product, to determine the *dsr* sequences and perform the phylogenetic analysis.

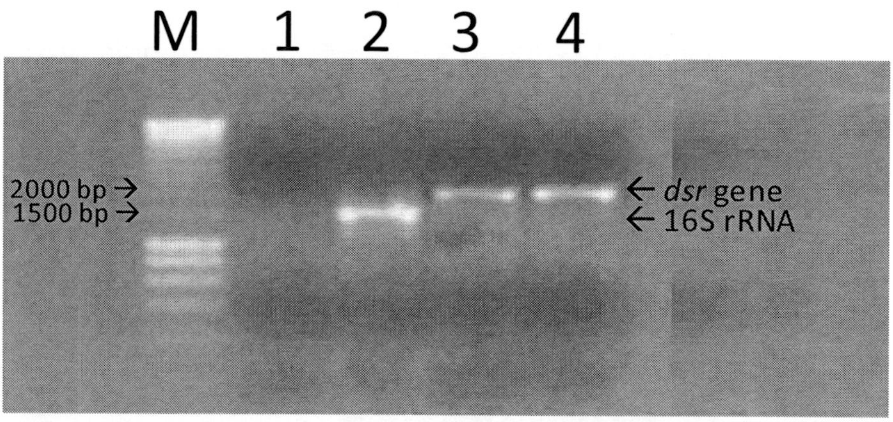

Figure 4. PCR products on an agarose gel (1%, v/w) stained with ethidium bromide. Note: M, molecular ladder (100 bp, BioLine, USA), 1, negative control, 2, amplification of a product corresponding to 1500 pb (16S rRNA) and 3-4 amplification of product corresponding to 1900 bp (*dsr* gene, per duplicate).

Effect of the Substrate on Microbial Sulfate Reduction (MSR) and H_2S (Sulfide) Formation in Different Sediment Samples

In order to prepare inocula for further biodegradation experiments of TCE and biotransformation of metals, a set of three different samples of sediments taken from different location in the vents area were cultivated with lactate and VFAs at 37°C. The selection of the substrates was based on the fact that lactate is probably the most preferred substrate for SRB, whereas acetate, propionate and butyrate are frequently found in natural environments and used in this type of research to evaluate the activity of SRB in marine sediments (Kleikemper et al., 2002). The sediments were incubated using the same procedure mentioned in the former section and none of these samples was used before for any TCE biodegradation experiments and to the best of our knowledge they can be considered as coming from a pristine environment. For these long incubation periods, both, substrate (either lactate or VFAs) and sulfate were added approximately every three weeks to the cultures at which time mass balances were calculated.

Results on sulfate reduction and accumulation of hydrogen sulfide (H_2S) for each one of the samples are shown in Table 2 (for *sample 1*), Table 3 (for *sample 2*) and Table 4 (for *sample 3*), respectively. Although all the three samples were incubated approximately the same time (230 days), the difference in the incubation time from one table to another indicates that the data presented are the maximum values obtained in 230 days.

Table 2. Maximum percentages of sulfate reduction (MSR) and conversion to H_2S for *Sample 1* incubated at 37°C

Incubation time (days)	Treatment	Sediment	Temperature (°C)	% SO_4^{-2} reduced	Sulfide concentration (mg/L)	% SO_4^{-2} converted to Sulfide
0-213	No substrate	*Sample 1*	37	70.6±18.3	130±4	11.5±3.7
0-213	VFAs	*Sample 1*	37	79.2±7.3	139±17	13.5±0.7
0-213	Lactate	*Sample 1*	37	91.5±7.4	155±24	13.3±0.80

The values presented in the table correspond to the mean and standard deviation of the maximum data obtained during the incubation period.

Table 2 shows that *sample 1* yields higher percentage of MSR with lactate (91.5±7.4%) than with VFAs (79.2±7.3%) but the recovery as H_2S is low and almost similar in both cases (approximately 13% of the reduced sulfate was converted to H_2S). Also, when no substrate is added to the medium there is a relatively high percentage of MSR, and the recovery as H_2S is comparable to the obtained with the two added substrates. Percentages of MSR and percentages of sulfate reduced *and* converted to H_2S are very close between the sample incubated without substrate and the one incubated with VFAs. These results may suggest: first, that the natural occurring organic matter in the sediments contains short chain volatile fatty acids as well because treatments with lactate yielded higher sulfate reduction and second other microorganisms besides SRB may predominate in the cultures as sulfide production remained low, indicating that dissimilatory SRB may not be outnumbering other microorganisms.

Table 3 shows the results for sediment *sample 2.* The results suggest that even though VFAs yielded higher percentage of MSR (61±16.7%), lactate promoted higher conversion of the sulfate reduced *to* H_2S. For lactate, MSR was of 62.7±9.2%, along with 85.4±32.8% of sulfate *converted to* H_2S, against only 19±5.4% of sulfate converted to H_2S with VFAs as substrate.

Figure 5 shows two different periods during the incubation of the samples. The Panel A shows an increase of H_2S concentration (mg/L), as sulfide in the liquid during the first days of incubation of the samples, corresponding to MSR establishment (days 1-45). Panel B shows a time period that shows the decrease in sulfate (%) and Panel C shows the corresponding H_2S concentration (mg/L) for that same period. Panels B and C start on day 38 (days 38-78), some days after sulfate addition.

Table 3 Maximum percentages of sulfate reduction (MSR) and conversion for *Sample 2* incubated at 37°C

Incubation time (days)	Treatment	Sediment	Temperature (°C)	% SO_4^{-2} reduction	Sulfide concentration (mg/L)	% SO_4^{-2} converted to Sulfide
0-173	No substrate	*Sample 2*	37	65.4±9.6	0	0
0-173	VFAs	*Sample 2*	37	61±16.7	58.99±15.44	19±5.4
0-173	Lactate	*Sample 2*	37	62.7±9.2	80.21±21.59	85.4±32.8

The values presented in the table correspond to the mean and standard deviation of the maximum data obtained.

Table 4 presents the results obtained with *sample 3* as inoculum. This sample yielded results in a relatively shorter period of time than the one for *sample 1* or *sample 2.* The incubation without substrate achieved 64.09±19.10% of MSR but H_2S was not produced. The samples amended with VFAs and lactate achieved almost percentages of MSR, whereas lactate yielded the highest conversion of sulfate to H_2S (sulfide), almost all the sulfate reduced (101±37.9% of it) was converted to H_2S even though the percentage of MSR was 42.5±31.4%. The "over production" of sulfide shown in Table 4 for the incubation with lactate may be due to the accumulation of it in the microcosms and the fluctuations between the liquid and gas phases, anyhow, at its lowest, conversion of sulfate reduced to H_2S was approximately 72%, which is the highest percentage achieved out of the three sediment samples.

Table 4 Maximum percentages of sulfate reduction (MSR) and conversion for *Sample 3* incubated at 37°C

Incubation time (days)	Treatment	Sediment	Temperature (°C)	% SO_4^{-2} reduction	Sulfide concentration (mg/L)	% SO_4^{-2} converted to Sulfide
0-111	No substrate	*Sample 3*	37	64±19	0	0
0-111	VFAs	*Sample 3*	37	44±21	68.4±0.5	63.2±27.5
0-111	Lactate	*Sample 3*	37	42.5±31	143.7±0.7	109.9±38

The values presented in the table correspond to the mean and standard deviation of the maximum data obtained.

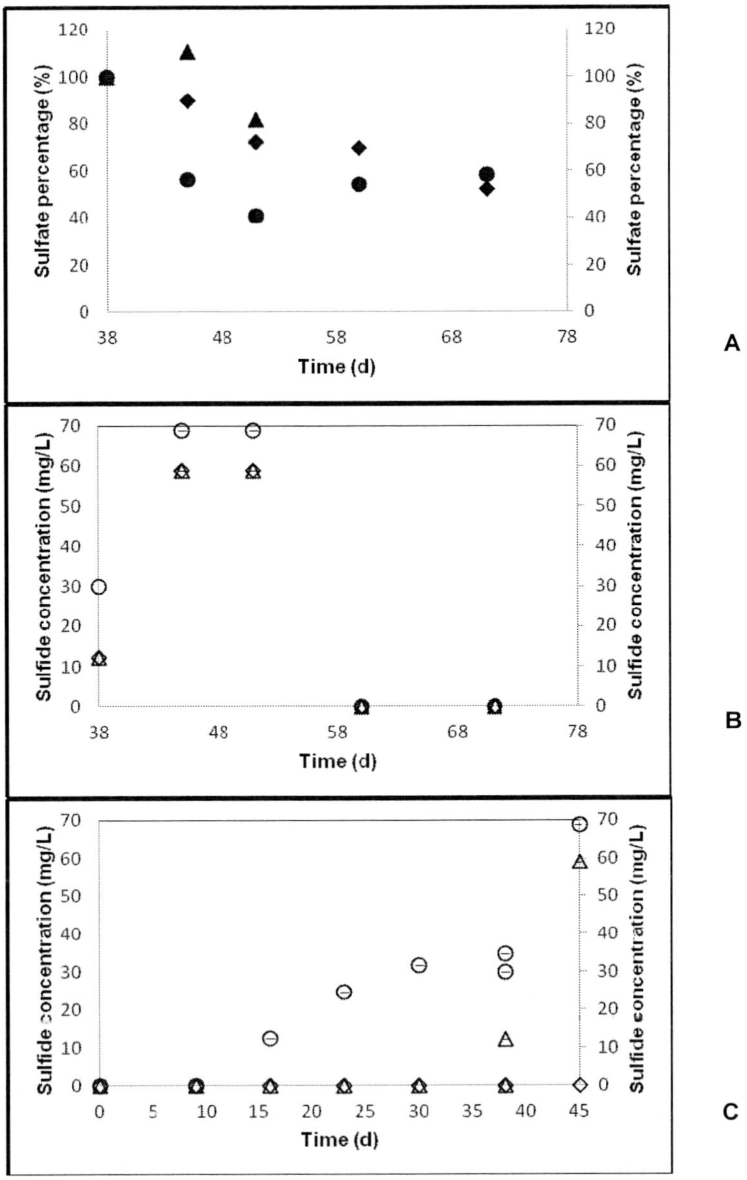

Figure 5. Sulfate percentages and sulfide concentration over time for *Sample 2*. Panel A. ♦ Sulfide no substrate; ▲ Sulfide VFAs; ● Sulfide Lactate (0-45 days). Panel B. ♦ Sulfate no substrate; ▲ Sulfate VFAs; ● Sulfate Lactate (38-78 days). Panel C. ♦ Sulfide no substrate; ▲ Sulfide VFAs; ● Sulfide Lactate (38-78 days).

Figure 6 shows the sulfate percentage in the cultures (Panel A) and the sulfide concentration (mg/L) (Panel B) for a time span in which (H_2S) sulfide concentration was high.

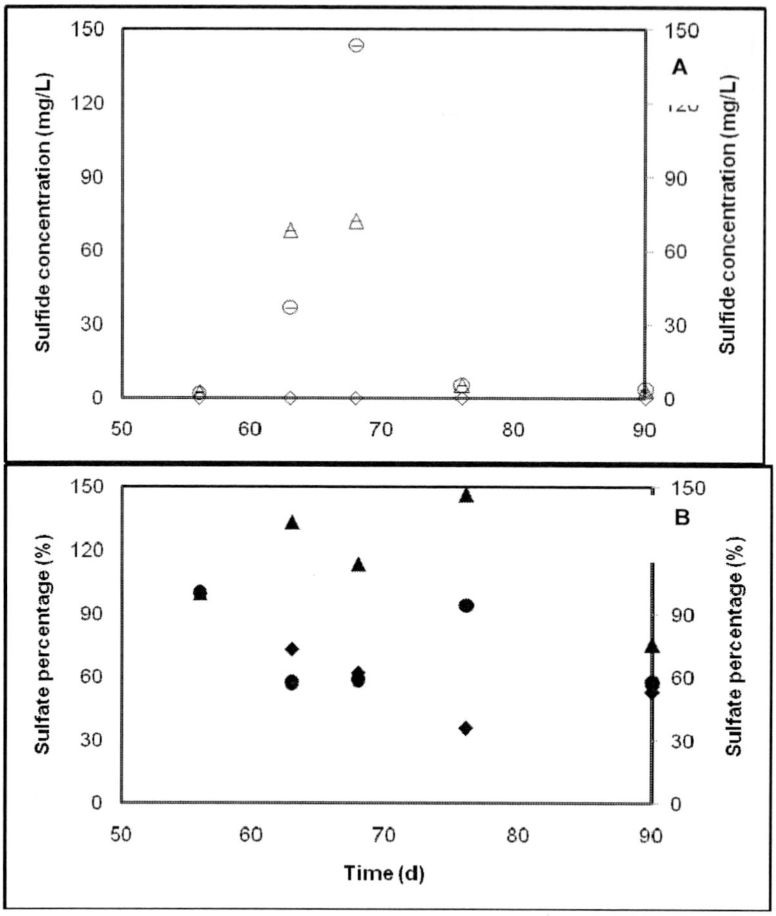

Figure 6. Sulfide concentration and sulfate percentages over time for *Sample 3.* Panel A. ♦ Sulfate no substrate; ▲ Sulfate VFAs; ● Sulfate Lactate. Panel B. ♦ Sulfide no substrate; ▲ Sulfide VFAs; ● Sulfide Lactate.

From results of this section, the following can be remarked: lactate in general was a better substrate than VFAs for the all the three samples, it is not surprising since lactate is one of the preferred substrates for SRB (Rabus et al., 2006; Barton and Fauque, 2009). The difference in the results of the three samples was expected, *samples* 1 and 2 were collected at temperatures between 60 and 70 °C, whereas *sample 3* was collected at a site where the temperature of the sediments was between 70 and 80°C. The fact that they were capable of reducing sulfate at a much lower temperature (37°C) is not surprising since it has been widely proven that certain microorganisms considered as extremophiles, inhabiting subzero or hot temperature environments yield higher amounts of certain proteins or polymers when cultivated at very different temperatures than the ones in which they originally inhabit (Abildgaard et al., 2004; Alam et al., 2005; Finke and Jørgensen, 2008). This demonstrates that microorganisms can adapt to the temperature and yield results (in this case SR) under the culture conditions after some time. Despite the relatively long incubation periods of time showed (total incubation time), as in the case of *sample 2*, it was observed that MSR took place in all samples approximately three weeks after incubation started, which may obey to an increase in the biomass in the sediments.

In order to explain the absence of sulfide in some cases as in *Samples 2* and *3*, or the low conversion of sulfate to sulfide in some cases such as in *Sample 1*, we certainly would need to rely on a 16S rRNA analysis to know which SRB may be present in the microbial communities, however, based on the previous results reported in this chapter it is possible that the consortia present SRB, sulfur reducers and fermentative bacteria, in sight of the high MSR obtained paired to a relatively low H_2S formation. Other possibility is that the intermediates of MSR are not soluble and therefore, bacteria cannot perform the transformation, this also will need to be confirmed using chemical analysis for the samples.

CONCLUSION

Taken the results as a whole, it can be said that hydrothermal vents sediments located in Punta Mita, Nayarit are a source of microorganisms that can be exploited for biodegradation applications due to the physiology that they exhibit. The enrichments prepared with these sediments under sulfate reducing conditions can biodegrade TCE and reduce sulfate as a result of what seems to be a syntrophic association involving fermentative and sulfur reducing bacteria. Also, in absence of TCE and long term incubation periods of some sediment samples with sulfate, it was proved that indeed dissimilatory SR is being carried out specifically by SRB, which may confirm the syntrophies among the several microorganisms. In addition, in some samples incubated also without TCE, the role of fermentative bacteria and other, so far unknown microorganisms, suggest that the microbial communities may be more complex. Despite the SR carried out in these cultures, H_2S was formed at lower ratios than expected, which may indicate that other metabolisms are involved and /or some sulfur species formed during SR are not available to the microorganisms. Further experimental work with these sediments will allow us to determine the composition of the microbial communities and gain more knowledge on the carbon and sulfur cycling in this kind of environment.

Finally, to the best of our knowledge, our work is the first reporting SR, biodegradation tests and insights on the microbial communities composition of the hydrothermal vents located in Punta Mita, Nayarit, Mexico.

ACKNOWLEDGMENTS

The authors are grateful for the financial support provided by Instituto de Ciencia y Tecnología del D.F (ICYT, D.F-Grant PICS-08-79).

REFERENCES

Abildgaard, L., Ramsing, N.B., Finster, K. (2004). Characterization of the marine propionate-degrading, sulfate-reducing bacterium *Desulfofaba fastidiosa* sp. nov. and reclassification of *Desulfomusa hansenii* as *Desulfofaba hansenii* comb. Nov. (2004). International *Journal of Systematic and Evolutionary Microbiology*. 54:393-399.

Alam, S.I., Dube, S., Reddy, G.S.M., Bhattacharya, B.K., Shivaji, S., Singh, L. (2005). Purification and characterisation of extracellular protease produced by *Clostridium* sp. from Schirmacher oasis, Antarctica. *Enzyme and Microbiology Technology*, 36:824-831.

Amann, R. I., W. Ludwig, and K. H. Schlefer.(1995). Phylogenetic identification and in situ detection of individual microbial cell without cultivation. *Microbiology Reviews* 59:143-169.

Barton, L.L. and Fauque, G.D. (2009). Chapter 2. Biochemistry, physiology and biotechnology of sulfate-reducing bacteria. *Advances in Applied Microbiology*. 68:41-98.

Beller, H. R., A. M. Spormann, P. K. Sharma, J. R. Cole, and M. Reinhard. (1996). Isolation and characterization of novel toluene-degrading sulfate-reducing bacterium. *Applied and Environmental Microbiology* 62:1188-1196.

Benner, S. G., Blowes, D. W., Ptacek, C. J., Mayer, K. V. (2002). Rates of sulfate reduction and metal sulfide precipitation in a permeable reaction barrier. *Applied Geochemistry*. 17:301–320.

Bhushan, B.; Halasz, A.; Hawari, J. (2006). Effect of iron (III), humic acids and anthraquinone-2,6-disulfonate on biodegradation of cyclic nitramines by *Clostridium* sp. EDB2. *Journal of Applied Microbiology*. 100:555-563.

Boonchayaanant, B., Nayak, D., Du. X., Criddle, C.S. (2009). Uranium reduction and resistance to reoxidation under iron-reducing and sulfate-reducing conditions. *Water Research*. 43: 4652-4664

Boopathy, R., Kulpa, C. F.,Wilson, M. (1993). Metabolism of 2,4,6-trinitrotoluene (TNT) by Desulfovibrio sp. (B strain). *Applied Microbiology and Biotechnology*. 39: 270–275.

Chang, Y.C., Hatsu, M., Jung, K., Yoo, Y.S., Takamizawa, K. (2000). Isolation and characterization of a tetrachloroethylene dechlorinating bacterium, *Clostridium bifermentans* DPH- 1. *Journal of Bioscience and Bioengineering*. 89:489-491.

Chang, Y. J., Peacock, A. D., Long, P. E., Stephen, J. R., Mckinley, J. P., Macnaughton, S. J.,Anwar Hussain, A. K. M., Saxton, A. M., White, D. C. (2001). Diversity and characterization of sulfate-reducing bacteria in groundwater at a uranium mill tailings site. *Applied and Environmental Microbiology*. 67: 3149–3160.

Cottrell, M. T., and S. C. Cary. (1999). Diversity of dissimilatory sulfite reductase genes of bacteria associated with the deep-sea hydrothermal vent polychaete annelid Alvinella pompejana. *Applied and Environmental Microbiology* 65:1127-1132.

Dennie, D., Gladu, I., Lepine, F., Villemur, R., Bisaillon, J. G., Beaudel, R. (1998). Spectrum of the reductive dehalogenation activity of *Desulfitobacterium frappieri* PCP-1. *Applied and Environmental Microbiology*. 64: 4603–4606.

Dhillon, A., A. Teske, J. Dillon, D. A. Stahl, and M. L. Sogin. (2003). Molecular characterization of Sulfate-reducing bacteria in the Guaymas Basin. *Applied and Environmental Microbiology* 69:2765-2772.

Dilling, W., and Cypionka, H. (1990). Aerobic respiration in sulfate-reducing bacteria. *FEMS Microbiology Letters*. 71:123–127.

Dojka, M. A., P. Hogenholtz, S. K. Haack, and N. R. Pace. (1998). Microbial diversity in hydrocarbon and chlorinated solvent-contaminated aquifer undergoing intrinsic bioremediation. *Applied and Environmental Microbiology* 64:3869-3877.

Doshi, S. M. (2006). Bioremediation of acid mine drainage using sulfate-reducing bacteria. US Environmental Protection Agency Office of Solid Waste and Emergency Response/ Office of Superfund Remediation and Technology Innovation. Washington, DC. *www.epa.gov.*

Drzyzga, O., Gerritse, J., Dijk, J. A., Elissen, H., Gottschal, J. C. (2001). Coexistence of a sulphate-reducing *Desulfovibrio* species and the dehalorespiring *Desulfitobacterium frappieri* TCE1 in defined chemostat cultures grown with various combinations of sulphate and tetrachloroethene. *Environmental Microbiology*. 3: 92–99.

Finke, N. and Jørgensen, B.B. (2008). Response of fermentation and sulfate reduction to experimental temperature changes in temperate and Arctic marine sediments. *The ISME Journal*. 2:815-829.

Galushko, A., D. Minz, B. Schink, andF. Widdel. (1999). Anaerobic degradation of naphthalene by pure culture of a novel type marine sulfate-reducing bacterium. *Environmental Microbiology* 1:415-420.

Gerritse, J., Drzyzga, O., Kloestra, G., Keijmel, M., Wiersum, L. P., Hutson, R., Collins, M. D., Gottschal, J. C. (1999). Influence of different electron donors and acceptors on dehalorespiration of tetrachloroethene by *Desulfitobacterium frappieri* TCE1. *Applied and Environmental Microbiology*. 65: 5212–5221.

Guerrero-Barajas, C., and E. I. García-Peña. (2010). Evaluation of enrichments of sulfate reducing bacteria from pristine hydrothermal vents sediments as potential inoculum for reducing trichloroethylene. *World Journal of Microbiology and Biotechnology*. 26:21-32.

Guerrero-Barajas, C., Garibay-Orijel, C., Rosas-Rocha, L.E. (2011). Sulfate reduction and trichloroethylene biodegradation by a marine microbial community from hydrothermal vents sediments. *International Biodeterioration and Biodegradation*. 65:116-123

Harms, G., Zengler, K., Rabus, R., Aeckersberg, F., Minz, D., Rossello-Mora, R., Widdel, F. (1999). Anaerobic oxidation of o-xylene, m-xylene, and homologous alkylbenzenes by new types of sulfate-reducing bacteria. *Applied and Environmental Microbiology*. 65: 999–1004.

Hockin, S. L., and Gadd, G. M. (2007). Bioremediation of metals and metalloids by precipitation and cellular binding. In *''Sulphate-Reducing Bacteria: Environmental and Engineered Systems''* (L. L. Barton and W. A. Hamilton, Eds.), pp. 405–434. Cambridge University Press, Cambridge, UK.

Holliger, C., Wohlfarth , G., Diekert, G.(1999). Reductive dechlorination in the energy metabolismof anaerobic bacteria. *FEMS Microbiology Reviews*. 22: 383-398.

Hugenholtz, P., C. Pitulle, K. L. Herhberger, and N. R. Pace. (1998). Novel division level bacterial diversity in Yellowstone hot spring. *Journal of Bacteriology* 180:366-376.

Jimenez-Perez, J. R., I. Y. Young, and L. J. Kerkhof. (2001). Molecular characterization of sulfate reducing bacteria in anaerobic hydrocarbon degrading consortia and pure culture using dissimilatory sulfite reductase (dsr AB) genes. *FEMS Microbiology Ecology* 35:145-150.

Kallmeyer, J.and Boetius, A. (2004). Effects of temperature and pressure on sulfate reduction and anaerobic oxidation of methane in hydrothermal sediments of Guaymas Basin. *Applied and Environmental Microbiology*. 70: 1231-1233.

Kazumi, J., Caldwell, M.E., Suflita, J.M., Lovley. D.R.,Young, L.Y. (1997) Anaerobic degradation of benzene in diverse anoxic environments. *Environmental Science and Technology*. 31: 813–818.

Kleikemper, J., Pelz, O., Schorth, M.H., Zeyer, J. (2002). Sulfate-reducing bacterial community response to carbon source amendments in contaminated aquifer microcosms. *FEMS Microbiology Ecology*. 42:109-118Klein, M., M. Friedrich, A. J.

Klein, M., M. Friedrich, A. J. Rogers, P. Hugenholtz, S. Fishbain, H. Abicht, L. L. Blackall, D. A. Stahl, and M. Wagner. (2001). Multiple lateral transfers of dissimilatory sulfite reductase genes between major lineages of sulfate-reducing prokaryotes. *J. bacterial.* 183:6028-6035.

Kleinsteuber, S., Schleinitz, K.M., Breitfeld, J.,Harms, H.,Richnow, H.H., Vogt, C.(2008). Molecular characterization of bacterial communities mineralizing benzene under sulfate-reducing conditions. *FEMS Microbiology Ecology*. 66:143-157.

Knöller, K., Vogt, C., Richnow, H. H., Weise, S. M. (2006). Sulfur and oxygen isotope fractionation during benzene, toluene, ethylbenzene, and xylene degradation by sulfate reducing bacteria. *Environmental Science and Technology* 40: 3879–3885.

Lane, D. J. (1991). 16S/23S rRNA sequencing. pp. 115-175. *In* Stackebrandt, E. and Goodfellow (ed.), *Nucleic acid techniques in bacterial systematics*. John Wiley & Sons, Nueva York, N.Y.

Loeffler, F. E., Tiedje, J. M.,Sanford, R. A. (1999). Fraction of electrons consumed in electron acceptor reduction and hydrogen thresholds as indicators of halorespiratory physiology. *Applied and Environmental Microbiology*. 65: 4049–4056.

Lovley, D. R. (1995). Microbial reduction of iron, manganese, and other metals. *Advances in Agronomy* 54: 175–231.

Martins, M., Faleiro, M.L., Barros, R.J., Veríssimo, A.R., Barreiros, M.A., Costa, M.C. (2009). Characterization and activity studies of highly heavy metal resistant sulphate-reducing bacteria to be used in acid mine drainage decontamination. *Journal of Hazardous Materials*. 166:706-713.

Meriah Arias, Y. and Tebo, B.M. (2003). Cr(VI) Reduction by sulfidogenic and nonsulfidogenic microbial consortia. *Applied and Environmental Microbiology*. 69: 1847-1853

Minz, D., S. Fishbain, S. J. Green, G. Muyzer, Y. Cohen, B. E. Rittman, and D. A. Stahl. (1999). Unexpected population distribution in a microbial mat community: sulfate-reducing bacteria localized to the highly oxic chemocline in contrast to eukaryotic preference for ammonia. *Applied and Environmental Microbiology* 65:4659-4665

Miralles, G., Grossi, V., Acquaviva, M., Duran, R., Bertrand, J.C., Cuny, P. (2007). Alkane biodegradation and dynamics of phylogenetic subgroups of sulfate-reducing bacteria in an anoxic coastal marine sediment artificially contaminated with oil. *Chemosphere*.68:1327-1334.

Mohn, W.W. and Tiedje, J.M. (1992). Microbial reductive dehalogenation. *Microbiology Reviews*. 56: 482-507.

Musat, F. and Widdel, F. (2008) Anaerobic degradation of benzene by a marine sulfate-reducing enrichment culture, and cell hybridization of the dominant phylotype. *Environmental Microbiology*.10: 10–19.

Ni, S., Fredrickson, J.K., Xun, L. (1995) Purification and characterization of a novel 3-chlorobenzoate-reductive dehalogenase from the cytoplasmic membrane of *Desulfomonile tiedjei* DCB-1. *Journal of Bacteriology*. 177: 5135-5139.

Nonaka, H., Keresztes, G., Shinoda, Y., Ikenaga, Y., Abe, M., Naito, K., Inatomi, K.,Furukawa, K., Inui, M.,Yukawa, H. (2006). Complete genome sequence of the dehalorespiring bacterium *Desulfitobacterium hafniense* Y51 and comparison with *Dehalococcoides ethenogenes* 195. *Journal of Bacteriology*. 188: 2262–2274.

Okeke, B., Chang, Y.C., Hatsu, M., Suzuki, T., Takamizawa, K. (2001). Purification, cloning, and sequencing of an enzyme mediating the reductive dechlorination of tetrachloroethylene (PCE) from *Clostridium bifermentans* DPH-1. *Canadian Journal of Microbiology*, 47:448-456.

Pace, N. R. (1997). A molecular view of microbial diversity and the biosfere. *Science*. 276:734-740.

Phelps, C.D., Kerkhof, L.J.,Young, L.Y. (1998). Molecular characterization of a sulfate-reducing consortium which mineralizes benzene. FEMS Microbiology Ecology. 27: 269–279.

Purdy, K.J., Nedwell, D.B., Embley, T.M. (2003). Analysis of the sulfate-reducing bacterial and methanogenic archaeal populations in contrasting Antarctic sediments. *Applied and Environmental Microbiology*. 69: 3181-3191.

Rabus, R., Nordhaus, R., Ludwig, W.,Widdel, F. (1993). Complete oxidation of toluene under strictly anoxic conditions by a new sulfate-reducing bacterium. *Applied and Environmental Microbiology*. 59: 1444–1451.

Rabus, R., Hansen, T. A., and Widdel, F. (2006). Dissimilatory sulfate- and sulfur-reducing prokaryotes. In ‘‘The Prokaryotes’’ (M. Dworkin, S. Falkow, E. Rosenberg, K.-H. Schleifer, E. Stackebrandt, Eds.), Vol. 2, *Ecophysiology and Biochemistry*. pp. 659–768. Springer, Berlin.

Risatti, J. B., W. C. Capman, and D. A. Stahl. (1994). Community structure of a microbial mat: the phylogenetic dimension. *Proc. Natl- Acad. Sci. USA* 91:10173-10177.

Rueter, P., R. Rabus, H. Wilkes, F. Aeckersberg, F. A. Rainey, H. W. Jannasch, and F. Widdel. (1994). Anaerobic oxidation of hydrocarbons in crude oil by new types of sulfite reducing bacteria. *Nature*, 372:455-458

Selesi, D., and Meckenstock, R.U. (2009). Anaerobic degradation of the aromatic hydrocarbon biphenyl by a sulfate-reducing enrichment culture. *FEMS Microbiology Ecology*. 68:86-93.

Shen, Y., and Buick, R. (2004). The antiquity of microbial sulphate reduction. *Earth-Science Reviews*. 64: 243–272.

Simonton, S., Dimsha, M., Thomson, B., Barton, L. L., Cathey, G. (2000). Long-term stability of metals immobilized by microbial reduction. In Proceedings of the 1996 HSRC/WERE Joint Conference on the Environment. *http://www.engg.ksu.edu/HSRC/00P{roceed/thomson.pdf.*

So, S. C. M., and L. Y. Young. (1999). Isolation and characterization of sulfate-reducing bacterium that anaerobically degrades alkanes. *Applied and Environmental Microbiology* 65:2969-2976.

Stewart, D.I., Burke, I.T., Hughes-Berry, D.B., Whittleston, R.A. (2010). Microbially mediated chromate reduction in soil contaminated by highly alkaline leachate from chromium containing waste. *Ecological Engineering*. 36:211-221.

Suyama, A., Iwakiri, R., Kai, K., Tokunaga, T., Sera, N.,Furukawa, K. (2001). Isolation and characterization of *Desulfitobacterium* sp. strain Y51 capable of efficient dehalogenation of tetrachlorothene and polychloroethenes. *Bioscience Biotechnology and Biochemistry*. 65: 1474–1481.

Takahashi,Y., Suto, K., Inoue, C. (2010). Polysulfide reduction by Clostridium relatives isolated from sulfate-reducing enrichment cultures. *Journal of Bioscience and Bioengineering*. 109: 372-380.

Thomson, B. M., Simonton, D. S., Barton, L. L. (2001). Stability of arsenic and selenium immobilization by in situ microbial reduction. In Proceedings of the 1996 HSRC/WERE Joint Conference on the Environment, *http://www.engg.ksu.edu/HSRC/01Proceed.*

Tsukagoshi, N., Ezaki, S., Uenaka, T., Susuki, N., Kurane, R. (2006). Isolation and transcriptional analysis of novel tetrachloroethene reductive dehalogenase gene from *Desulfitobacterium* sp. strain KBC1. *Applied Microbiology and Biotechnology*. 69: 543–553.

Quistad, S.D. and Valentine, D.L. (2011). Anaerobic propane oxidation in marine hydrocarbon seep sediments. *Geochimica et Cosmochimica Acta*. 75:2159-2169.

Wagner, M., A. J. Roger, J. L. Flax, G. A. Brusseau, and D. A. Stahl. (1998). Phylogeny of dissimilatory sulfite reductase supports an early origin of sulfate respiration. *Journal of Bacteriology* 180:2975-2982.

Widdel, F., and T. A. Hansen. (1992). The dissimilatory sulfate- and sulfur-reducing bacteria, p. 583-624. *In* A. Balows, H. G. Truper, M. Dworkin, W. Harder, and K.-H. Schleifer (ed.), *The prokaryotes, 2nd ed.* Springer, New York, N.Y.

Zhang, X., and L. Y. Young. (1997). Carboxylation as an initial reaction in the anaerobic metabolism of naphthalene and phenanthrene by sulfidogenic consortia. *Applied and Environmental Microbiology* 63:4759-4764.

In: Biotechnology
Editors: L. G. Torres Bustillos and I. Garcia Peña
ISBN 978-1-62081-071-2

Chapter 11

HYDROGEN BONDING IN PHARMACY: OXALIC ACID DERIVATIVES AS A CASE OF STUDY

Francisco J. Martínez-Martínez*[1]*, Efrén V. García-Báez*[2]*,
***Juan Saulo González-González*[1]*, and Itzia I. Padilla-Martínez*[*2]**
[1]Facultad de Ciencias Químicas, Universidad de Colima, México
[2]Departamento de Ciencias Básicas,
Unidad Profesional Interdisciplinaria de Biotecnología IPN, Mexico

ABSTRACT

Herein, an overview of the work that our group has developed using oxalamate/oxalamide derivatives is presented. First, we deal with general aspects of oxalic acid in the context of pharmacy, followed by a brief description of structural features of these molecules supported on multinuclear NMR spectroscopy, IR, voltamperometric studies and x-ray diffraction. Results indicate the robustness, reliability and versatility of these systems both in the solid and solution, characteristics that make them suitable for further tests as co-crystallizers of pharmaceutical compounds and molecular gelling agents.

INTRODUCTION

Oxalic acid or ethanodioic acid is a colorless transparent solid that melts at 187 °C; it is soluble in water, alcohol and ether. Oxalic acid is mainly used as an analytic reactive (Arantes *et al.*, 2009) and as a component of cleaning solutions. Oxalic acid has been the subject of intense studies in inorganic chemistry (Gao *et al.*, 2000), organic chemistry (Ghorbani-Vaghei *et al.*, 2010), biochemistry (Lombard *et al.*, 2011) and biotechnology (Kim *et al.*, 2011) because of its importance in biological systems.

[*] ipadillamar@ipn.mx

Due to its great economic importance, oxalic acid is industrially produced by chemical processes (salts of formic acid with treatment of sulfuric acid) and biotechnology, such as the oxidation of carbohydrates by nitric acid. More recently, new microbiological methods have been used for large scale production, using *Cyanobacteriae* or *Aspergillium Niger* (Cameselle, *et al.*, 1998).

Oxalic acid is widely distributed in animals and plants, such as fruits and vegetables, and has been found as a metabolic product of the glycoxylic or ascorbic acid (Caliskan, 2000). In humans, the increased urinary levels of the ion oxalate leads to the formation of calculus (calcium oxalate salts) in the urinary system. Due to this ability to form complexes with metals, oxalic acid is used as chelating agent of metallic ions in the anti-pollution industry and as household cleaning solutions. This ability was explored by pharmaceutical sciences and lead to the discovery of one of the compounds used in the treatment of cancer, the oxalylplatin, (Figure 1), which is a derivative of *cis*-platinum coordination compounds. Its mechanism of action is similar to the alkylating agents (Ehrsson *et al.*, 2010 and Wheate *et al.*, 2010).

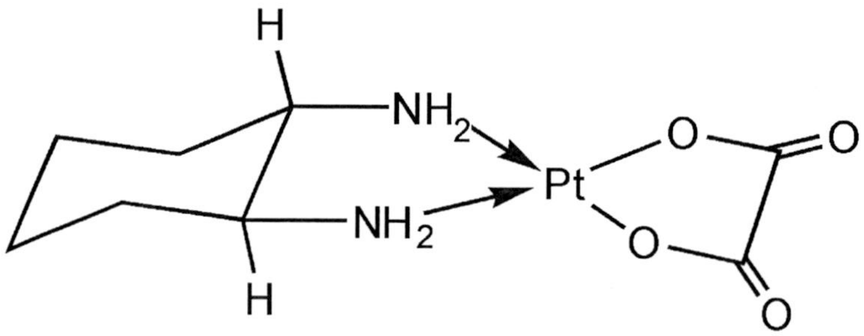

Figure 1. Oxalylplatin complex used in cancer treatment.

This capacity as a chelating agent has been exploited in the formation of coordination complexes where the ligand oxalate, which is the deprotonated form of oxalic acid, binds to a metallic nucleus (Benory *at al.*, 1985; Baggio *et al.*, 1997; Coronado *et al.*, 2008) (Figure 2).

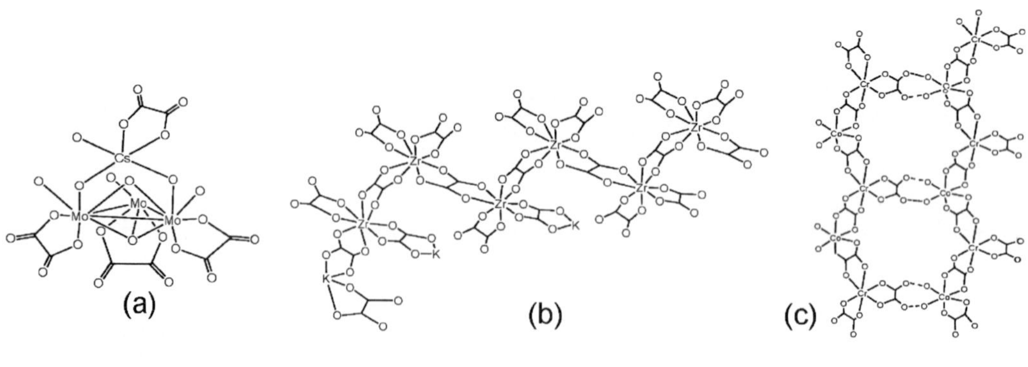

Figure 2. Oxalate polynuclear complexes.

Another use is in the co-crystallization of pharmaceutically active molecules, where oxalic acid gives stability in processing, formulation, packaging and storage of unstable compounds. An example is caffeine, it is susceptible to moisture but when caffeine is mixed with oxalic acid, the resulting co-crystal acquires a high degree of stability to relative moisture (Trask *et al.*, 2005). (Figure 3).

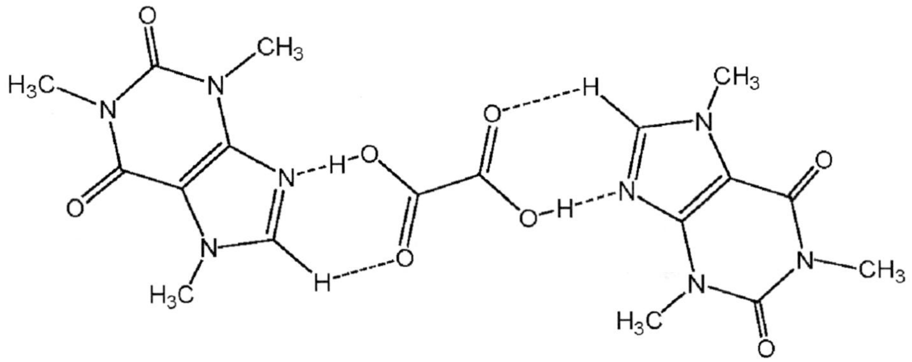

Figure 3. Molecular structure of caffeine-oxalic acid co-crystal.

The acidic hydrogen atoms and carbonyl groups confer to the oxalic acid, the capability to form intra- and inter-molecular hydrogen bonds (Derissen and Smith, 1974) (Figure 4.) Hydrogen bond (D-H····A) is defined as an attractive interaction in which one electropositive H atom is shared between two electronegative species D (hydrogen donor) and A (hydrogen acceptor). This interaction must meet four basic characteristics: electrostatic (acid/base), polarization (hard/soft), van der Waals (dispersion/repulsion) and covalency (charge transfer).

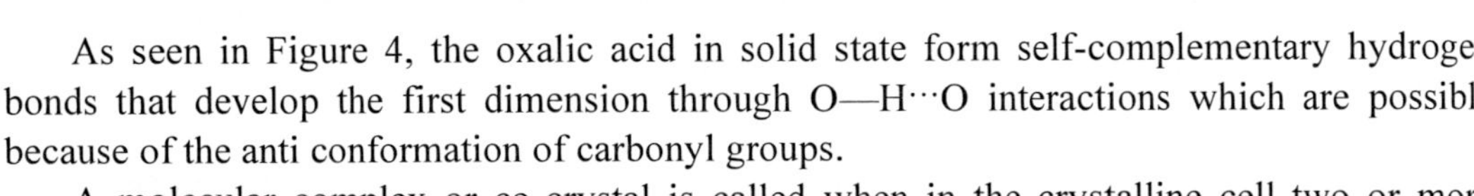

Figure 4. Intermolecular hydrogen bonding in oxalic acid.

As seen in Figure 4, the oxalic acid in solid state form self-complementary hydrogen bonds that develop the first dimension through O—H···O interactions which are possible because of the anti conformation of carbonyl groups.

A molecular complex or co-crystal is called when in the crystalline cell two or more different molecules are found, as in the above example with caffeine-oxalic-acid. Thus, oxalic acid has a great ability as co-crystallizing agent by means of hydrogen bonds as a consequence of its chemical structure. The molecular complexes between oxalic acid and different compounds (Vishweshwar *et al.*, 2003; Aakeröy *et al.*, 2006) are shown in Figure 5, where the preferred arrangement of the carbonyl groups is the anti conformation.

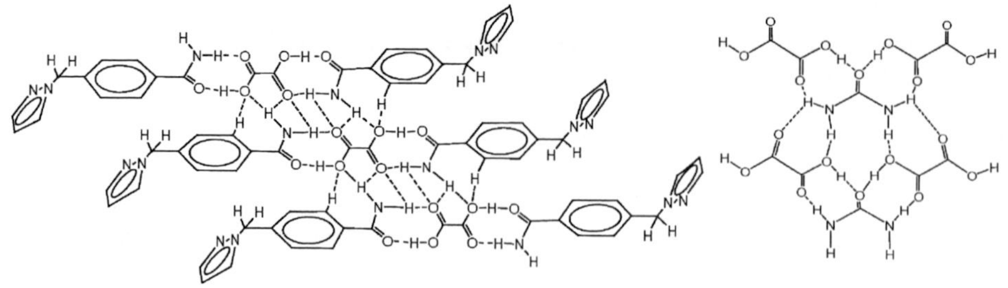

Figure 5. Molecular crystals of 4-pyrazol-1-ylmethyl-benzamide (lef) and urea (right) with oxalic acid. The anti conformation between oxalic acid carbonyls is worthy to note.

As shown in Figure 6 (Harkema and Brake, 1978), carbonyl groups in oxalic acid also can adopt the *syn* conformation in co-crystals.

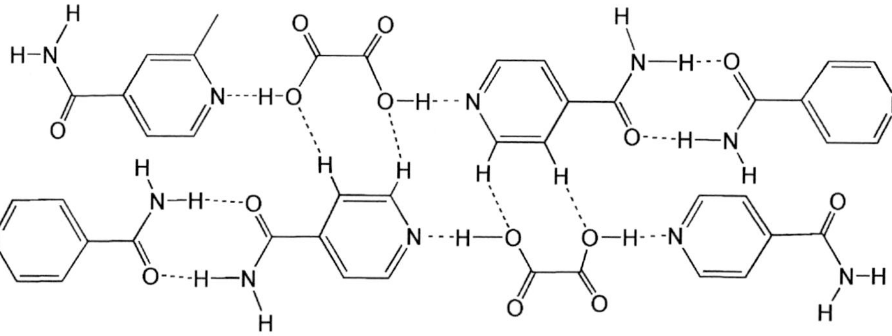

Figure 6. Isonicotinamide-oxalic acid molecular crystals. The *syn* conformation between oxalic acid carbonyls is worthy to note.

Oxalamates and oxalamides are common oxalic acid derivatives. Oxalamates are formed by the substitution of both OH groups, in the oxalic acid, by an ester and an amide groups. Whereas in oxalamides, both hydroxyl groups have been substituted by amide groups (Figure 7).

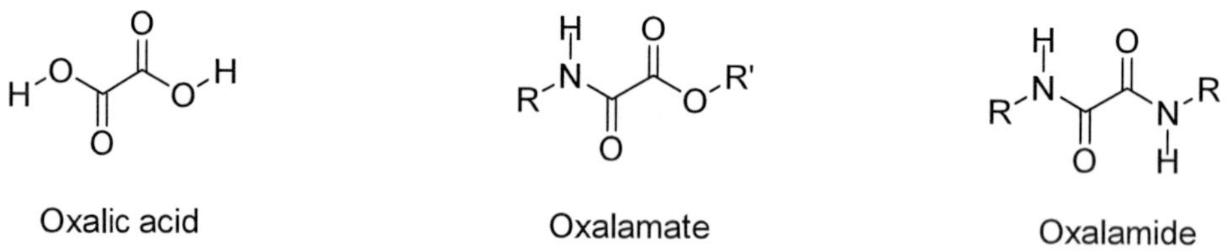

Figure 7. Oxalic acid and its derivatives.

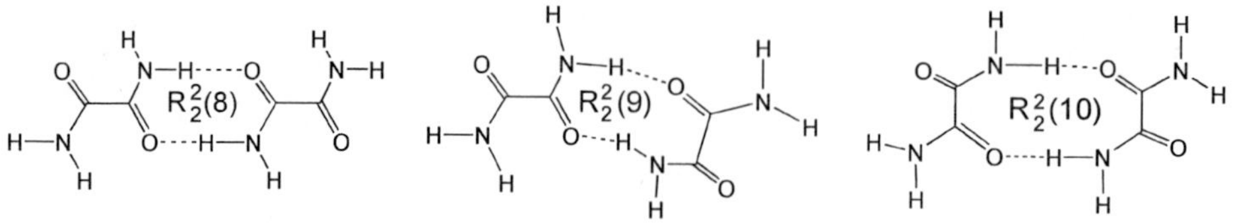

Figure 8. Three possible arrangements for hydrogen bonding in oxalamides.

Oxalamides have been used in the study of intra- and inter-molecular hydrogen bonds. The preferred conformation of the carbonyl groups in oxalamides is the anti conformation. Oxalamides have also been used as building blocks in supramolecular chemistry and crystal engineering (Nguyen, *et al.*, 1998, 2001). Three stable arrangements are possible for hydrogen bonding: eight, nine or ten membered rings (Figure 8).

Oxalamides are good chelating agents because of their N-H and C=O groups behave as Lewis acids and bases. Several complexes have been reported, such as those shown in Figure 9, between oxalamide and copper (Gao *et al.*, 2000), and, oxalamide and ruthenium (Majid *et al.*, 2004).

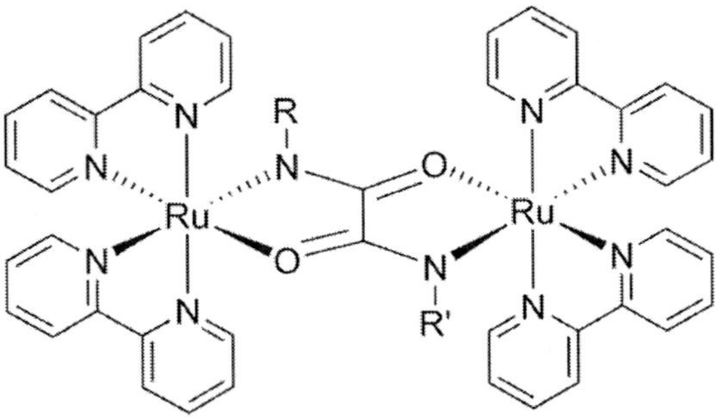

Figure 9. Dinuclear oxalamide–ruthenium complex.

Oxalamides are also capable to form supramolecular gels. Gels are soft materials with low mobility that exhibit mechanical properties characteristic of a solid. Most gels incorporate solvent molecules into a three-dimensional network of organized aggregates consisting of gellating agents, usually polymeric. On the other hand, supramolecular gels are dimensionally controlled assemblies comprising low-molecular-weight molecules held together by non-covalent interactions. Because of the low energy characteristic of such interactions, components can be readily replaced with alternatives, promising material systems for drug delivery and tissue engineering (Maeda, 2008). Supramolecular gels that consist of molecules sensitive to a certain stimulus can be modulated and controlled by that external stimulus. The structures and properties of molecular gels are tunable by hydrogen bonding and by electrostatic interactions between anion stimuli and the gelator building components. In this sense, an oxalamide-substituted anthraquinone, which gelates alcohols and aromatic solvents through stacking between the gelator planes has been reported (Frkanec and Žinić, 2010) (Figure 10 left). The xylene-gel of this molecule is transformed into a solution by the addition of fluoride salts whereas the corresponding ethanol-gel is preserved upon the salt addition. The fluoride anion interrupts the hydrogen bonding between the gelator molecules in xylene, but in ethanol, fluoride anion is solvated enough to interact with the gelling agent.

Bis(amino acid)- and bis(amino alcohol)-oxalamide gelling-agents, shown in Figure 10 right (Frkanec and Žinić, 2010), form molecular gels mostly in a consequence of strong and directional intermolecular hydrogen bonding provided by oxalamide units and lack of molecular symmetry due to the presence chiral centers. These gelators exhibit ambidextrous gelation properties, being capable to form gels with apolar and also highly polar solvent systems, and tending to organize into bilayers or inverse bilayers in hydrogel or organic solvent gel assemblies. The results of spectroscopic structural studies on these molecules are related to organizations found in the crystal structures of selected gelling agents, confirming a similar arrangement in gel assemblies and in the solid state.

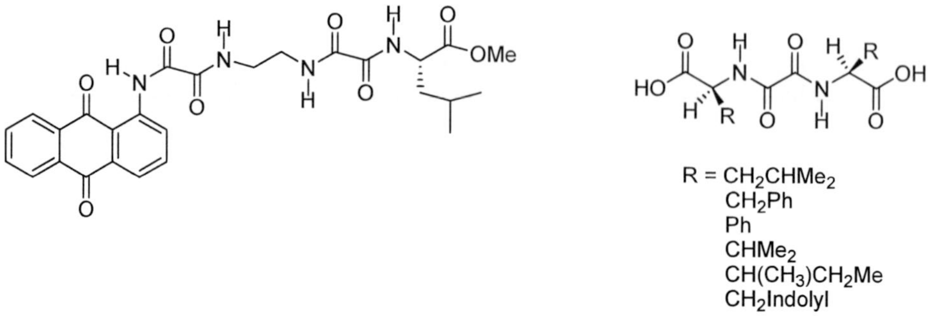

Figure 10. Some oxalamides reported as molecular gelling agents of organic solvents.

Molecular gels of these oxalamide derivatives can be transcripted into silica using sol-gel polymerizations (Šijaković-Vujičić *et al.*, 2007). Thus, TEM (Transmission Electron Microscopy) images of silica nanotubes and twisted silica ribbons have been observed from the corresponding template gels.

In this contribution, an overview of the work that our group has developed using oxalic acid derivatives, as model compounds for the study of hydrogen bond, is presented. The aim of this review is to show the suitability of these molecules for molecular association particularly with pharmaceutical compounds.

Methodology

All chemicals and solvents were of reagent grade and used as received. All new compounds were prepared and fully characterized as described in the corresponding original reference. The crystal structures of new compounds were deposited in the Cambridge Crystallographic Data Center. Theoretical calculations were carried out using the DFT/PBEPBE method with the 6-31+G (3df, 3pd) basis set. All calculations were done using the Gaussian98 program.

Results and Discussion

Oxalamate and Oxalamide Synthesis

The most frequent method herein used for the synthesis of ethyl oxalamates, is the condensation reaction of a primary amine with ethyl clorooxalacetate in tetrahydrofuran (THF) at 0 °C, in the presence of triethylamine as HCl scavenger. The subsequent reaction with other amine forms the corresponding non-symmetric bis-oxalamide (Scheme 1).

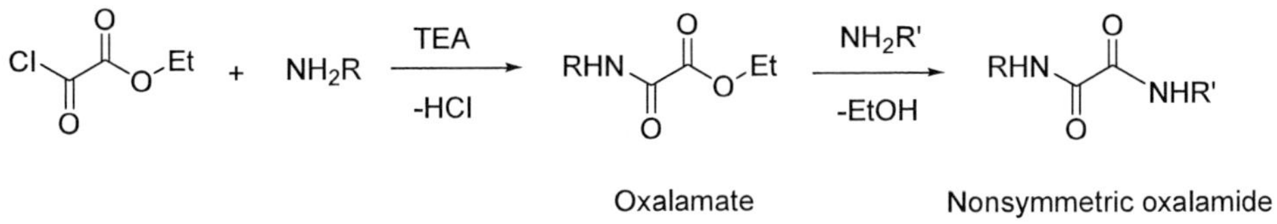

Scheme 1.

In similar form, symmetric bis-oxalamides were synthesized (Scheme 2) by the condensation reaction between oxalyl dichloride and two mol equivalents of the corresponding amine.

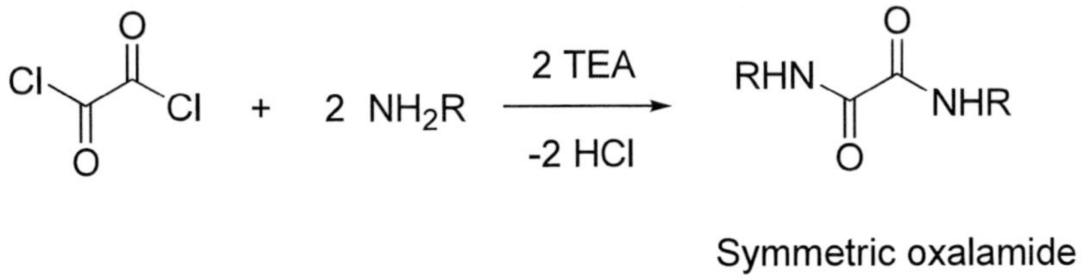

Scheme 2.

Hydrogen Bonding Trends of Oxalamate and Oxalamide Derivatives in Solution by NMR

^{1}H NMR. The study of the conformation in solution has been performed by ^{1}H NMR spectroscopy. Particularly, the temperature dependence of amide hydrogen chemical shift ($\Delta\delta$(NH)/ΔT) was used to stablish intramolecular hydrogen bonding interactions. It is accepted that □$\Delta\delta$(NH)/ΔT value, in proton accepting solvents like DMSO-d6, is a good indicator of proton mobility. $\Delta\delta$(NH)/ΔT values larger than 4 ppb K^{-1} are attributed to solvated NH groups whereas smaller values than 3 ppb K^{-1} are attributed to NH groups strongly involved in intramolecular hydrogen bonding and thus are not available for solvent interactions (Kesler, 1982). A series of oxalamides were synthesized with the aim to establish the role of the hydroxy group in hydrogen bonding, using the above criteria (Martínez-Martínez *et al.*, 1993). Thus, a $\Delta\delta$(NH)/ΔT value of -0.5 ppb K^{-1} was measured for compound **1** and -5.2 ppb K^{-1} for compound **2**. The large difference between these $\Delta\delta$(NH)/ΔT values, alowed us to propose the participation of the hydroxy oxygen atom in a three-center hydrogen (THB) bond in compound **1** and a simple one in compound **2**.

1

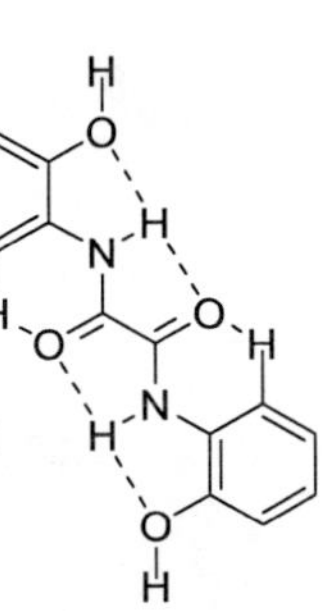

2

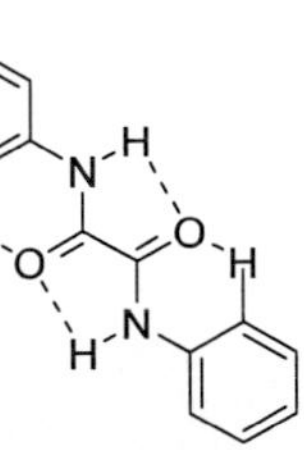

The $\Delta\delta/\Delta T$ values have been shown to be a very useful tool to establish THB in solution in compounds **3-5** for NH^A. Although the similar chemical environment of NH^A and NH^B hydrogen atoms in oxalamides **3**–**5**, the last is not involved in THB interaction in solution ($\Delta\delta/\Delta T$ higher than 3.0 ppb K^{-1}), because its participation in the tautomeric equilibrium, Scheme 3, between forms **I** and **II** (Padilla-Martínez *et al.*, 2001).

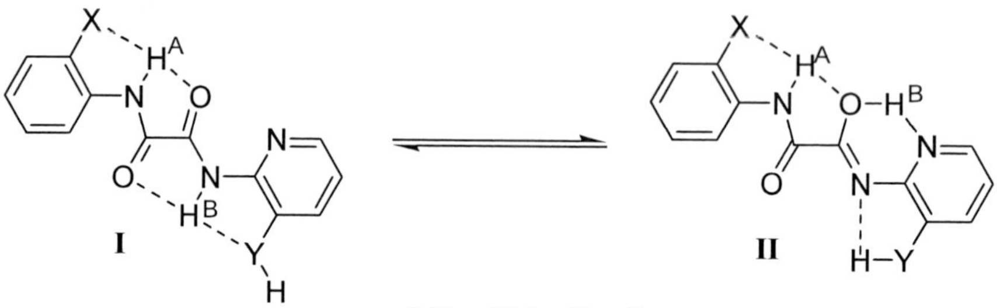

3 X = OMe, Y = O
4 X = OMe, Y = NH
5 X = NO_2, Y = NH

Scheme 3.

The three-center hydrogen bond (THB) is one of the two modes of bifurcated hydrogen bonding which is known to occur. It can be characterized as that configuration where a hydrogen atom covalently bonded to an electronegative atom is hydrogen bonded to another two lying in or close to the plane defined by them. This kind of interaction frequently occurs in the crystal structures of organic molecules, and biomolecules such as aminoacids, proteins and DNA. In spite of its wide occurrence, much less is known about the energy involved in this interaction. In fact, there is a controversy about the energetic superiority of THB over regular or two-centered HB. To address this issue, we measured the thermodynamic properties and cooperativity involved in (THB) using aromatic *ortho*-A substituted amides, oxalamates and bisoxalamides (A = H, OMe, F, CH_2OH, NO_2, $COCH_3$) as model molecules. The *ortho*-substituent was varied from electron-donating (A = OMe) to electro-withdrawing capabilities (A = F, NO_2, $COCH_3$). Amides were used as model compounds for regular HB, oxalamates for THB and bis-oxalamides to test cooperativity in THB (Gomez-Castro *et al.*, 2008).

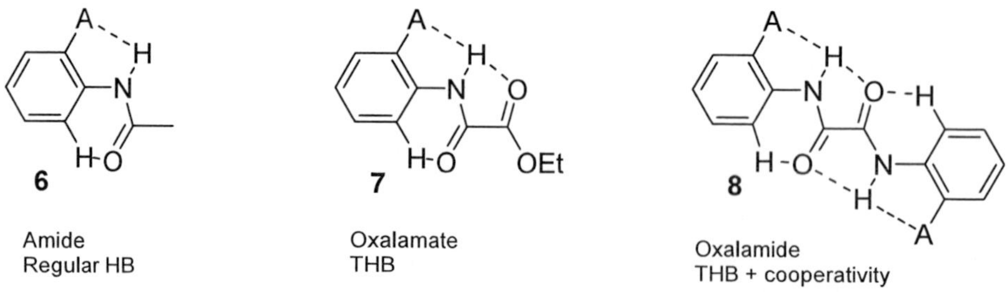

$\Delta H°$ and $\Delta S°$ associated with the disruption of intramolecular hydrogen bonding by solvent were estimated using the temperature dependence data of the N-H chemical shift ($\Delta\delta/\Delta T$). The results suggest that the influence of the A group is more important when electron-withdrawing, increasing both the enthalpy and entropy with an important contribution from conformational changes. The data allowed the estimation of the $Ph{=}NH^+$ rotational barrier of 14.0 kJ mol^{-1} in the amide and 16.7-18.0 kJ mol^{-1} in oxalyl moiety. Correlations between ΔH° and ΔS° with NH temperature gradients predicted an enthalpy change of 18.7(1.0) and 24.4(1.7) kJ mol^{-1} for the energy required to break a full THB bond (A···H···O=C) and entropy differences between the non-hydrogen bonded and hydrogen bonded state of 42.0(4.7) and 61.9(11) J mol^{-1} K^{-1} in oxalamate and bis-oxalamide series, respectively. These results are in agreement with the participation of cooperative effects in THB.

^{13}C NMR. The ^{13}C chemical shifts of oxalic carbon atoms are in the characteristic range for ester or amide carbonyls, disregarding π conjugations through OC—CO bond. The amidic carbon chemical shifts in oxalamides appear at lower frequencies than analogous amides by 8-10 ppm because of the electro-withdrawing effect of the neighboring carbonyl group (Martínez-Martínez *et al.*, 1993), whereas the ester carbonyl appears at higher frequencies than the amide carbonyl in oxalamates. This data is in agreement with two independent carbonyls for the oxalyl moiety.

^{15}N NMR. The ^{15}N chemical shifts of several symmetrical and non-symetrical bis-oxalamides derived from anilines, were measured (Martínez-Martínez *et al.*, 1998). This parameter is linearly dependent on the hydrogen atom mobility according the equation: □$\Delta\delta(^1H)/\Delta T = 106.9(\pm 9.5) + 0.40(\pm 0.04)\ \delta(^{15}N)$. They appear at low frequencies (-263 to -266 ppm) for those N atoms bonded to H atoms involved in intramolecular three-center hydrogen bonds. Whereas NH atoms with more mobility appear at higher frequencies (-258 to -260 ppm). This spectroscopy was used to differentiate the ^{1}H chemical shift of OH and NH in compound **1** (Martínez-Martínez *et al.*, 1993). Usually the chemical shift of OH is at higher frequencies than NH, but in compound **1,** this last appears at a higher frequency due to its involvment in three-center hydrogen bonding.

Hydrogen Bonding by Cyclic Voltammetry

The role of a single (amide) or three-center (oxalamide) intramolecular hydrogen bond on the intermolecular hydrogen bonding interaction with the weakly basic quinone chloranyl (Q^{2-}) was evaluated by cyclic voltammetry (Gómez *et al.*, 2004). Both amide/oxalamide (DH) form hydrogen bonds, stabilizing those reduction intermediates species [(DH)Q^{2-}] from further reduction (Scheme 4).

The association process between the semiquinone and the amides is not important. However, the dianion is associated according to different thermodynamic and kinetic conditions. The voltammetric behavior of these systems was studied by the simulation of two consecutive electron transfer steps with two successive association processes coupled to the dianion. The association process [(DH)Q^{2-}] presents a wide range of equilibrium constant values: K/M^{-1} = 150-170 (amide), 9-10 (oxalamide) when A = H and 5-7 x 10^4 (amide), 1.1-1.2x10^3 (oxalamide) when A = OMe, in agreement with low values when THB is present. The same trend is observed for the kinetic constant k (M^{-1} s^{-1}): 5x10^6 (amide), 1x10^4 (oxalamide) when A = H and 1.0x10^{10}, 1.9x10^5 (oxalamide) when A = OMe. According to the amide structure, the interference of the electron transfer kinetics on the association processes, increases with the extent of the interaction. This result is explained by considering the acidity of the hydrogen donor groups in the amide and the participation of two intermolecular hydrogen bonds. When the proton donor group in the amide is located between a carbonyl and a methoxy group, the kinetics of the association process is lowered by the effect of a structural reorganization of the length and angle bonds.

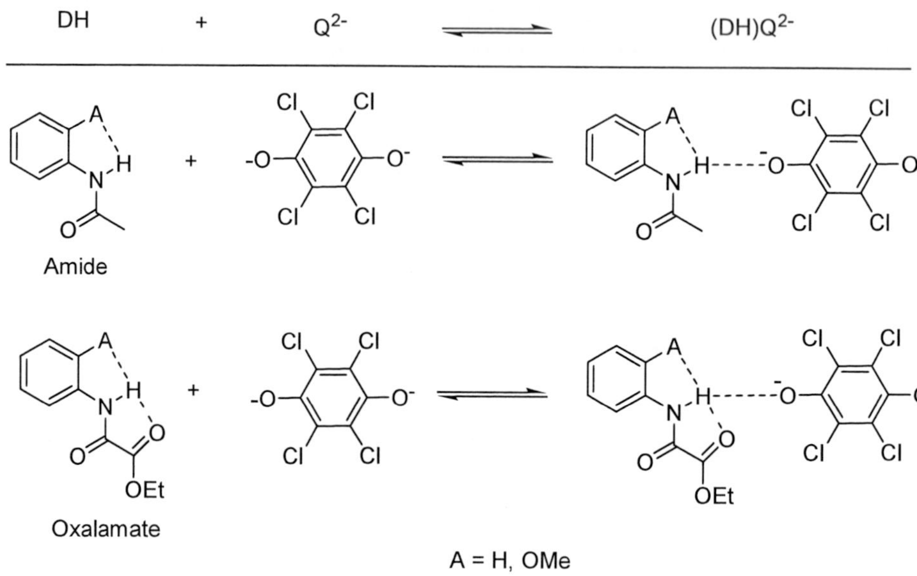

Scheme 4.

Hydrogen Bonding Patterns in Oxalamate and Oxalamide Derivatives in Solid State

Oxalamates are composed by an amide and an ester group directly connected to each other whereas oxalamides are mono-substituted diamides of oxalic acid. Both represent construction units with high in-plane hydrogen bonding potential; they are self-complementary and capable of unidirectional hydrogen bonding. In many cases, their self-assembly produces robust one-dimensional hydrogen-bonded chains, although, depending on the substitution and presence of additional hydrogen bonding units, a variety of more complex structures might be observed. The NH and amide carbonyl are usually in a *trans* disposition, as also are amide and ester carbonyls or both carbonyl amides, in oxalamate and oxalamide, respectively, thus the most common conformation is *trans-anti* (*t,a*) in oxalamates and *trans-anti-trans* (*t,a,t*) in oxalamides, as shown for diethyl *N,N'*-1,4-cyclohexyldioxalamate (Martínez-Martínez *et al.*, 2005) (Figure 11 left).

In the case of dioxalamates derived from tertiary amides, that is without amide NH groups, the oxalyl carbonyls can be syn or perpendicular to each other, as shown in Figure 11 right, for diethyl piperazine-1,4-diyldioxalamate (Martínez-Martínez *et al.*, 2004).

Oxalamates. They can be associated forming ten-membered rings ($R^2_2(10)$) or four-membered chains ($C(4)$) (Figure 12). In the first case, both amide and ester carbonyls participate as hydrogen acceptors whereas in the last case, only the amide carbonyl acts as an acceptor.

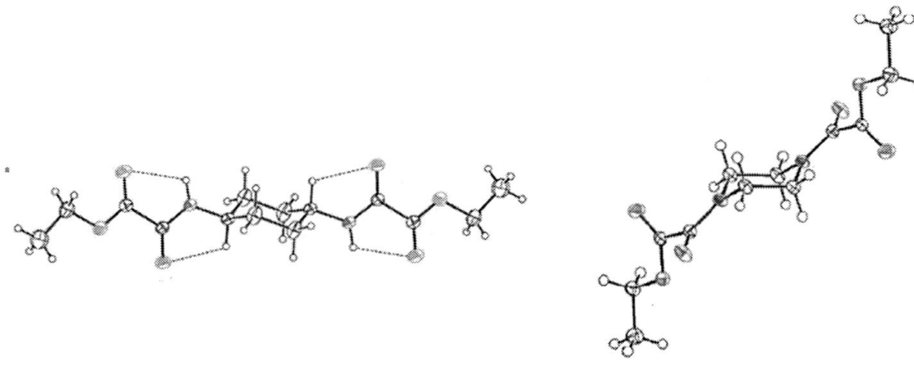

Figure 11. Molecular X-ray structures of secondary (left) and tertiary (right) oxalamates.

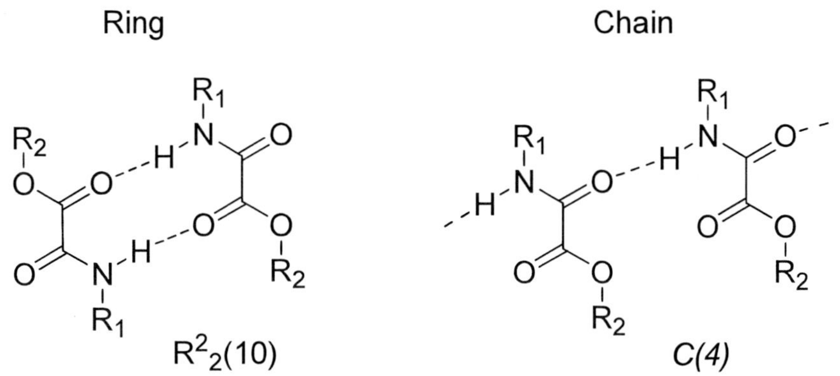

Figure 12. Supramolecular motifs generally found in the crystal lattice of oxalamates.

Ethyl oxalamates derived from non-substituted anilines form $R^1_2(6)$ and $R^2_2(10)$ ring motifs (García-Báez, Gómez-Castro *et al.*, 2003) (Figure 13). When the aromatic ring bears a hydrogen bonding acceptor group (A) in the *ortho*-position, the amide hydrogen is engaged in a three-center hydrogen bond and thus the crystal is composed by independent molecules or weakly bonded ones with the participation of C—H donors to form $R^2_1(5)$ ring motifs (García-Báez, Martínez-Martínez *et al.*, 2003).

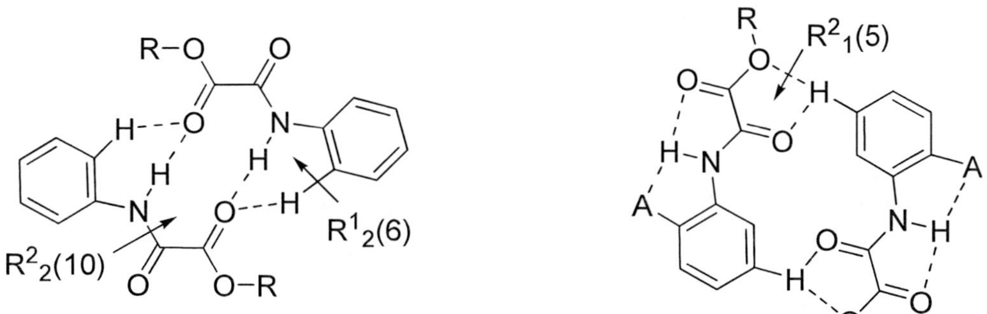

Figure 13. Self association patterns found in the crystal lattice of phenyl-oxalamates.

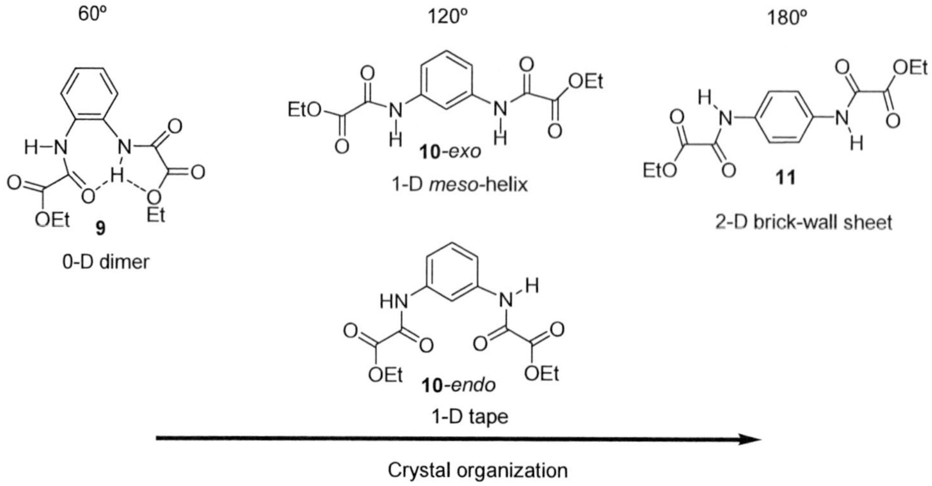

Figure 14. The organization of the crystal lattice depends on the topology of the substitution in plenylenene-dioxalamates.

The hydrogen bonding self assembly of ethyl dioxalamates derived from 1,2-, 1,3- and 1,4-dianilines, compounds 9-11, is directed by the topology of the substitution in the phenyl ring (Muñoz *et al.*, 2010). The complexity or the organization in the crystal lattice increases from zero, one- and two-dimensional as long as the two ethyl oxalamate arms become apart (Figure 14).

Two conformational polymorphs of di-ethyloxalamate 10 have been isolated; both are organized in one-dimension in the solid state, conformer 10-*exo* as a *meso*-helical structure and 10-*endo* as tapes. The former conformer is only 1.0 kcal mol^{-1} more stable than the last one (Padilla-Martínez *et al.*, 2005) thus both polymorphs can equally be obtained. The conformation was controlled through the bulkiness and hydrogen bonding properties of the substituents in the phenyl ring, compounds 12-14, (González-González *et al.*, 2011). The supramolecular arrays in oxalamate derivatives of 1,3-diaminobenzene, 2-methyl-benzene-1,3-diamine and 2,4,6-trimethyl-benzene-1,3-diamine, Figure 15, are directed by self-complementary N—H···O=C (amide) hydrogen bonding interactions, whose organization in the crystal depends on the twist of the oxalamate arms, meanwhile in oxalamate derivatives of 5-tert-butyl-2,6-diamineanisol with an *exo*(ap)–*exo*(ap) conformation, the supramolecular arrays are directed by π-stacking, dipolar carbonyl–carbonyl interactions and C–H···O soft contacts.

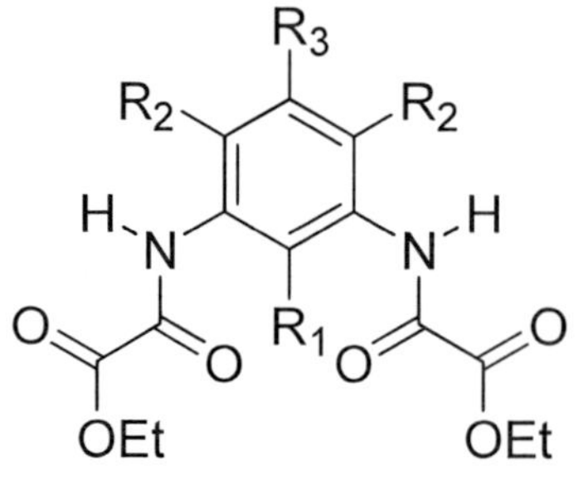

12 $R_1 = CH_3$, $R_2 = R_3 = H$
13 $R_1 = R_2 = CH_3$, $R_3 = H$
14 $R_1 = OMe$, $R_2 = H$, $R_3 = t$-Bu

Figure 15. Studied dioxalamates with different torsion angles.

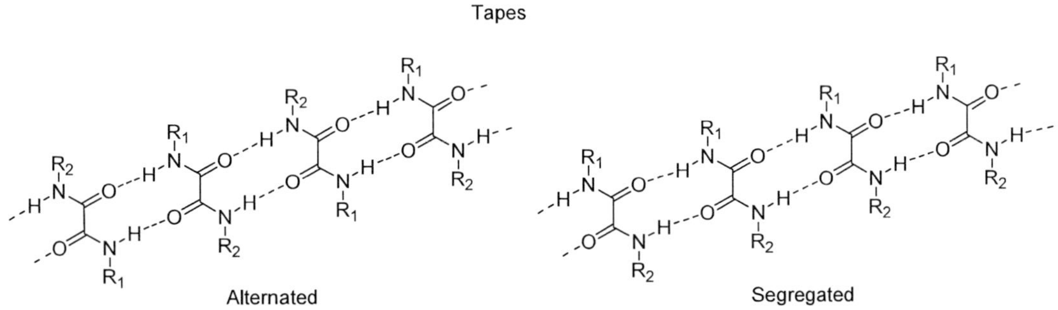

Figure 16. Tapes formed by self-complementary hydrogen bonding in aliphatic oxalamides.

Oxalamides. Those derived from primary amines associate by self-complementary N—H···O=C hydrogen bonding to form $R^2_2(10)$ motifs that propagate as alternated or segregated tapes typical of aliphatic oxalamides (Nguyen *et al.*, 2001) (Figure 16).

Diaryl oxalamides 8 are planar, the crystal lattice is constituted by discrete independent molecules when intramolecular three-center hydrogen bond is formed (Padilla-Martínez *et al.*, 2001). If planarity is disrupted by the use of substituents to twist the phenyl ring from the oxalamide plane, molecular crystals are readily formed with two moles of pentafluorophenol through O—H···O=C and N—H···O hydrogen bonding (Piotrkowska *et al.*, 2007).

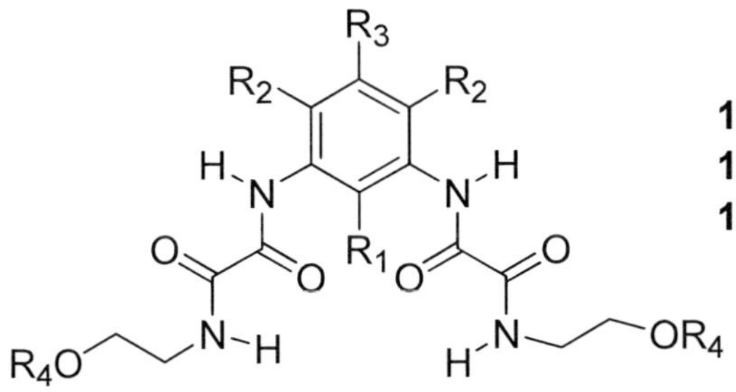

15 $R_1 = CH_3$, $R_2 = R_3 = H$, $R_4 = H$
16 $R_1 = R_2 = CH_3$, $R_3 = H$, $R_4 = CH_2CH_2OH$
17 R_1 = OMe, $R_2 = H$, R_3 = *t*-Bu, $R_4 = H$

Figure 17. Studied phenylene dioxalamides with different torsion angles.

The conformation of oxalamides 15-17, Figure 17, was modulated by the steric requirements of the aryl substituents (González-González *et al.*, 2011). There is a clear trend, in the solid state, to adopt an *exo*(ap)–*exo*(ap) conformation by increasing the steric requirements of the C2-substituent. Further substitution in C4 and C6 leads to both oxalyl arms almost perpendicularly positioned to the phenyl ring resulting in a very reliable conformation also observed in diamides derived from 2,4,6-trimethyl-benzene-1,3-diamine. Carbonyl amide groups are also differentially used for hydrogen bonding in oxalamides, NH/O=C amide dislike interactions (different amide type) are mainly used for intramolecular HB in oxalamides 15–17 and for supramolecular synthons only when also involved in intramolecular three-center hydrogen bonding. NH/O=C amide-like interactions (same amide type) are exclusively used for intermolecular hydrogen bonding in compound 15.

It is worthy to mention that *N,N'*-(1,3-(2,4,6-trimethyl)-phenyl)-bis-(*N*2-(2-(2-hydroxyethoxy)ethyl)oxalamide) 16 adopts the form of a supramolecular *meso*-helix, which is, to the best of our knowledge, the first published example of helical 1,3-phenyl-dioxalamide (Figure 18).

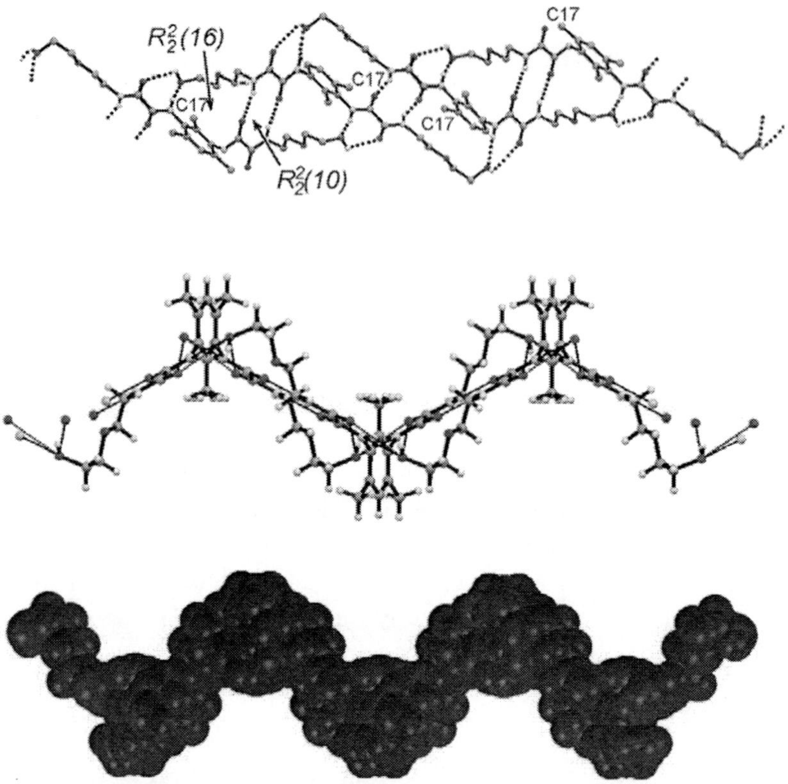

Figure 18. Supramolecular *meso*-helix formed by *N,N'*-(1,3-(2,4,6-trimethyl)-phenyl)-*bis*-(*N*2-(2-(2-hydroxyethoxy)ethyl)oxalamide) **16**. Hydrogen bonding ring motifs are highlighted in the top view and *meso*-helix in middle and bottom (van der Waals).

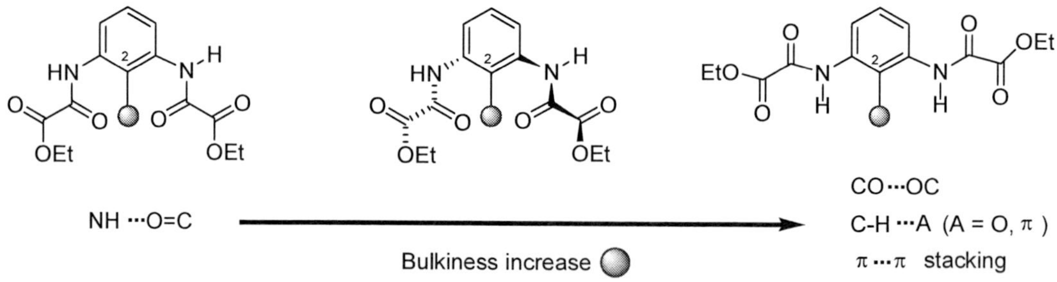

Figure 19. The participation of harder NH···O=C amide interactions in the crystal network structure diminishes, as long as the conformation goes from *endo*(syn)-*endo*(syn) (left) to *exo*(ap)-*exo*(ap) (right) to allow the softer C=O···C=O, C–H···A (A = O, π) and π-stacking interactions to lead.

Finally, the concurrence of varied interactions such as C=O/C=O, C–H/A (A = O, π) and π-stacking, forms crystal networks of comparable robustness as those formed by hydrogen bonding, in compounds **14** and **17**. These compounds are characterized by full in plane *exo*(ap)-*exo*(ap) conformation which is favorable for softer dipolar interactions to occur. (Figure 19)

CONCLUSIONS

The oxalic acid derivatives, herein studied, are good models for the study of intramolecular three-center hydrogen bonds. The entalpy and entropy associated with THB dissociation in DMSO solution, as well as the kinetics of association with chloranyl, strongly support the concept of cooperativity in such systems. In the solid state, 1,3-phenyl di-oxalamide/di-oxalamate compounds have shown to form robust and reliable supramolecular architectures by the concurrence of N—H···O hydrogen bonds and a variety of soft interaction such as C=O···C=O, C–H···A (A = O, π) and π-stacking. The structure in solution shows a close similarity with that observed in the solid state.

Because of its robustness, reliability and versatility, the oxalic acid derivatives are good candidates to be tested as co-crystallizing agents of active pharmaceutical ingredients in order to improve their stability. Pharmaceutical co-crystallizing agents represent an emerging field in the pharmaceutical industry; due to one of the main problems facing this industry which is the low stability of some drugs in the solid state.

Another area in which the oxalamate/oxalamide compounds represent an opportunity of development, due to its ability to form coordination bonds with metal ions, is developing new alkylating agents such as TAED (tetraacetylethylenediamine) which is already commercially sold. Likewise, this alkylating property can be exploited in the design of drugs that act as metalloenzyme inhibitors, that the oxalamidic fragment of the molecule can trap the metal ions which the enzyme uses as co-factor, causing it not to be able to perform its function. Also, due to its ability to form gels, the oxalamate/oxalamide compounds offer a potential for topical drug administration, drug delivery or tissue engineering.

Finally, according to the results that we obtained in the X-ray diffraction studies, the oxalamate/oxalamide compounds can adopt conformations which forms cavities, depending on the phenylenediamine used. These cavities have the ability to form hydrogen bonding interactions, which enables them to interact in a non-covalent way with other molecules that might come into those cavities. This led our research to the study of host-guest interactions, such as that carried out between an enzyme and its active site, in which the oxalamate/oxalamide cavities serves as receptors to dihydroxybenzenes with the purpose to study systems such as the dopamine beta-hydroxylase which is an enzyme that converts dopamine to norepinephrine.

ACKNOWLEDGEMENTS

This work was supported by CONACYT grant 83378, SIP-IPN and Ramón Alvarez-Buylla Funds Universidad de Colima grant 764/11.

REFERENCES

Aakeröy, C.B., Desper, J., Scott, B.M.T. (2006). Balancing supramolecular reagents for the reliable formation of co-crystals. *Chem. Comm.* 1445-1447.

Arantes, V., Qian, Y.H., Milagres, A.M.F., Jellison, J., Goodell, B. (2009). Effect of pH and oxalic acid on the reduction of Fe^{3+} by a biomimetic chelator and on Fe^{3+} desorption/adsorption onto wood: Implications for brown-rot decay. *Int. Biodet. Biodeg.* 63: 478-483.

Baggio, R., Garland, M.T., Perec, M. (1997). Preparation and x-ray crystal structure of the polymeric zirconium(IV) oxalate complex $[K_2\{Zr(C_2O_4)_3\}\cdot H_2C_2O_4\cdot H_2O]_n$. *Inorg. Chem.* 36: 737–739.

Benory, E., Bino, A., Gibson, D., Cotton, F.A., Dori, Z. (1985). Oxalato complexes of the trinuclear aquo ion of molybdenum(IV), $[Mo_3O_4]^{4+}$. *Inorg. Chim. Acta* 99: 137-142.

Caliskan, M. (2000). The metabolism of oxalic acid. *Turk. J. Zool.* 24: 103-106.

Cameselle, C., Bohlmann, J.T., Nuñez, M.J., Leman J.M. (1998). Oxalic acid production by *Aspergillus niger*. *Bioprocess Engineering* 19: 247-252.

Coronado, E., Galán-Mascarós, J.R., Martí-Gastaldo, C. (2008). Single chain magnets. *J. Am. Chem. Soc. 130*: 14987–14989.

Derissen, J.L., Smith, P.H. (1974). Refinement of the crystal structures of anhydrous alfa- and beta-oxalic acids. *Acta Crystallogr.* B30: 2240-2242.

Ehrsson, H., Wallin, I., Yachnin, J. (2002). Pharmacokinetics of oxalylplatin in humans. *Medical Oncology* 19: 261-265.

Frkanec, L., Žinić, M. (2010). Chiral bis(amino acid)- and bis(amino alcohol)-oxalamide gelators. Gelation properties, self-assembly motifs and chirality effects. *Chem. Commun.* 46: 522–537.

Gao, E.Q., Bu, W.M., Yang, G.M., Liao, D.Z., Jiang, Z.H., Yan, S.P. Wang, G.L. (2000). Crystal structure, redox and spectral properties of copper(II) complexes with macrocyclic ligands incorporating both oxamido and imine groups. *Chem. Soc. Dalton Trans.* 1431–1436.

García-Báez, E.V., Gómez-Castro, C.Z., Höpfl, H., Martínez-Martínez, F.J., Padilla-Martínez, I.I. (2003). Ethyl N-phenyloxamate. *Acta Crystallogr.* C59: 0541-0543.

García-Báez, E.V., Martínez-Martínez, F.J., Höpfl, H., Padilla-Martínez, I.I. (2003). π-Stacking Interactions and C—H···X (X = O, aryl) Hydrogen bonding as directing features of the supramolecular self-association in 3-carboxy and 3-amido coumarin derivatives. *Cryst. Growth Des.* 3: 35-45.

Ghorbani-Vaghei, R., Veisi, H., Keypour, H., Dehghani-Firouzabadi, A.A. (2010). Practical and efficient synthesis of bis(indolyl)methanes in water, and synthesis of di-, tri-, and tetra(bis-indolyl)methanes under thermal conditions catalyzed by oxalic acid dehydrate. *Molecular Diversity* 14: 87-96.

Gómez, M., Gómez-Castro, C.Z., Padilla-Martínez, I.I., Martínez-Martínez, F.J., González, F. J. (2004). Hydrogen bonding effects on the association process between chloranil and a series of amides. *J. Electroanal. Chem.* 567: 269-276.

Gómez-Castro, C.Z., Padilla-Martínez, I.I., Martínez-Martínez, F.J., García-Báez, E. V. (2008). Thermodynamic characterization of three centered hydrogen bond using o-aromatic amides, oxalamates and bis-oxalamides as model compounds. *Arkivoc* (v) 227-244.

González-González, J.S., Martínez-Martínez, F.J., Peraza-Campos, A.L., Rosales-Hoz, M.J., García-Báez, E.V., Padilla-Martínez, I.I. (2011). Supramolecular architectures of conformationally controlled 1,3-phenyldioxalamic molecular clefts through hydrogen bonding and steric restraints. *CrystEngComm* 13: 4748–4761.

Harkema, S., Brake, J.H.M.T. (1978). Structure and thermal expansion of urea-oxalic acid (1:1). *Acta Crystallogr.* B35: 1011-1013.

Kesler, H. (1982). Conformation and biological activity of cyclic peptides. *Angew. Chem. Int. Ed. Engl.* 21: 512.

Kim, H.Y., Lee, J.W., Jeffries, T.W., Choi, I.G. (2011). Response surface optimization of oxalic acid pretreatment of yellow poplar (*Liriodendron tulipifera*) for production of glucose and xylose monosaccarides. *Biores. Technol.* 102: 1440-1446.

Lombard, M.C., N'Da, D.D., Breytenbach, J.C., Smith, P.J., Lategan C.A. (2011). Synthesis, in vitro anti-malarial and cytotoxicity of artemisinin-aminoquinoline hybrids. *Bioorg. Med. Chem. Let.* 21: 1683-1686.

Maeda, H (2008). Anion-Responsive Supramolecular Gels. *Chem. Eur. J.* 14: 11274–11282.

Majid, F., Rastegar, E.K., Todd, Tang, H., Wang, Z.Y. (2004). A new class of near-infrared electrochromic oxamide-based dinuclear ruthenium complexes. *Org. Let.*, *6:* 4519–4522.

Martínez-Martínez, F.J., Ariza-Castolo, A., Tlahuext, H., Tlahuextl, M., Contreras, R. (1993). ^{1}H, ^{13}C, ^{15}N, 2D and variable temperature NMR study of the role of hydrogen bonding in the structure and conformation of oxamide derivatives. *J. Chem. Soc., Perkin Trans* 2: 1481-1485.

Martínez-Martínez, F.J., Padilla-Martínez, I.I., Brito, M.A., Geniz, E.D., Rojas, R C., Saavedra, J.B.R., Höpfl, H., Tlahuextl, M., Contreras, R. (1998). Three-center intramolecular hydrogen bonding in oxamide derivatives. NMR and X-ray-diffraction study. *J. Chem. Soc.*, Perkin Trans : 401-406.

Martínez-Martínez, F.J., Rojas-Pérez, E.E., García-Báez, E.V., Höpfl, H., Padilla-Martínez, I.I. (2004). Diethyl piperazine-1,4-diyldioxalate. *Acta Crystallogr.* C60: o699-o701.

Martínez-Martínez, F.J., Maya-Lugardo, P., García-Báez, E.V., Höpfl, H., Hernández-Díaz, J., Padilla-Martínez. I.I. (2005). Diethyl *N,N'*-cyclohexane-1,4-diyldioxalamate. *Acta Crystallogr.* E61, o2994–o2996.

Muñoz, M.C., Blay, G., Fernández, I., Pedro, J.R., Carrasco, R., Castellano, M., Ruiz-García R., Cano, J. (2010). Topological control in the hydrogen bond-directed self-assembly of ortho-, meta-, and para-phenylene-substituted dioxamic acid diethyl esters. *CrystEngComm*, 12: 2473-2484.

Nguyen, T.L., Scott, A., Dinkelmeyer, B., Fowler, F. W., Lauher, J. W. (1998). Design of molecular solids: utility of the hydroxyl functionality as a predictable design element. *New J. Chem.* 129-135.

Nguyen T.L., Fowler, F.W., Lauher J.W. (2001). Commensurate and Incommensurate Hydrogen Bonds. An Exercise in Crystal Engineering. *J. Am. Chem. Soc.* 123: 11057–11064.

Padilla-Martínez, I.I., Martínez-Martínez, F.J., García-Báez, E.V., Torres-Valencia, J.M., Rojas-Lima, S., Höpfl, H. (2001). Further insight into three-center hydrogen bonding. Participation in tautomeric equilibria of heterocyclic amides. *J. Chem. Soc., Perkin Trans.* 2: 1817–1823.

Padilla-Martínez, I.I., Martínez-Martínez, F.J., Guillén-Hernández, C.I., Chaparro-Huerta, M., Cabrera-Pérez, L.C., Gómez-Castro, C.Z., López-Romero, B.A., García-Báez, E.V. (2005). Switching from twisted to planar oxalamide molecular clefts through intramolecular three-center hydrogen bonding. *Arkivok*, 401-415.

Piotrkowska, B., Gdaniec, M., Milewska, M.J., Polonski, T. (2007). Aryl-perfluoroaryl stacking interactions, hydrogen bonding and steric effects in controlling the structure of supramolecular assemblies of *N,N'*-diaryloxalamides. *CrystEngComm*, 9: 868-872.

Šijaković-Vujičić, N., Ljubešić, N., Žinić, M. (2007). Transcription of Gel Assemblies of Bola type bis(oxalamide)-dicarboxylic acid and -diester gelators into silica nanotubes and ribbons under catalyzed and non-catalyzed conditions. *Croat. Chem. Acta* 80: 591-598.

Trask A.V., Motherwell, W.D.S., Jones W. (2005). Pharmaceutical co-crystallization: Engineering a remedy for caffeine hydration. *Cryst. Growth Des*. 5: 1013-1021.

Vishweshwar, P., Nangia, A., Lynch, V.M. (2003). Molecular complexes of homologous alkanedicarboxylic acids with isonicotinamide: x-ray crystal structures, hydrogen bond synthons and melting point alternation. *Cryst. Growth Des.* 3: 783-790.

Wheate, N.J., Walker, S., Craig, G.E., Oun, R. (2010). The status of platinum anti-cancer drugs in the clinic and in clinical trials. *Dalton Trans*. 39: 8113-8127.

In: Biotechnology
Editors: L. G. Torres Bustillos and I. Garcia Peña
ISBN 978-1-62081-071-2

Chapter 12

FRACTAL AND MULTIFRACTAL STUDY OF HEARTBEAT INTERVAL TIME SERIES ANALYSIS IN YOUNG PATIENTS WITH METABOLIC SYNDROME

A. Muñoz Diosdado * ***and G. Gálvez Coyt***
Basic Sciences Department, UPIBI-IPN, Mexico

ABSTRACT

A wider range of complex structures and systems of interest have in recent years been quantitatively characterized using the idea of a fractal dimension, in particular, the physiological systems which have extraordinary complexity. The non-stationarity and non-linearity of signals generated by living organisms defy traditional mechanistic approaches based on homeostasis and conventional biostatistical methodologies. This complexity has fueled growing interest in applying concepts and techniques from statistical physics, including chaos theory, to a wide range of biomedical problems from molecular to organismic levels. The data obtained when studying a complex system often appears as a time series, in principle it seems to be that they lack useful information. Nevertheless, a more detailed analysis with methods and techniques originated in statistical physics and non-linear dynamics can show the existence of correlations that are used to characterize complex systems.

In this work, we have three objectives; first, we introduce methods from non-linear dynamics used to analyze time series: multifractal analysis, detrended fluctuation analysis (DFA) and the method of the Higuchi's fractal dimension. Second, we describe the construction of a database focused on the study of heart variability of young subjects with metabolic syndrome (MS). The syndrome consists of having 3 or more of the following problems: central obesity, alterations in the metabolism of the glucose (with or without Diabetes Mellitus and resistance to the insulin), arterial hypertension, dyslipidemia with diminished HDL cholesterol and hypertriglyceridemia. The database contains 24 hour electrocardiogram records taken with a Holter and their corresponding RR interval time series. The database currently has 90 subjects data, approximately one third are healthy young and the remaining are young people diagnosed with MS. Third, we examine the RR time series of these young university students (17 to 24 years old) with MS. Starting

* amunozdiosdado@gmail.com

from the analysis of the heartbeat time series we used the DFA method, the Higuchi's fractal dimension method and the multifractal analysis to locate the possible presence of heart problems. The results show that although the young persons have metabolic syndrome, the majority do not present alterations in the heart dynamics. However, there were cases where the used fractal parameter values differ significantly from the healthy people values. This suggests carrying out a more detailed exam and a plan for those young persons in order to change their feeding habits and their physical activity routines.

INTRODUCTION

Fractal and Multifractal Analysis

Several years ago, a considerable interest started in describing objects with extremely strange forms, such as coastlines lengths, galaxies, snowflakes or mathematical sets such as the Cantor and Koch curves (Harte, 2001). This tends to happen when the set is very irregular and the characteristics of the system at certain magnification level are essentially the same as those at any other magnification level; the irregularity is repeated at ever finer levels and thus to infinity. Such sets are said to be self-similar. A way of describing the size of these sets is calculating the fractal dimension. For example, the dimension of a coastline may be larger than one but smaller than two, indicating that it is not a simple line and has features that allow filling part of the plane. In the same way, the dimension of the area of the surface of a snowflake can be larger than two but less than three, indicating that its surface is more complex than the regular geometric forms and partially fills the volume. The fractal dimension can be used as a quantitative descriptor of the morphology of the structure and may occasionally suggest how the structure has been formed. When the fractal is geometric, the fractal dimension can be calculated by using simple algorithms (such as box counting). Some physiological structures such as the mammal lungs are almost deterministic fractals (Harte, 2001). Stochastic fractals can be generated by preserving the stochastic properties in all magnifications. In this case, as the magnification is changed, the structure is not identical, it is only similar. Lightning is an example of a stochastic fractal with statistical similarity on many different scales (Dewey, 1997).

Although many physical systems may seem self-similar over a magnification range, there are lower and high lengths, below and above which the fractal behavior no longer exists, i.e. physical structures cannot be infinitely repeated. It might seem that the fractal's ability to model a physical phenomenon is limited. Technically, it is only when the iteration is going to infinity when the real fractal is formed, but this is not a major constraint because with a finite number of iterations we can approach to the real fractal in such a way that the distinction between this structure (prefactal) and the real fractal is not important. Fractal models are used to predict scaling laws on several magnitude orders of an experimental parameter, the model breaks down when the scale is as small as the basic unit or as large as the entire structure. That is, there must be well defined limits for any fractal model.

Frequently, we are not interested in fractals *per se*, but rather in the measures that support such objects. These measures are often probability measures. For example, the occurrence of earthquakes can have fractal features. It is known that earthquakes occur in the faults which are essentially a fracture in the Earth's crust. A fracture in a three-dimensional object can have a dimension of two. However, considering the situation in which small faults unfolds from larger faults and in these small faults we can find small faults, it is expected that this object to have a dimension larger than two but less than three. However, within this network of faults, there are certain areas which are more active than others, i.e., have a larger probability that they likely produce an earthquake. Then we can think that the set of possible places where an earthquake can occur is also a fractal set, but in this set there is a probability measure that describes the possible occurrence of an event. Usually this probability distribution is extremely irregular, even it is probable that we do not have a probability density. Phenomena of this type have multifractal features.

An important point to consider is the fact that Mandelbrot and Van Ness (Mandelbrot and Van Ness, 1968, Malamud and Turcotte, 1999) extended the concept of geometrical self-similarity to time series, which was made in the context of self affine time series. A two-dimensional object has self-affinity if it is statistically self-similar when the two axes are scaled differently. The classic example of a self-affine time series is Brownian motion.

The starting point of the multifractal analysis is a singular measure. Such measures can be created, for example, iterating a given map as the logistic or Henon map and building a probability distribution of the visitation relative frequency in different regions. Another way is to consider a multiplicative process, where at every stage of the construction, a unit interval is divided in intervals of different lengths, each of which contains a given probability. Such a measure is qualitatively similar to those that have been found in experiments or as model results used to simulate, for example, a sand pile or an earthquake. If the measure is a complicated singular distribution then we expect to use self-similarity properties to be able to describe it. To do this, we do a partition of the measure subject to certain restrictions. For example, you can choose a partition (cover) with boxes of equal length. The total probability (integrated measure) in each one of these boxes would be a set of numbers; a function of these numbers is expected to be minimized. On the other hand, boxes of varying sizes could be chosen to make a partition in the measure, where the variation in the size is decided in a way that minimizes any function of the measure in each of these boxes. Depending on the chosen partition and the function that is minimizing, different exponents that reflect properties of the scaling measure can be obtained. For example, the box counting dimension, the Rényi and Hausdorff dimensions, and others (Gammel, 1994).

All methods used to study multifractals have advantages and disadvantages, our analysis used the method proposed by Chhabra and Jensen (Chhabra and Jensen, 1989, Chhabra et al., 1989) because it is a practical method, efficient and above all, accurate for the direct calculation of multifractal spectrum. The time series can be considered as a singular measure $P(x)$ if the time series is normalized. The curve $f(\alpha)$ is estimated by covering the measure with boxes of equal length $L = 2^{-n}$ and calculating the probabilities $P_i(L)$ in each of the boxes. Then we construct the one-parametric family of normalized measures $\mu_i(q, L)$:

$$\mu_i(q, L) = \frac{[P_i(L)]^q}{\sum_j [P_j(L)]^q}$$

The fractal dimension can be obtained from

$$f(q) = \lim_{L\to 0} \frac{\sum_i \mu_i(q, L) log \mu_i(q, L)}{log L}$$

and the average power of the measure or Hölder exponent

$$\alpha_q = \frac{log P_i(L)}{log L}$$

is obtained from

$$\alpha(q) = \lim_{L \to 0} \frac{\sum_i \mu_i(q, L) log P_i(q, L)}{log L}$$

Finally, for each value of q, the numerators in the right side of the equations are evaluated for boxes of decreasing size (increasing N). $f(q)$ and $\alpha(q)$ are extracted from the slopes of the lines that are the graphs of the terms of the numerators versus $logL$, the graph of $f(q)$ vs $\alpha(q)$ is the multifractal spectrum. The parameter q provides a kind of microscope in the sense that allows exploring different regions of the singular measure. For $q > 1$, the more singular regions of P are amplified, while for $q<1$ are stressed the less singular regions, and for $q=1$ the measure $\mu(1)$ reproduces the original measure. The above equations give a relationship between the fractal dimension f and the average power of singularities α as implicit functions of the parameter q.

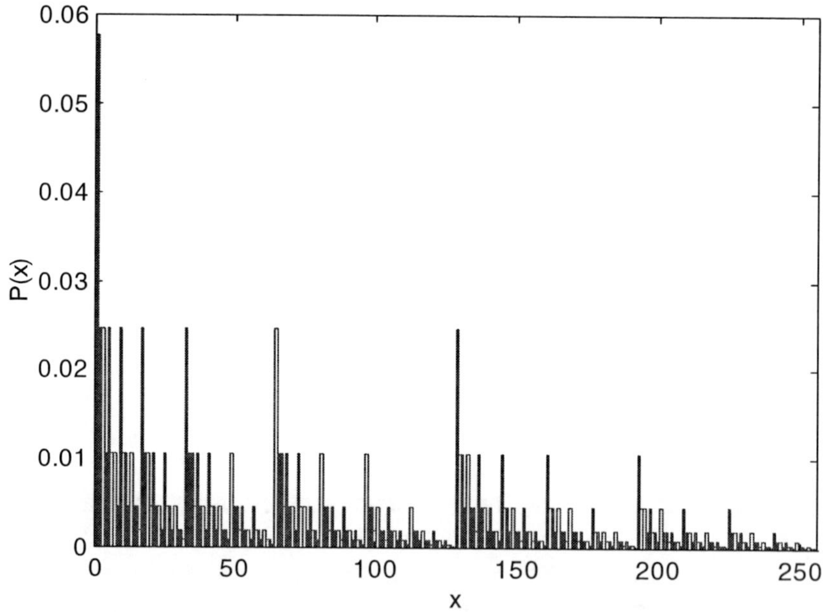

Figure 1. The measure $M(x)$ for the binary multiplicative process after $n = 8$ generations with $p_1 = 0.7$ and $p_2 = 0.3$.

These fundamental concepts of multifractal formalism can be illustrated by applying them to the measure that is generated by the multiplicative process defined by the division of the unit interval into two pieces of half the original size. The multiplicative process divides the population in every part of the same shape, so there are four parts in the next step. As this process is iterated, it produces more and more short segments that contain less and less of the total measure. In Figure 1 is displayed in a bar graph, one step in the construction of multifractal by using the multiplicative process after 8 generations. After n generations there are $N = 2^n$ cells sequentially labeled for the index $i = 0,..,N-1$. The longitude of the i-th cell is $L_n = 2^{-n}$, and the measure, or fraction of the population is $P_i = p_1^k p_2^{n-k}$, where k is the number of 0's in the representation of the binary fraction of the number $x = i/2^n$. Cells describing the population distribution cover the line completely and contain all the measure, in other words, each and every member in the population.

The values of α and $f(\alpha)$ were obtained according to the algorithm already described. The multifractal spectrum is the graph $f(\alpha)$ vs α as the one subsequently shown in the Figure 2; it is a graphic that reminds a kind of parabolic curve that opens downward. The α_0 value is the α value where the spectrum has its maximum. The bandwidth of the spectrum (also called degree of multifractality), estimates the α range where $f(\alpha) \geq 0$, and is obtained by extrapolating the adjusted curve to zero; then the width of the multifractal spectrum is defined as $\Delta\alpha = \alpha_{max} - \alpha_{min}$ where $f(\alpha_{max}) = f(\alpha_{min}) = 0$.

In the series with data corresponding to the binary multiplicative process, the values obtained for the values of α_{min}, α_{max} and α_0 are practically the same that the theory predicts [Feder J. 1988, Guzmán-Vargas et al., 2004, Muñoz-Diosdado et al., 2008).

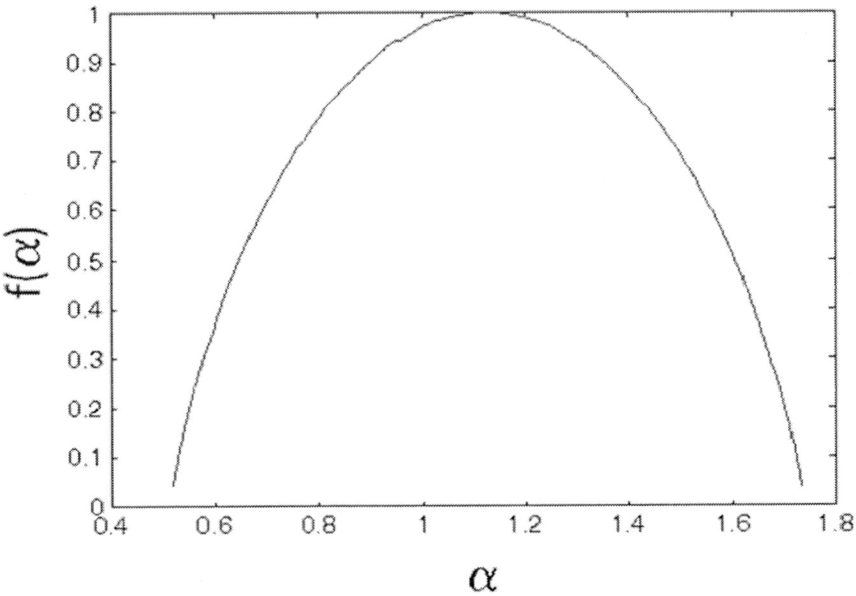

Figure 2. Multifractal spectra of the measure obtained with the binary multiplicative process with $p_1 = 0.7$ and $p_2 = 0.3$ in the 8th generation, it means, only 256 data in the time series.

As a different way to visualize the multifractal spectrum, there are color panels where each color corresponds to a fractal dimension (Ivanov et al., 2002, Del Río Correa et al., 2002, Stanley et al., 1999). The method is to explore the time series following a procedure very similar to the spectral analysis of the light from a star, using a color panel which is a kind of "optical" multifractal spectrum, a wider range of colors indicates a high degree of multifractality, i.e. multifractal spectra are wider; they have a more polychromatic appearance while the panels corresponding to time series that tend to monofractality tend to be more monochromatic, indicating loss of multifractality.

Time Series with Fractal Methodologies

It is not enough with the multifractal analysis for characterizing completely a time series, especially if we want to study correlations in the time series, it is necessary to resort to techniques of fractal analysis, in particular, we will describe methods frequently used as spectral power, Higuchi's method and detrended fluctuations analysis or DFA-method.

Spectral Power

The spectral power density is a well established method to investigate the temporary fluctuations of the time series. A time series can be described in the dominion of time as $x(t)$ or in the dominion of frequency in terms of amplitude $X(f,T)$ where f is the frequency. This amplitude can be calculated by means of the Fourier transform applied to $x(t)$ in the interval $0 < t < T$.

$$X(f,T) = \int_0^T x(t)e^{2\pi i f t}\, dt$$

The quantity $|X(f,t)|^2 df$ is the contribution to the total energy of $x(t)$ of these components with frequencies between f and $f + df$. The spectral power [14] is defined as:

$$S(f) = \lim_{T\to\infty} \frac{|X(f,T)|^2}{T}$$

Here, $X(f,T)$ is the Fourier transform of the time series $x(t)$, T is the monitoring total time with $T = n/\Delta t$, where Δt is the sampling time.

For self-affine series, the spectral power behaves as a power law with frequency given by $S(f) = f^{-\beta}$. First, $S(f)$ is calculated by using the fast Fourier transform (FFT), and the spectral exponent β is estimated by the slope of the adjusted least squares line that fits to $logS(f)$ vs $log(f)$ and, according to Malamud and Turcotte [4], β characterizes temporal fluctuations of the time series, in this sense, the β value is strongly related to the type of correlations in a time series. For example, the white noise has $\beta \approx 0$; the Brownian motion, the typical example of short-range correlated noise of short-range has $\beta \approx 2$; and the $1/f$ noise, which exhibits long-range correlations has $\beta \approx 1$. The β value is a measure of persistence or antipersistence in a time series (Malamud and Turcotte, 1999). It is well known that the power spectrum can be considered as the cosine transform for the Fourier transform of the correlation function as it is established by the Weiner-Khintchine theorem (Bergé et al., 1984). In this sense, also the β value is strongly related to the type of correlations in a time series. As an example, Figure 3 shows the power spectra of some self-affine time series generated with the method proposed by Malamud and Turcotte (1999), in this case a noise Gaussian with $\beta = 2$ and 10000 points, the slope of the fitted line gives the value of the spectral exponent $\beta = 2.0002$ which is very close to the expected value and corresponds to a Brownian noise.

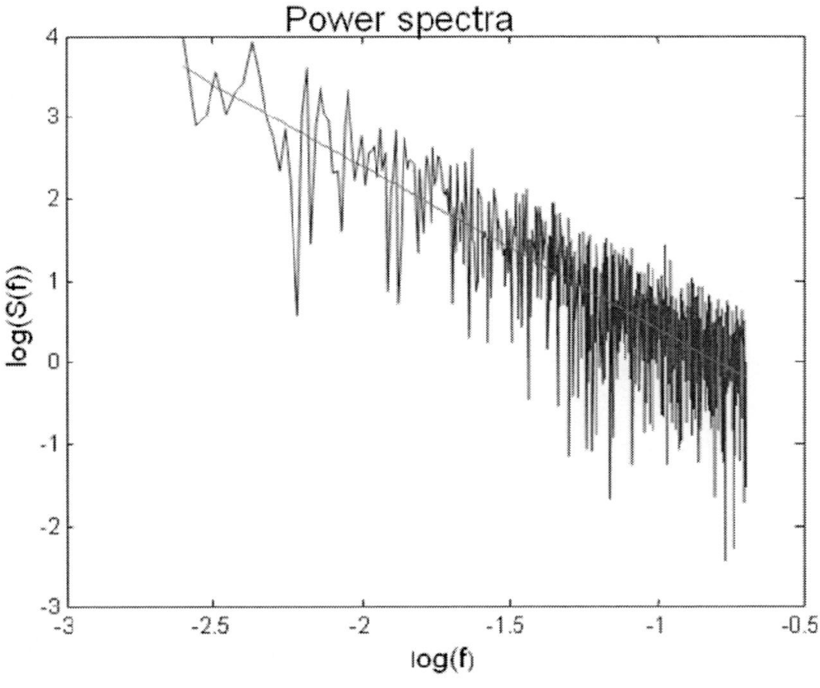

Figure 3. Evaluation of the spectral exponent for a Gaussian noise of Brownian type.

The Method of the Higuchi's Fractal Dimension

Fractal dimension of self-similar objects in the plane is defined in terms of the isotropic distribution of its parts, which can be scaled by a single scale factor. This property changes for self-affine fractals because its spatial distribution is not isotropic and the scaling factor is different for each direction (Turcotte, 1997). Higuchi (Higuchi, 1988, Higuchi, 1990) proposed a method for calculating the fractal dimension of self-affine curves in terms of the slope of a straight line that adjusts the length of the curve versus the time interval (lag) on a log-log graph. The method consists of considering a finite set of data taken at a regular interval,

$$v(1), v(2), v(3), \dots, v(N)$$

From these series, new series are constructed v_k^m, defined as

$$v_k^m : v(m), v(m+k), v(m+2k), \dots, v\left(m + \left[\frac{N-k}{k}\right] \cdot k\right)$$

with $m = 1,2,3,\dots,k$, where [ ] denotes the Gaussian notation and m, k are integers indicating the initial time and the time interval, respectively. For a time interval equal to k, k new sets of time series are obtained. For example, for $k = 3$ and $N = 100$, the three series that are obtained by this process would be: $v_3^1 : v(1), v(4), v(7), \dots, v(97)$,

$v_3^2: v(2), v(5), v(8), \dots, v(98),$

$v_3^3: v(3), v(6), v(9), \dots, v(99)$. The curve length v_k^m, is defined as:

$$L_m(k) = \frac{1}{k}\left[\left(\sum_{i=1}^{\frac{N-m}{k}} |v(m+ik) - v(m+(i-1)k)|\right)\frac{N-1}{\left[\frac{N-m}{k}\right]k}\right] \quad (8)$$

The term $\left[\frac{N-m}{k}\right]kk$ is a normalization factor. Then, the curve length for the interval k, is given by $\langle L(k)\rangle$, the average value over k sets $L_m(k)$. Finally, if

$$\langle L(k)\rangle \propto k^{-D},$$

then the curve is fractal with dimension D (Higuchi, 1988). In the case of self-affine curves this fractal dimension is related to the spectral exponent β by means of $\beta = 5 - 2D$, where if D is in the range $1 < D < 2$ then $1 < \beta < 3$ (Higuchi, 1988). Higuchi showed that this method provides an accurate estimate of the fractal dimension of even a small data number. Higuchi developed its method as an alternative to spectral analysis because although there is a relationship between D and β, the standard deviation of the obtained fractal dimension by using FFT is greater than that which is obtained by calculating the fractal dimension with this method. As the FFT method required making averages of power spectra to obtain a stable range, it will be required many of these averages to obtain values as precise and stable as those afforded by this technique. In addition, Higuchi method has allowed defining clearly the two or more regions in which the graph of $log L_m(k)$ vs $log\ k$ is divided when there are changes in the correlation dynamics, i.e. the points that divide different scaling regions with different values of fractal dimension D (Higuchi, 1988, Higuchi, 1990, Telesca et al., 2003). If the Higuchi method is applied to the same series for which we calculated the spectral power in Figure 2, it is the graph shown in Figure 4, as we can see, the adjustment in this case gives us a Higuchi's fractal dimension of $D = 1.5075$, if we use the relation $\beta = 5 - 2D$ we obtain a value of $\beta = 1.985$, very close to the $\beta = 2$ value that should be given as a series of Brownian type.

Detrended Fluctuation Analysis

Peng et al., 1993 introduced a new method to detect long-term correlations, the detrended fluctuation analysis (DFA). This method is based on the classic random walk variations and has been used to detect long-range correlations in highly heterogeneous sequences of DNA and other physiological signals (Peng et al., 1993, Peng et al., 1994). The DFA method has some advantages over conventional methods (Varotsos et al., 2002), because it allows the detection of long-range correlations in time series highly non-stationary and avoids spurious detection of apparent long-term correlations that are an artifact of non-estacionarity and can also help to identify different states of the system according to their different scaling behaviors (Hu et al., 2001, Chen et al., 2002).

In the DFA method, we begin with a time series of total length *N*, first, this series is integrated $y(k) = \sum_{i=1}^{k}[B(i) - B_{ave}]$. Second, the series is divided into boxes of equal length *n*. In each of these boxes, a least squares line (or a polynomial curve of order *k*) is fitted to the data (this is called the local trend $y_n(k)$, in each box). For *k=1* we have the DFA-1 method, $k = 2$ corresponds to DFA-2 method, etc. Then, the points $y_n(k)$ (of the fit) are subtracted from the integrated series $y(k)$ in each box (Figure 5). For a given box of size *n* the mean quadratic root of the fluctuation $F(n) = \sqrt{\frac{1}{N}\sum_{k=1}^{N}[y(k) - y_n(k)]^2}$ is calculated. This calculation is repeated on all time scales (box sizes), from $n = minbox$ to $n = maxbox$, to obtain a relationship between *F(n)* and *n*. Typically, *F(n)* would increase with the size of the box. A linear relationship between *F(n)* and *n*. Typically, *F(n)* would increase with the box size. A linear relationship in a log-log graph indicates the presence of a scaling power law (fractal):

$$F(n) \sim n^{\gamma}$$

Under such conditions, fluctuations can be characterized by the scaling exponent or correlation exponent γ, i.e., the slope of the line that adjusts $log(F(n)$ versus $log(n)$. The case $\gamma = 1/2$ represents the absence of long-range correlations (white noise), if $\gamma < 1/2$ the signal is anti-correlated, if $\gamma > 1/2$ there are positive correlations in the signal, if $\gamma = 1$ it means $1/f$ noise and $\gamma = 1.5$ represents a Brownian motion (Karasik et al., 2002). The exponent γ γ is related to the spectral exponent β by $\beta = 2\gamma - 1$ (Iyengar et al., 1996), then with the fractal dimension of Higuchi by $\gamma = 3 - D$. Figure 6 shows how the DFA method is applied to the same series of Brownian type to which it was applied spectral power and the Higuchi's method, in this case it was obtained $\gamma = 1.485$ a value very close to the value of $\gamma = 1.5$ that is the expected value for Brownian noise.

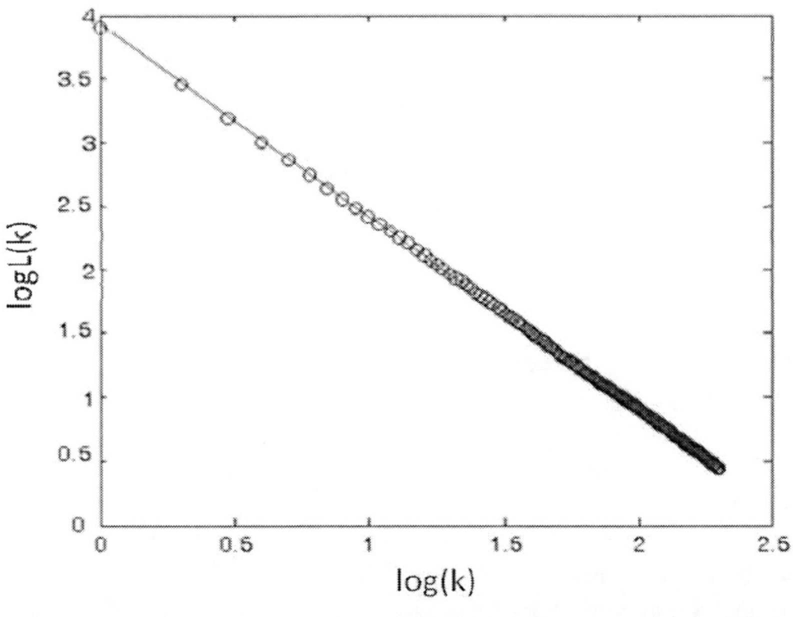

Figure 4. Evaluation of the Higuchi's fractal dimension of one Gaussian Brownian noise.

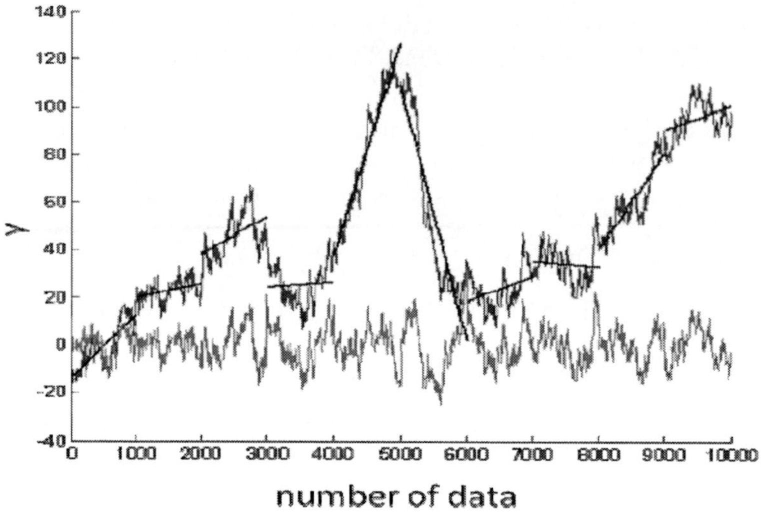

Figure 5. Elimination of the trend in the DFA1 method. As it can be seen, it is chosen a size box, then a fit to a line in each box is done and when subtracting the tendency to the original data it is obtained the lower curve, which has no trends.

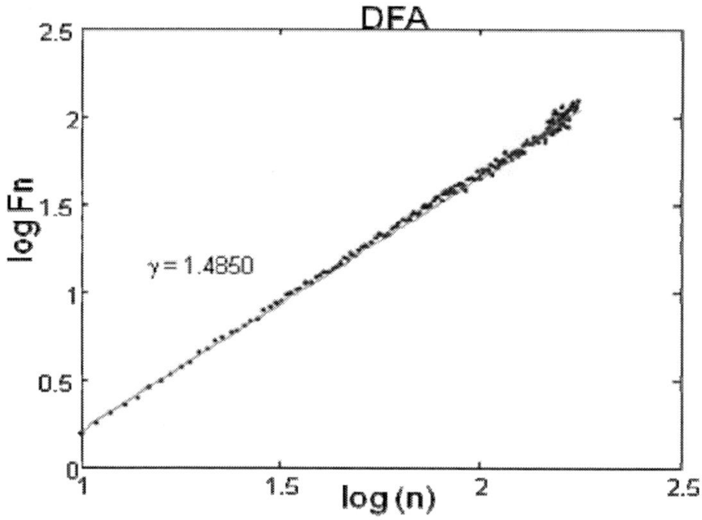

Figure 6. Evaluation of the DFA exponent for a Brownian Gaussian noise.

The Construction of the Database

Currently the clinical diagnoses have become totally dependent on the ability to measure and analyze physiological signals, such as the electrocardiogram. The advantage is that now it is possible to handle massive volumes of information that corresponds to the monitoring for several hours (even days). Physiological signals represent processes that are non-linear and non-stationary by nature (Glass and Mackey, 1988, Service, 1999), although the traditional techniques of analysis assume them linear and stationary. It has been established that such data sets may contain information that traditional methods are unable to find and that may have clinical value (Kresh, and Izrailtyan, 1998, Ivanov et al, 1999, Muñoz-Diosdado, 2009, Muñoz-Diosdado et al., 2009).

Researchers need access to databases of high quality, rigorously validated and that are standardized and not only with data of healthy subjects, but with different pathological conditions (Chhabra et al., 1989). Indeed, as it has been suggested on numerous occasions, both experimental data and clinical data are obtained with very high investment, they are analyzed only for those who collected them, and once analyzed (probably partially) they are filed or lost. There are important aspects to be taken into account when trying to build a database of physiological variables. The first thing is that it requires a significant effort to develop software for signal processing, for the analysis of time series and for the handling of such databases. Commercial software does not meet all expectations, much less when the quantities of information are very large. Researchers frequently invest many resources in developing software for a single project and practically nobody do the validation of algorithms and programs. Other researchers probably could not use the new tools by variations in the software used to implement algorithms; usually the publications did not specify details of software and programs. In biomedical research in general, and in cardiovascular research in particular, the variability in biomedical signals requires non-trivial computational techniques for its quantification and analysis. It is necessary to offer researchers resources for the exchange and dissemination of biomedical signals and algorithms.

In the opinion of those who constructed the "Research Resource for Complex Physiologic Signal" (Goldberger et al., 2000), an effort of this kind should contain three interrelated components: the bank's data, both analytical and computational analysis tools and a resource for the communication and dissemination of the results. The database must be a well characterized set of files that can be used by different researchers. That means that the collected data must be well documented, in such a way that the measurement conditions and the characteristics of the subject of measurement are known. Ideally, it will be good not to measure a single parameter but several variables. Data should be available via the internet and physically. The set of analysis techniques should be implemented in a software free or easily accessible. It must include both classical methods and new techniques derived from statistical physics, fractal and dynamic analysis, analysis of non-stationary processes, simulation of physiological signals, etc. It should be possible to exchange and disseminate information and analysis techniques, so that it can be considered, discussed, reviewed and evaluated in detail by any interested researcher. In this way it would be possible to detect errors in the data, in the algorithms or in the interpretation of the results, and its use in education and training of new researchers would be extremely relevant. The disposition of the researchers to share original databases can ensure that the data is validated when they are revised and re-analyzed. In addition, the data can be analyzed using new techniques. In fact, the deposit in an information bank of data obtained with government funding should be an obligation, previously taking all measures for the absolute protection of the anonymity of the subjects from whom such measurements were taken. When the databases are constructed with similar criteria, it will be possible to eliminate different interpretation of results due to different ways of taking data, different analytical techniques, and different computer implementations of the algorithms. In fact, there is a trend in the area of scientific publishing which provides that in the future, it will not be published articles based on the analysis of physiological signals, if they do not provide the original data for validation (Goldberger et al., 2000).

METHODOLOGY FOR THE MEASUREMENTS

This part of the chapter describes some steps that we have been taken to be able to integrate a database of 24 hours ECG recordings with a Holter, of healthy young persons and young people diagnosed with MS. Metabolic syndrome is a complex group of interrelated metabolic disorders, which have caused difficulties for their classification, including: obesity, resistance to insulin or diabetes, hypertension and Dyslipidemia (Moller, and Kaufman, 2005, Tapia, 2003). In this sequence of multifactor events, we want to define early alterations that translate endothelial dysfunction and therefore electrophysiological changes of the heart muscle. The aim is to find deviations in electrocardiographic records of young people with metabolic syndrome using non-linear dynamics, techniques that possible will allow us to establish whether there is a cardiovascular problem or a problem is developing. Therefore, we need to measure ECG from subjects with MS and healthy subjects in a control group.

It is well known that the ECG is a record of the electrical activity of the heart and in general is not limited to the conduction area, but to the whole heart. The heart rate variability (HRV) is the variation of the distance between the R's in the ECG intervals. It depends on fluctuations in the stimulation of the autonomous nervous system on the heart, and it is influenced by the age, the baroreceptor reflex, breathing, temperature, etc. Young adults have a high HRV, expression of the optimum degree of adjustment of the vagus nerve reflexes. The reduction of the HRV is associated with deterioration of vagal control and a sympathetic dominance.

We selected young students with ages ranging from 17 to 24 years, for three study groups according to the following selection criteria: Group 1-(with alterations in lipid metabolism) individuals of both sexes with BMI (body mass index) > 30 and index waist - hip IWH > 0.90 for men and 0.85 for women. Total serum cholesterol > 200 mg/dl, HDL < to 45 mg/dl, LDL > 150 and triglycerides > 150 mg/dl. Group 2-(with carbohydrates metabolism disturbances) individuals of both sexes with a family history of diabetes type 2 in first-line, blood glucose in fasting conditions > 110 mg/dl. Group 3 (controls) individuals of both sexes without disturbances in the metabolism of lipids, carbohydrates, or a family history of diabetes. For the three groups we applied as exclusion criteria: suffering from heart disease, congenital disorders of lipids and carrying out a systematic practice of sports. The 24-Hour ECG records were taken with a Fukuda Denshi Holter; model FM150 with a sampling frequency of 125 Hz. The disposable electrodes are also from Fukuda (TEH-177MDT). The subjects previously signed an informed consent and performed a brief clinical survey, a record of blood pressure and anthropometric assessment and a sample of blood was taken for clinical laboratory studies (hematic biometry and blood biochemistry).

We have 90 ECG 24 hours Holter records, approximately each group referred above has 30 subjects. The digitized records have a 16 byte format, so the signal can be easily opened using MATLAB or any software for the analysis of massive data sets. At this first stage, we are working with the analysis of interbeat time series, we write programs to determine the ECG R points, and our results were compared with the results provided by the software of the Holter Fukuda FM150.

The real ECG's present problems for the proper determination of R points, for example, in Fig. 7 we show an ECG of one minute duration of a person with an arrhythmia, and in Fig. 8 it can be seen in detail a part of the ECG. It is notorious the displacement of the so-called

ECG basal line, which is the reason for what many computational algorithms do not properly find the R points.

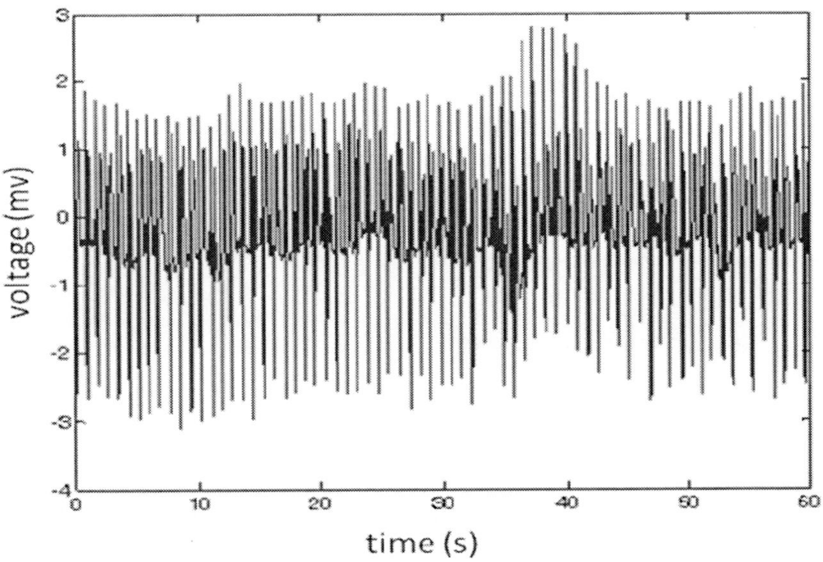

Figure 7.One minute duration of one ECG record.

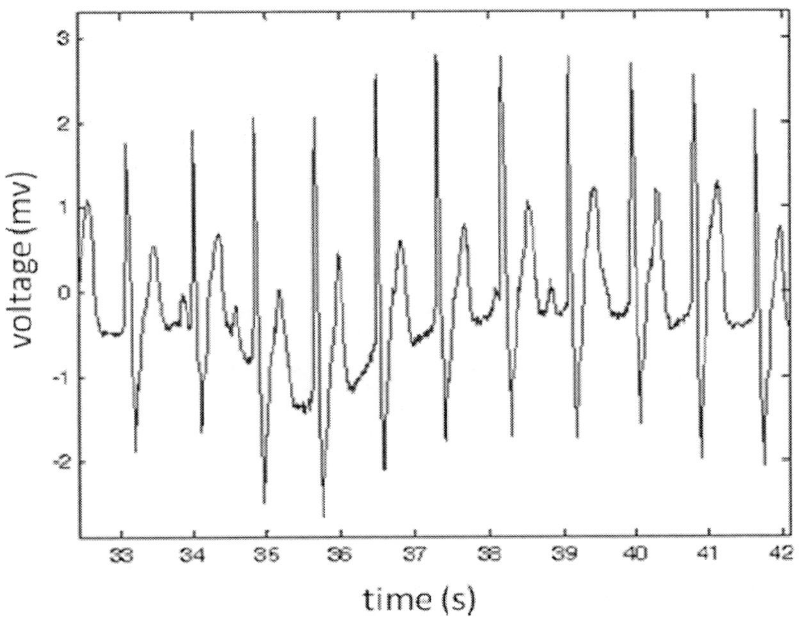

Figure 8. An approximation to the ECG registration of Figure 7. Note the displacement of the basal line.

The signal that modulates the displacement of the basal line can be considered as a very low frequency signal compared to the signal of interest. An adaptive filter is used here to remove this interference; Figure 9 shows a diagram of this filter. The adaptive filter has a single input coefficient and the input is always 1. Under these conditions, the adaptive system will be able to model the mean value of the ECG. The application of the adaptive filter to the ECG of the Figure 8 can be seen in Figure 10. On the other hand, it can be seen that the elimination of the basal line does not change the location of the signal R peaks.

To determine the R peaks, a criteria combination is used in this work because the application of a single criterion does not adequately locate all the peaks of the ECG. The first used criterion is precisely the signal amplitude which is the maximum close to the peak and the second uses the fact that in the R peak the signal has a noticeable change of slope (or first derivative) from positive to negative at that point. In Figure 11, we show the graph of the signal change or first derivative with respect to the time, for a segment of the signal from Figure 8.

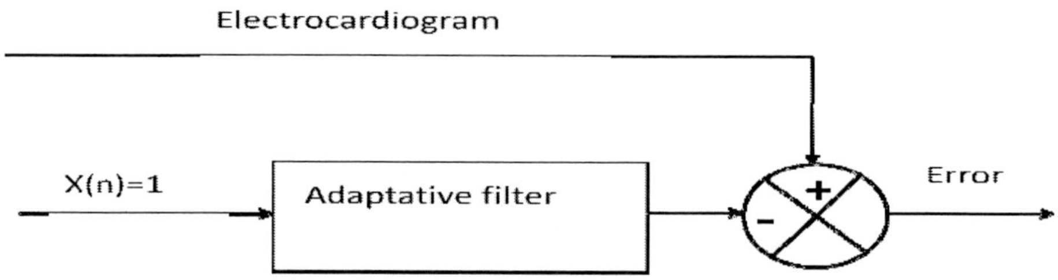

Figure 9. Diagram of the proposed adaptive filter to eliminate the variations of the basal line.

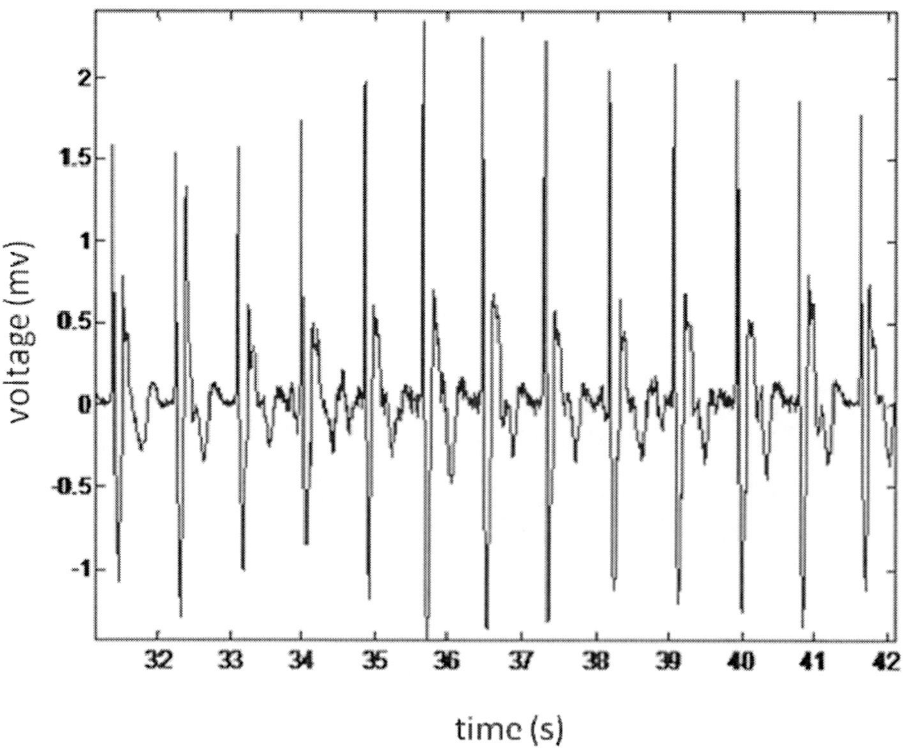

Figure 10. Correction of the basal line using the described filter.

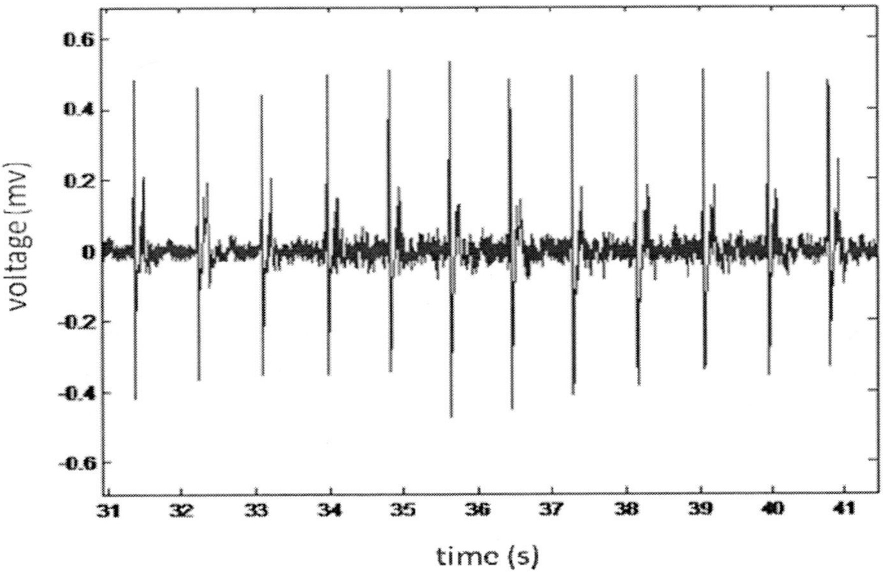

Figure 11. Behavior change (first derivative) of a segment of the ECG signal from Figure 8.

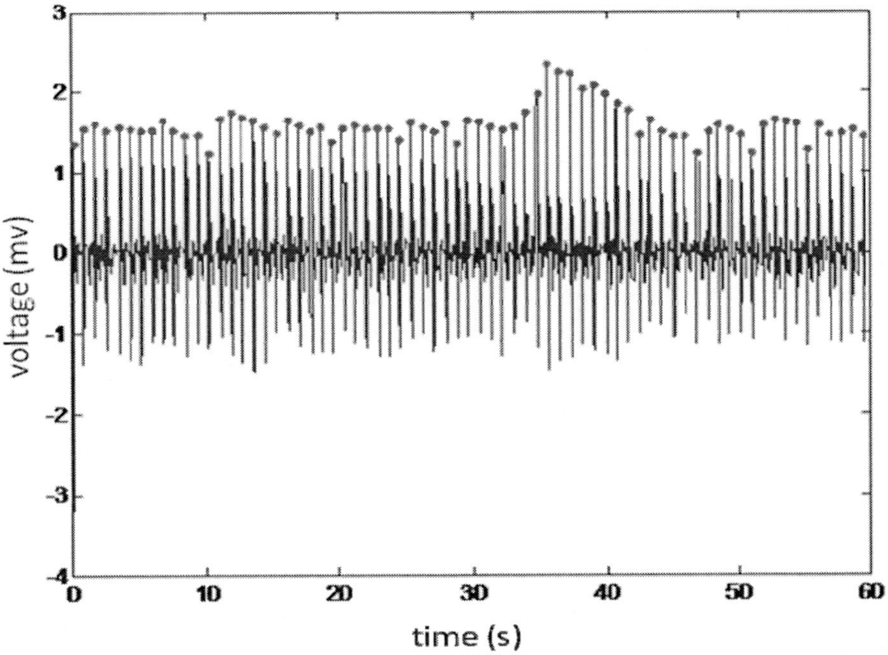

Figure 12. Implementation of the three criteria for the location of R peaks. The R peaks are indicated by dots.

The joint implementation of the two above-mentioned criteria requires a proper selection of some parameters as the signal amplitude, the slope maximum change, etc., which need to be adjusted depending on the particular problem. It is clear that it is necessary to apply other criteria that would eliminate the unwanted peaks. This third criterion uses the fact that the interbeat time must have a minimum value, and then applying this criterion with the proper selection of the necessary parameters gives the result shown in Figure 12, which shows the correct location of the R peaks, which are highlighted with points. So far we have nearly 90 records and it was planned to continue measuring, now people with the syndrome but at different ages. We have the ECG of all of them, the interbeat time series and the programs to evaluate fractal dimensions, multifractal spectra analysis, detrended fluctuation analysis, spectral power, etc.

The analysis techniques have been programmed in Matlab and the management of the database can be done in Access or Excel. The database is organized into different categories; for example, in the present case, the base has data of individuals in the control group and young people with MS. The database has the data of the patient; results of the biochemical tests, the analysis results are stored also in the database. We analyzed total RR time series, and segments corresponding to when the subject is awake or when the subject is asleep. The database has also the register of the activities during the 24 hours that the patient was connected to the Holter.

METABOLIC SYNDROME

In a previous work (Muñoz-Diosdado et al., 2010), we examined the presence of metabolic syndrome (MS) in young students (17-24 years). About 13.9% of 2,947 students were diagnosed with metabolic syndrome. For the purpose of this work, we took a sample of healthy students and a sample of students with metabolic syndrome were selected to took them 24-hour Holter records and from these records were extracted the RR time series (tachograms). Nonl-inear analysis techniques were applied to the RR time series (tachograms) of the group of young healthy people and to the group of subjects with MS.

In 1998, the World Health Organization (WHO) presented its operational definition of metabolic syndrome, which requires evidence of resistance to insulin and less 3 or 4 factors from hypertension, hyperlipidemia, obesity and microalbuminaria (WHO consultation). The most widely accepted definition was currently proposed by the National Cholesterol Education Programme (NCEP) of the United States in 2001 (Expert panel on Detection, 2001), which require the presence of at least 3 of the 5 factors: circumference of the waist, high triglycerides, low HDL cholesterol, hypertension and intolerance to glucose. The main challenge to assess the impact of the metabolic syndrome in diabetes mellitus type II (DM2) or cardiovascular disease (CD), is the current transition in its definition. The definition proposed by the NCEP includes abnormal glucose in fasting as a surrogate variable to insulin resistance (Expert panel on Detection, 2001, Bonora et al., 2003) (See Table 1). It is known that the clinical features of the MS are associated with increased risk of cardiovascular disease, including an increased risk of coronary heart disease for a given level of cholesterol of low-density and premature death (Moller and Kaufman, 2005).

Current definitions for the diagnosis of MS are focused in the adult population, but the best approach to prevention is the early recognition, detecting the MS in children and adolescents is needed to estimate the magnitude of this problem among the young population (Rodríguez-Morán et al., 2004). In Mexico, the Ministry of Health reported diseases including MS among the three most important causes of death and among the 10 leading causes of hospital admission. A study in 15,607 Mexican subjects inhabiting 417 urban communities aged 20 to 69 years, found that MS has a prevalence of the 26.3%, according to the criteria of US NCEP ATP III and the 13.61% according to the WHO criteria (WHO consultation). In both cases, we Mexicans are classified as a population at high risk for obesity, diabetes, dyslipidemia, hypertension, and cardiac coronary disease.

Table 1. Definition of the Methabolic Syndrome by the World Health Organization (WHO) and the National Cholesterol National Program (NCEP) from USA

WHO	NCEP
Abnormal glucose in fasting or intolerance to glucose or DM2, with 2 or more of the following: Body mass index > 30 Waist/hip > 0.85m (women) or > 0.9m (men) Triglycerides > 150mg/dl and/or HDL cholesterol < 35 mg/dl (men) or < 40 mg/dl (women) Microalbuminuria: urinary excretion of albumin/creatinine 30 mg/g Blood pressure > 140/90 mm/Hg	Three or more of: Abdominal obesity: waist circumference > 88 cm (women) o 102 cm (men) Triglycerides > 150 mg/dl HDL cholesterol < 45 mg/dl (women) or < 35 mg/dl (men) Blood pressure > 130/85 mm/Hg Glucose in fasting > 110mg/dl.

In this work we will analyze correlations between the results of the biochemical tests to diagnose the MS with the results of the techniques of nonlinear dynamics (Del Río Correa, and Muñoz-Diosdado 2002, Ivanov et al., 1999, Bunde et al., 2002) applied to the analysis of 24-hour Holter ECG records of patients with MS, with the hope that can be identified in a timely manner in the young population with MS the possibility of developing heart disease.

Results

The study population was selected as described earlier in this work. First, we applied algorithms previously designed to locate the points R and build the tachograms. For all RR time series we calculated the multifractal spectrum and estimated the degree of multifractality. This information will allow us to characterize the complexity of the RR time series based on multifractal spectrum and its symmetry and it is expected that these variables can help to differentiate between healthy and not healthy dynamics. Multifractal analysis has been used for the study of time series of physiological variables (Del Río Correa, and Muñoz-Diosdado 2002, Ivanov et al., 1999, Bunde et al., 2002, Stanley, 2000, Humeau et al., 2009, Humeau et al., 2008, Muñoz-Diosdado et al., 2004, Aranda et al., 2006, Muñoz Diosdado et al., 2005, Muñoz-Diosdado, and Del Río-Correa, 2006), the analysis of the symmetry of the multifractal spectrum as an additional tool has been proposed to characterize the spectra (Muñoz-Diosdado and del Río-Correa, 2006).

In order to make quantitative statements about the multifractal spectrum different time series, it is common to calculate the width of the spectrum or degree of multifractality, $\Delta\alpha=\alpha_{max}$ - α_{min}, this quantity is a measure of the fractal exponents range in the time series; therefore, if $\Delta\alpha$ is big, the signal is richer in structure. It is possible to use $\Delta\alpha_r=\alpha_{max}$ - α_0 and $\Delta\alpha_l=\alpha_0$ - α_{min}, where α_0 is the point where $f(\alpha)$ has its maximum, as indicators of symmetry, if $\Delta\alpha_r = \Delta\alpha_l$ then the spectrum is symmetrical, but if $\Delta\alpha_r > \Delta\alpha_l$ the spectrum is right skewed and if $\Delta\alpha_r < \Delta\alpha_l$ then the spectrum is left skewed. Multifractal spectra have the look shown in Figure 13, which corresponds to a young healthy person; note that the spectrum is right skewed.

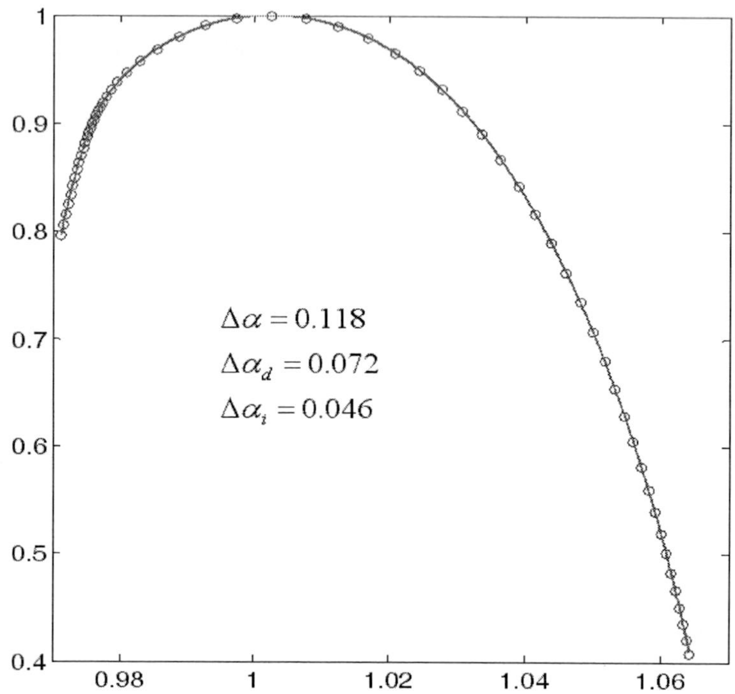

Figure 13. Multifractal spectrum of a healthy young person. The spectrum is right skewed.

Detecting the MS in young people is important. That is the reason why this study was carried out. The objective of this work is to see the possibility of recognizing precursors of heart problems in young people with metabolic syndrome, in fact, given that they are so young, a diagnosis made by a cardiologist probably will establish that they are healthy, in other words, there are no apparent problems. In fact, the cardiac problems appear with age or do not appear. However, would it be possible using the non-linear dynamic techniques to find any behavior that indicates some sort of deviation from the healthy dynamics? The multifractal spectrum of total tachograms of healthy subjects and those with MS was calculated for this purpose. The degree of multifractality and the symmetry of the spectra were calculated. We then selected two segments of these time series with two-hour duration, one of them when the subject is awake and another when the subject is asleep.

It has been reported (Ivanov et al., 1999) that wider spectra correspond to the RR time series of healthy individuals and narrow spectra correspond to people who have a heart problem. However, we found no significant differences in the analysis of the total tachograms total (24 hour's duration) in healthy young people and young people with MS (Figure 14). We studied the symmetry of multifractal spectra of the tachograms and we have found in a statistically significant way that the spectra of the healthy people tachograms are charged to the right (Figure 15), we used t- Student with a 0.05 significance level.

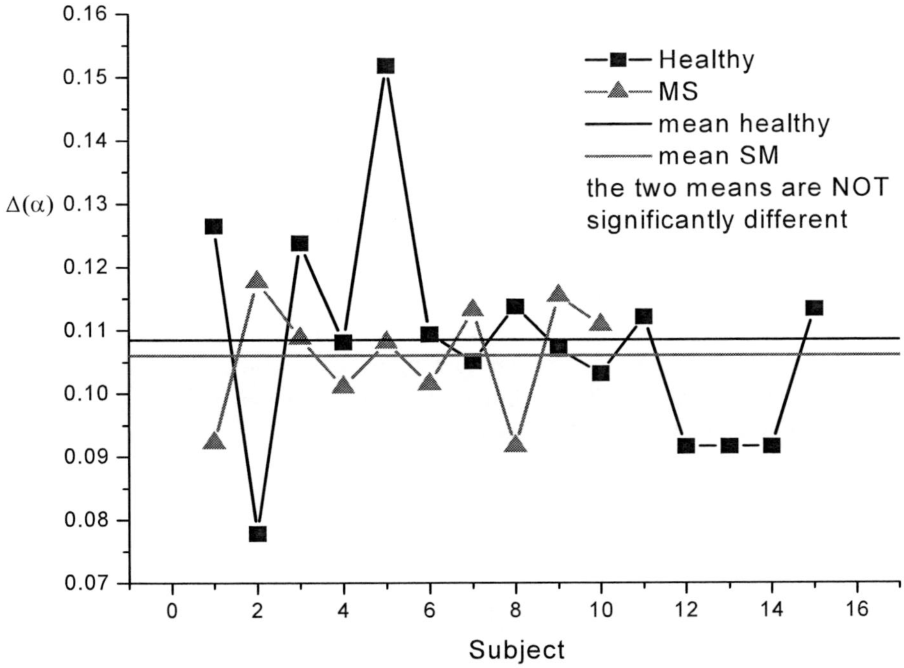

Figure 14. Multifractality degree of the tachogram of healthy subjects (square) and young people with MS (triangle). Although the average of the degree of multifractality of the tachograms is greater than the average of the degree of multifractality of the tachograms of young people with MS, the difference is not statistically significant.

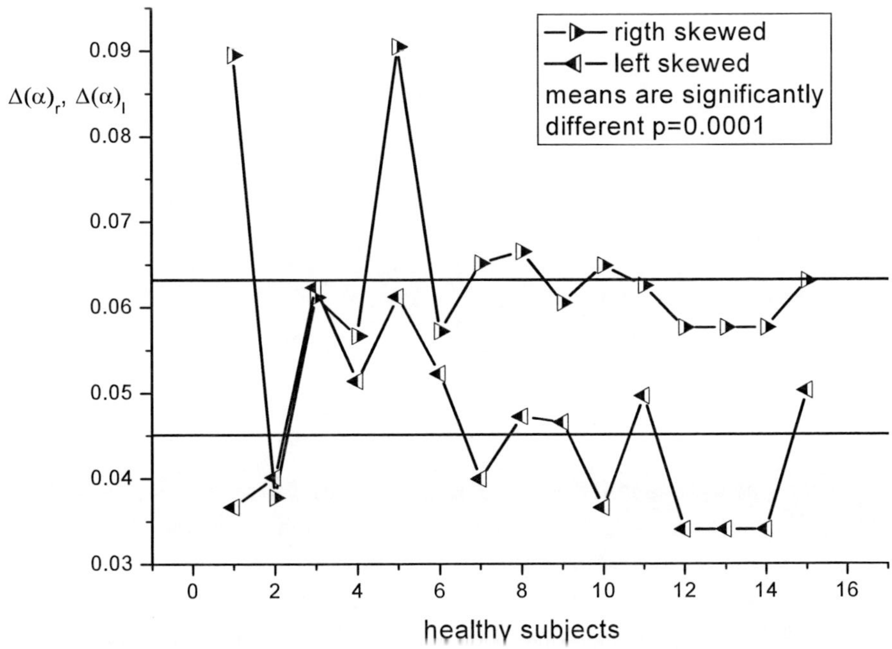

Figure 15. This figure shows that the spectra of the tachograms of the healthy people are usually right skewed.

In the group with MS, it happens the same; however, there is inversion in three spectra of tachograms of young people with MS (see Figure 16), which is abnormal, because the three spectra are either very symmetrical or definitively charged towards the left (Figures 17 and 18).

There is no significant difference in the degree of multifractality of the tachograms when the healthy subjects are awake or when they are asleep, meaning that the widths of the spectra of the healthy subjects are virtually the same regardless of whether they are awake or sleeping (Figure 19).

The situation is different when comparing spectra of young people with MS, the widths of the spectra of tachograms when they are asleep are different from the tachograms when they are awake (Figure 20). This fact seems to suggest some kind of disorder of the sleep of young people with metabolic syndrome, but it requires further investigation and a review of other cases to be more conclusive in this statement.

In short, although it is difficult that a young person has heart problems associated with the presence of the MS, it is possible to identify that there are deviations in the values of the parameters of the multifractal spectrum, particularly in the symmetry of those who have healthy young people, although only in some cases, this should lead to a more detailed analysis by cardiologists and should lead to a change in diet and physical activity habits.

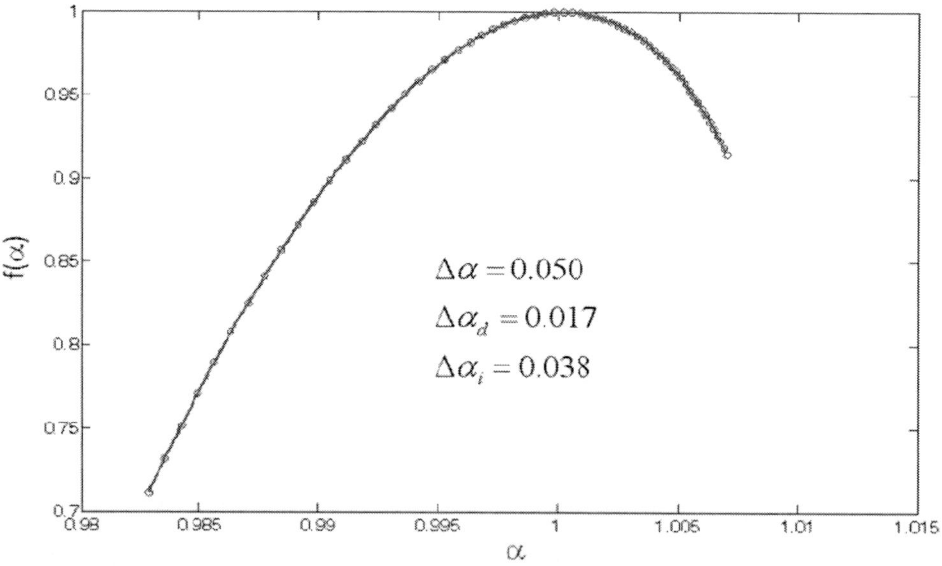

Figure 16. Multifractal spectrum of a young person with MS. The spectrum is left skewed.

The Higuchi's analysis applied to the RR time series for a healthy person is shown in Figure 21, and in the case of a person with MS, the graph is similar. Fractal dimensions obtained in each case are similar, the dimension of the person with MS being slightly larger. Figure 22 shows fractal dimensions for some of the studied cases. In this case it becomes evident that there are 4 young subjects with MS (triangles) that have a dimension above the average and tend to white noise (D = 2), which indicates a loss of correlation, especially for low scales.

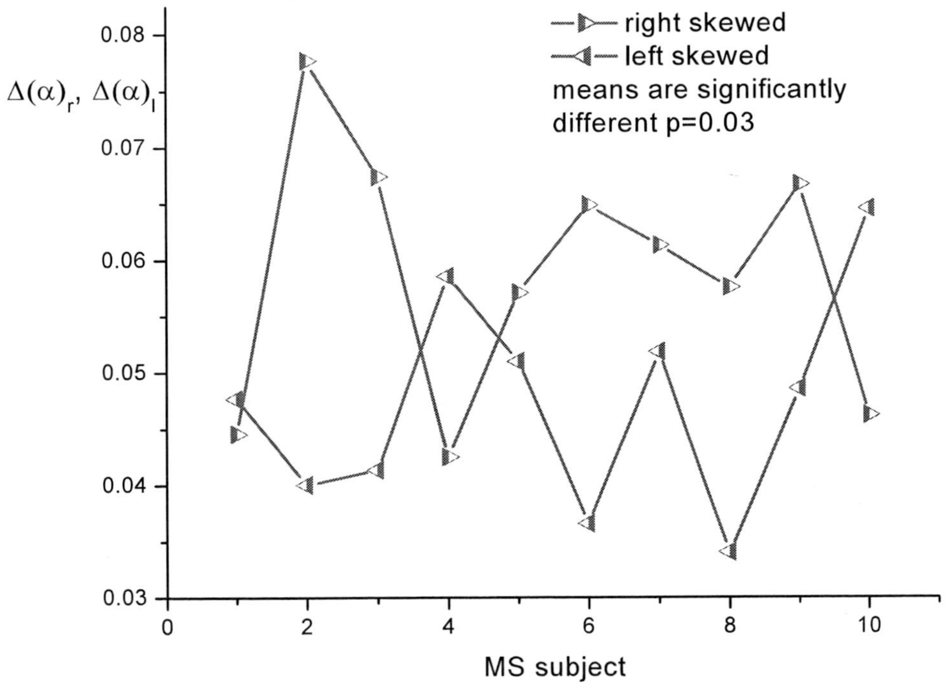

Figure 17. Three cases in which multifractal spectra of tachograms of young people with MS are clearly left skewed.

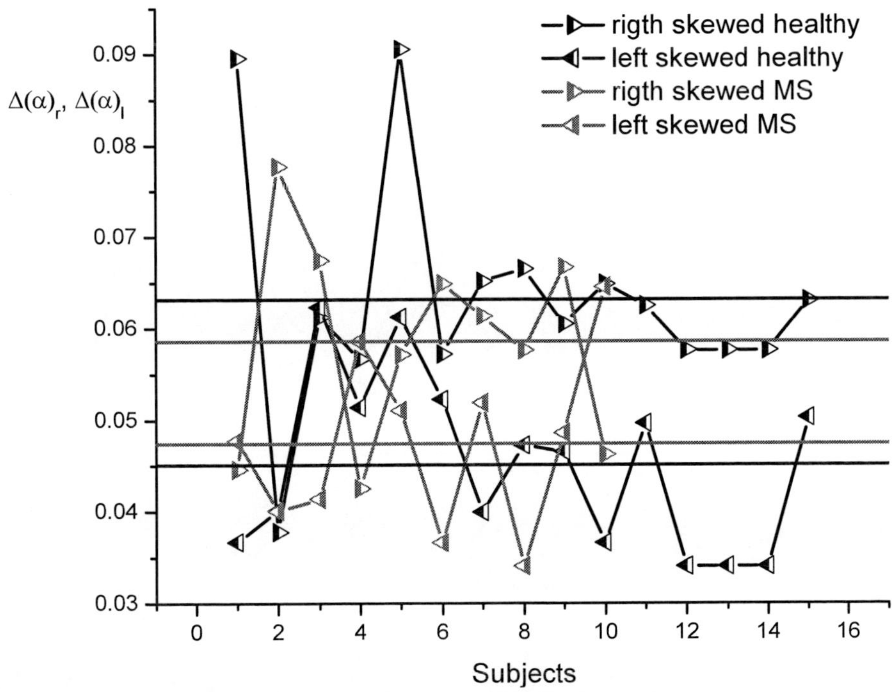

Figure 18. The analysis of the spectra symmetry confirms the observations made on above figures. There are certain anomalies in several tachograms of young people with MS.

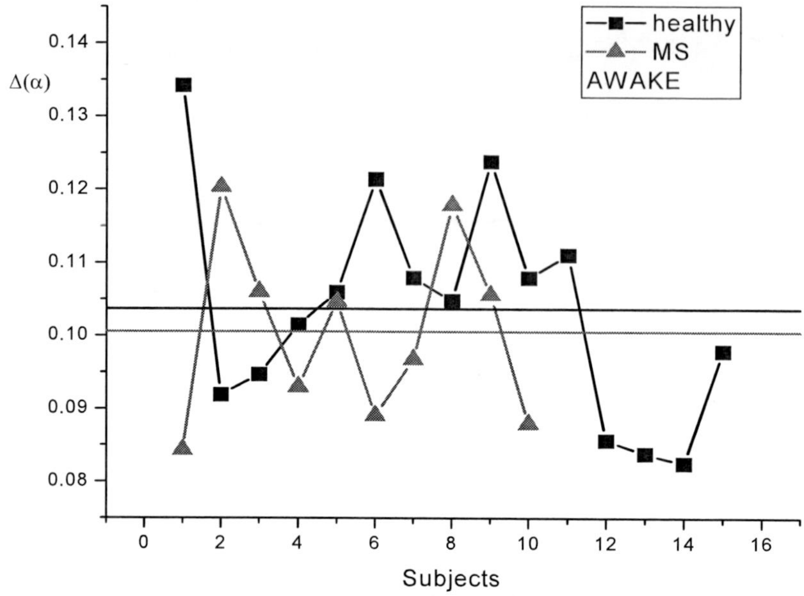

Figure 19. There is no difference between the spectra of the tachograms of young healthy when they are sleeping or awake.

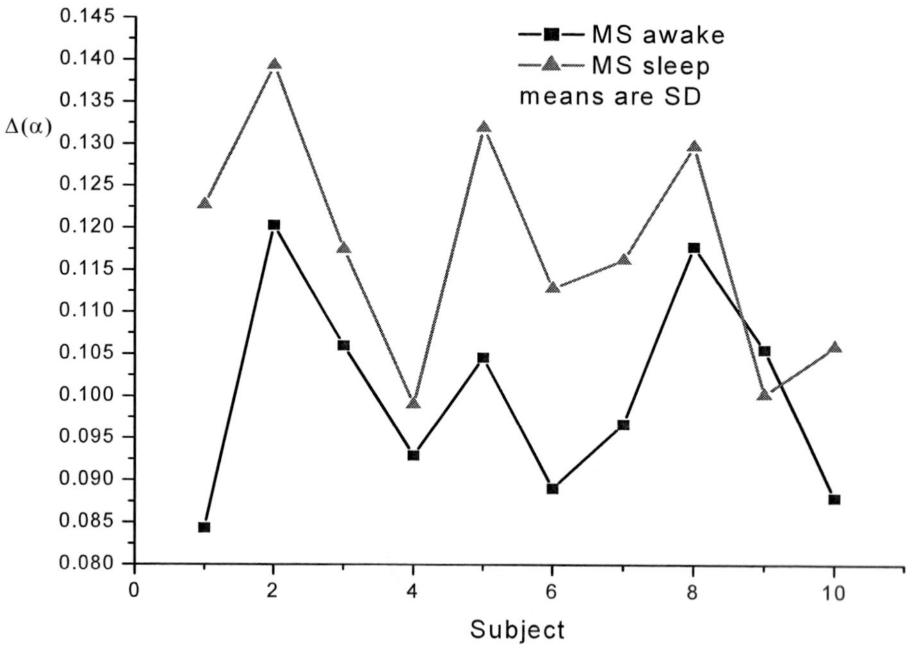

Figure 20. There exists difference between the spectra of the young people with MS tachograms when they are asleep or awake.

On the other hand, if we pay greater attention to Figure 21, (see the marked region in Figure 21) it may be noted that for a healthy person there is a small deflection down of the points on the adjusted straight line at low scales, and in the case of the subject with MS there is a slight oscillation about the adjusted line, and we used this difference to distinguish some MS subjects from healthy subjects, by defining the quadratic error:

$$E_r = \sum_{i=1}^{k} (y_i - y_r)^2$$

Where y_i is the point over the Higuchi's analysis graph and y_r the obtained value with the adjusted straight line.

Observing carefully once again Figure 21, it can be seen that for a healthy subject, the first point of the graph is below the line adjustment, on the other hand, for the MS subject, the first point is located very near and above the line adjustment, then, we add this idea as another form of discriminating healthy subjects from MS subjects.

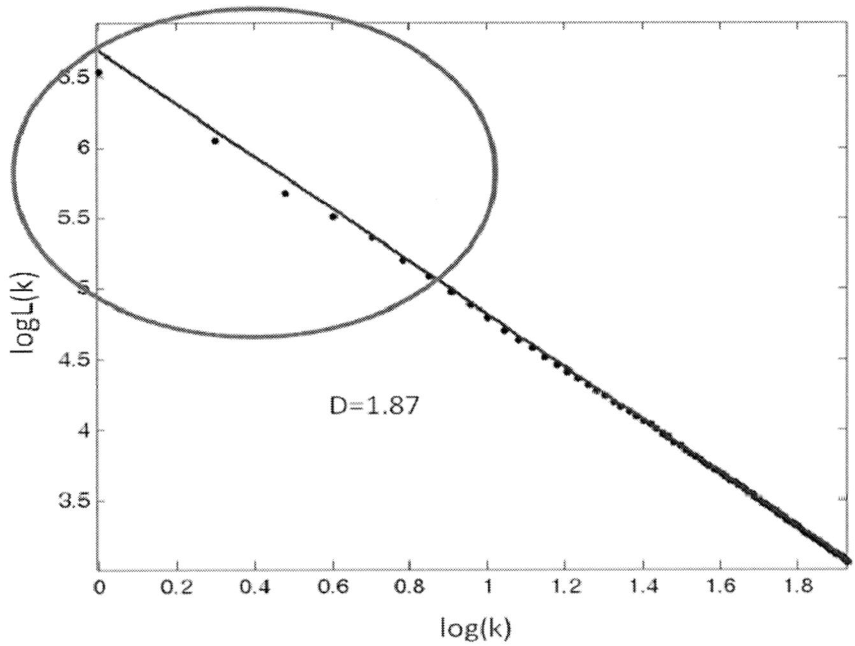

Figure 21. Higuchi's analysis of one healthy subject.

Figure 23 shows the graph obtained by applying the two criteria mentioned above; the quadratic error is plotted on the x axis and the distance to the first point on the y axis. The graph includes two rectangles that limit from the respective averages of the quadratic error and the distance to the first point in one and two standard deviations, respectively. There are six cases of people with MS that are outside the range of two standard deviations, which indicates that with a probability of 95%, these cases are outside the normal behavior.

Regarding the DFA method, it has been published (Guzmán-Vargas et al., 2008, Guzmán Vargas et al, 2009) that when RR time series of young, healthy people are analyzed, what we get on the graph to determine γ (exponent of correlations) is a straight line whose slope should be about $\gamma = 1$, old age and especially the cardiac failure have two effects: first, the value γ can be either diverted to 0.5 (white noise) or 1.5 (Brownian noise) and there are crossovers (changes in the dynamics of correlations) in such a way that two or more lines of adjustment are needed. In Figure 24, we show the typical behavior for a completely healthy young person, note the lack of crossovers and the γ value is close to 1. On the other hand, in Figure 25, we show a typical case of a DFA graph which could belong to a person with heart problems.

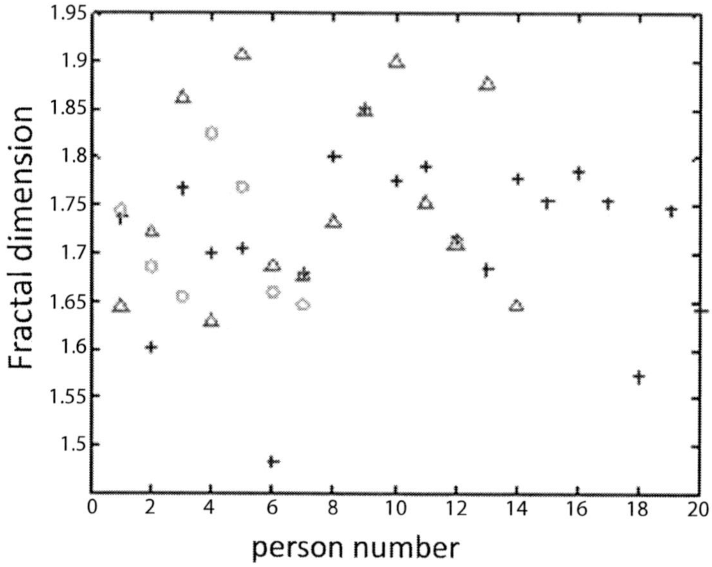

Figure 22. Fractal dimension of Higuchi for subjects with MS (triangles) and healthy subjects (crosses and circles).

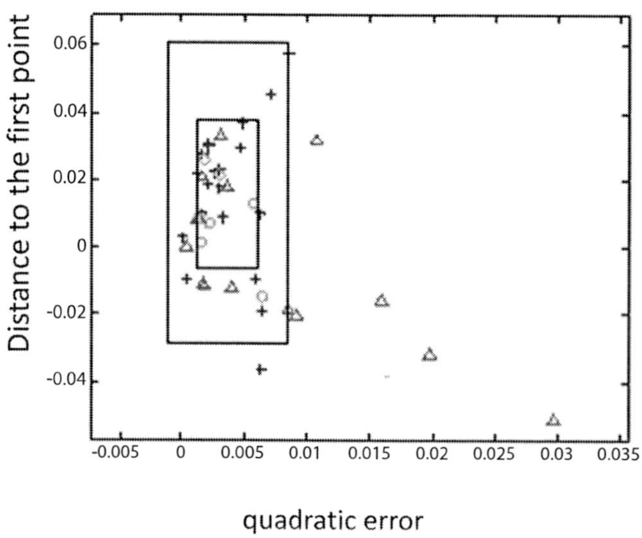

Figure 23. Graph of the criterion of distance from the first point for each of the cases analyzed and the quadratic error. Healthy (crosses and circles), MS (triangles).

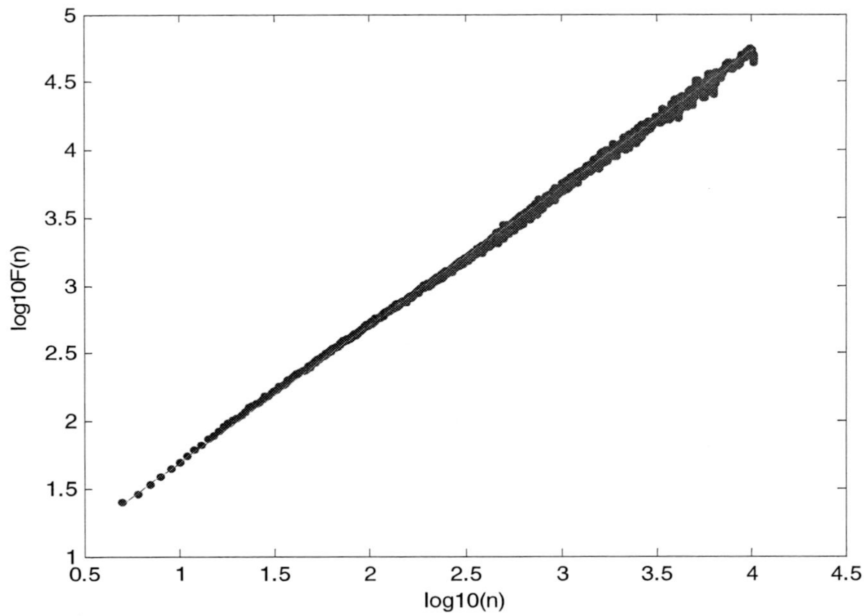

Figure 24. DFA graph for a RR time series of one healthy young subject. There are no crossovers and the slope corresponds to 1/f noise because $\gamma = 1.0135$.

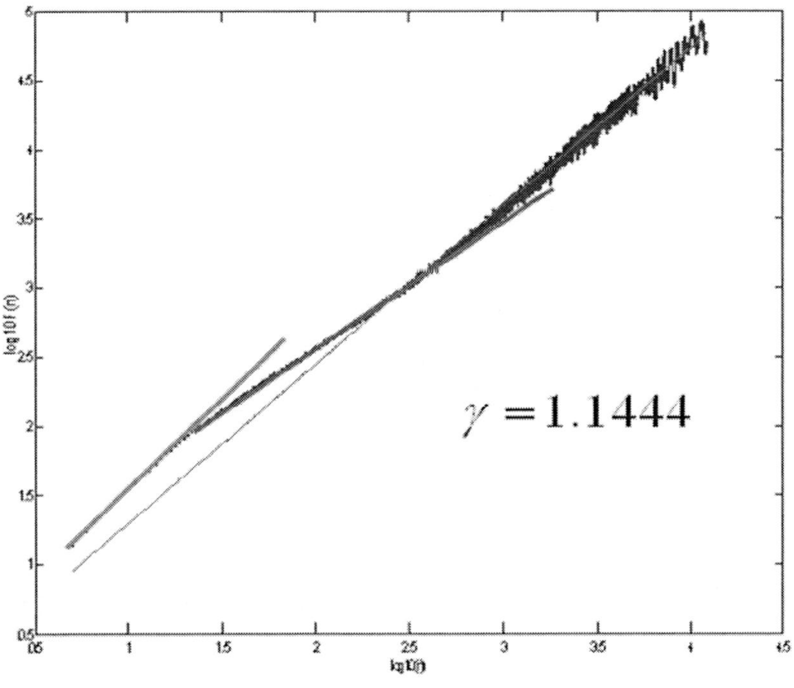

Figure 25. DFA graph for a RR time series of one subject with MS. There are two crossovers, so we calculated three γ values. The γ indicated is an average of the three values and it is close to 1/f noise.

However, as our database has only data of healthy young persons and young people with MS, the obtained values of γ are not away of γ = 1, i.e., we have not obtained evidence to show that some young people have RR time series that tend to white or Brownian noise. But we have obtained crossovers, one, two and even three crossovers. For example, if a graph shows two crossovers then we have to make three adjustments and determine three values of γ. However, we have not obtained γ values too far from the unitary value.

The problem we have faced is that even some of the subjects who were diagnosed as clinically healthy, show crossovers, this seems somewhat contradictory, and however, we believe that the situation of health and disease is not a binary issue. In other words, a completely healthy state is an ideal; in addition, if, for example, we assume that there is a single crossover then there are two values of γ, γ_1 and γ_2, if we evaluate the change $\Delta\gamma = \gamma_2 - \gamma_2$, then we would expect that if the person is healthy, the value of $\Delta\gamma$ should be small and if a person has cardiac problems then $\Delta\gamma$ should be greater.

This means that if we make a graph of the number of crossovers vs $\Delta\gamma$ (in the case of two or three crossovers we add the two or three $\Delta\gamma$) one would expect that people that might actually have some beginning of a heart problem was located above and to the right, in other words, with largest number of crossovers and a value of $\Delta\gamma$ larger than the others, it should also correspond to a subject with MS. But we must not lose sight that all subjects are young and that is why even though they have MS, the majority has the characteristics of the healthy. At the end it happened as expected, i.e. more deviant cases correspond precisely to some young people with MS, this is shown in Figure 26.

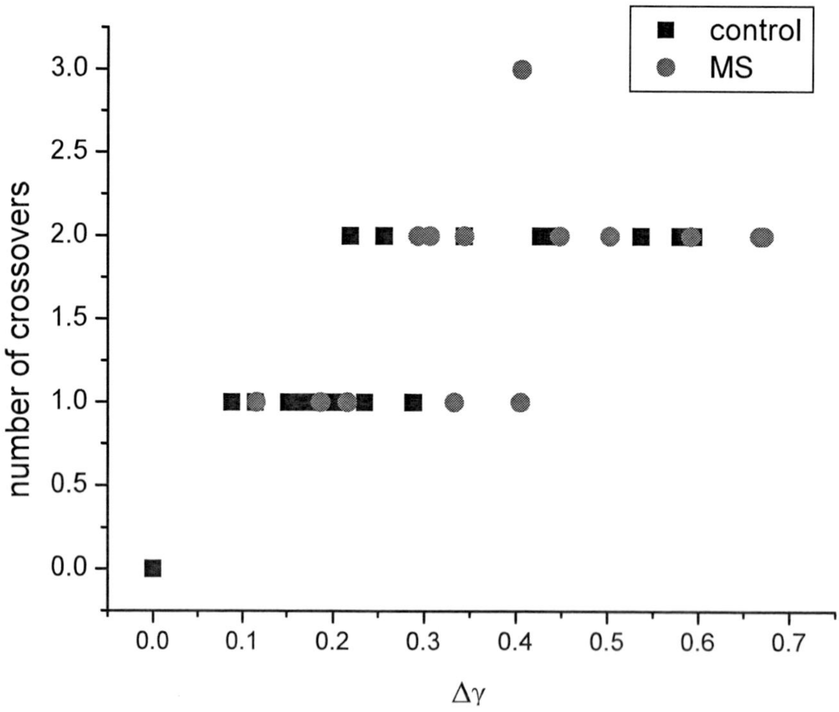

Figure 26. Graph with the number of crossovers vs $\Delta\gamma$ for healthy (squares) and MS (circles), the subjects with probable cardiac problems are above and to the right.

CONCLUSIONS

Metabolic syndrome is an important problem in Mexico; we need to change our eating habits and exercise more. In this sense, the problem is growing between young and children. It is possible that young subjects with MS have cardiac alterations, but it is difficult for the cardiologists to identify these deviations. We offer in this work an alternative to early detection of cardiac problems in healthy subjects with MS. This detection is based on the application of methodologies of non-linear dynamics to the analysis of RR time series obtained from 24 hour Holter ECG records. Multifractal analysis, detrended fluctuation analysis and Higuchi's fractal dimension were applied to the tachograms obtained from Holter records and we have found abnormalities in the spectra tachograms of the young with MS when compared with the young of the control group.

We describe the construction of a database of 24 hour Holter records and the corresponding intertbeat time series of healthy young persons and young people diagnosed with MS. With nearly 100 records, the base is still small but is intended to continue collecting data, the database will include a personal blog with the patient's main activities carried out during the day as well as the results of biochemical tests to diagnose their state of health. The database is intended to be free access as well as the programs to analyze signals with traditional techniques and techniques of non-linear dynamics.

In short, although it is unlikely that a young person has heart problems associated with the presence of MS, it is possible to identify, at least in some cases, behaviors different from healthy people, which should lead to a more detailed analysis from a specialist, and most importantly, this analysis should lead to actions to enable a change food and physical activity habits to restore normality.

ACKNOWLEDGMENTS

The authors wish to thank SIP, EDI and COFAA-IPN for partial support.

REFERENCES

Aranda, J.S., Salgado, E., Muñoz Diosdado, A. (2006). Multifractality in intracellular enzymatic reactions. *Journal of theoretical biology* 240(2): 209-217.

Bergé P., Pomeau Y., Vidal C. (1984). *Order within chaos.* USA: John Wiley & Sons.

Bonora, E., Kiechl, S., Willet, J., Oberhollenzer, F., Egger, G., Bonadonna, R.C., Muggeo, M. (2003). Metabolic syndrome: epidemiology and more extensive description. Cross-sectional data from the Bruneck Study. *International Journal of Obesity* 27: 1283-1289.

Bunde, A., Kropp J. and Schellnhuber, H.J. (2002). *The Science of Disasters,* Springer, New York.

Chen Z., Ivanov PCh., Hu K., Stanley H.E. (2002). Effect of non-stationarities on detrended fluctuation analysis. *Phys. Rev.* E 65: 041107.

Chhabra, A.B., Jensen, R.V. (1989). Direct determination of the $f(\alpha)$ singularity spectrum. *Physical Review Letters* 62: 1327-1330.

Chhabra, A.B., Jensen, R.V., Sreenivasan K.R. (1989). Extraction of underlying multiplicative processes from multifractals via the thermodynamic formalism. *Phys. Rev. A*. 40(8): 4593-4610.

Del Río Correa, J.L., Muñoz-Diosdado, A. (2002). Multifractality in physiological time series. *AIP Conference Proceedings* 630(1): 191-201.

Dewey, T.G. 1997. *Fractals in Molecular Biophysics*. New York: Oxford University Press.

Expert panel on Detection, Evaluation and treatment of High Blood Cholesterol in Adults. (2001). Executive summary of the third report of the National Cholesterol education Program (NCEP) Expert Panel on Detection, Evaluation and Treatment of High Blood Cholesterol in Adults (Adult Treatment Panel III). *JAMA* 285: 2486-2497.

Feder, J. (1988). *Fractals*. New York: Plenum Press.

Gammel, B.M. (1994). Kritisches Verhalten und Niederfrequenz-Anomalien beim Quanten-Hall-Effekt [PhD Thesis]. Technisque Universität München.

Glass, L., Mackey, M.C. (1988). *From Clocks to Chaos: The Rhythms of Life*. Princeton, NJ: Princeton University Press.

Goldberger, A.L., Amaral L.A.N., Glass, L., Hausdorff, J.M., Ivanov, P. Ch., Mark, R.G., Mietus, J.E., Moody, G.B., Peng, C.K., Stanley, H.E. (2000). PhysioBank, PhysioToolkit, and Physionet: Components of a New Research Resource for Complex Physiologic Signals, Circulation 101(23): e215-e220 *[http://circ.ahjournals.org/org/cgi/content/full/101/23/e215]*.

Guzmán-Vargas, L., Muñoz Diosdado, A., Angulo Brown, F. Influence of loss of time-constants repertoire in pathologic heartbeat dynamics. *Physica A* 348: 304-316.

Guzmán-Vargas, L., Muñoz Diosdado, A., Calleja Quevedo E., Angulo Brown F. (2008). Some fractal properties of human heartbeat dynamics. In *Statistical Mechanics Research. Edited by Byung-Soo Kim*, Nova Science Publishers, Inc. New York, 317-330.

Harte, D. (2001). *Multifractals, theory and applications*. USA: Chapman & Hall/CRC.

Higuchi, T. (1988). Approach to an irregular time series on the basis of the fractal theory. *Physica D* 31: 277-283.

Higuchi, T. (1990). Relationship between the fractal dimension and the power-law index for a time series: a numerical investigation. *Physica D* 46: 254-264.

Hu, K., Ivanov, P. Ch., Chen, Z., Carpena, P., Stanley, H.E. (2001). Effect of trends on detrended fluctuation analysis. *Phys. Rev. E* 64: 011114.

Humeau, A., Buard, B., Chapeau-Blondeau, F., Rousseau, D., Mahe, G., Abraham P. (2009). Multifractal analysis of central (electrocardiography) and peripheral (laser Doppler flowmetry) cardiovascular time series from healthy human subjects, *Physiological Measurement* 30 (7): 617-629.

Humeau, A., Chapeau–Blondeau, F., Rousseau, D., Rousseau, P., Trzepizur, W., Abraham P. (2008). Multifractality, sample entropy, and wavelet analyses for age-related changes in the peripheral cardiovascular system: Preliminary results, *Med. Phys*. 35 (2): 717-723.

Ivanov, P.Ch., Goldberger, A.L., Stanley H.E. (2002). Fractal and Multifractal Approaches in Physiology, in *The Science of Disasters*, Bunde A, Knopp J, Schellnhuber HJ. Eds. Germany: Springer-Verlag.

Ivanov, P.Ch., Nunez Amaral, L.A., Goldberger, A.L., Havlin, S., Rosemblum, M. G., Struzik, Z.R., Stanley, H.E. (1999). "Multifractality in human heartbeat dynamics", *Nature* (London) 399, 461-465.

Iyengar, N., Peng, C.K., Morin, R., Goldberger, A.L., Lipsitz, L.A. (1996). Age-related alterations in the fractal scaling of cardiac interbeat interval dynamics. *Am. J. Physiol.* 271: R1078-R1084.

Karasik, R., Ashkenazy, Y., Ivanov, P.Ch., Dvir, I., Lavie, P., Havlin, S. (2002). Correlation differences in heartbeat fluctuations during rest and exercise. *Phys. Rev. E* 66: 062902.

Kresh, J.Y., Izrailtyan, I. (1998). Evolution in functional complexity of heart rate dynamics: a measure of cardiac allograft adaptability, *Am J Physiol.* 275, R720–R727.

Malamud, B.D., Turcotte, D.L. (1999). Self-Affine Time Series: I Generation and Analyses, *Advances in Geophysics* 40: 1-90.

Mandelbrot, B.B., Van Ness, J.W. (1968). Fractional Brownian motions, fractional noises and applications. *SIAM Rev.* 10: 422-437.

Moller, D.E., Kaufman, K.D. (2005). Metabolic Syndrome: A clinical and molecular perspective, *Annu. Rev. Med.* 56: 45-62.

Moller, D.E., Kaufman, K.D. (2005). Metabolic Syndrome: A clinical Molecular Perspective. *Annu. Rev. Med.* 56: 45-62.

Muñoz Diosdado, A., Angulo Brown, F., Del Río Correa, J.L. (2005). Changes in multifractality with aging and heart failure in heartbeat interval time series in *Proc. 27th IEEE EMBS Annual International Conference, Shanghai, China.*

Muñoz-Diosdado, A., Del Río-Correa, J.L. (2006). Further Study of the Asymmetry for Multifractal Spectra of Heartbeat Time Series, in *Proc. 28th IEEE EMBS Annual International Conference, New York, USA.*

Muñoz Diosdado, A., Del Río Correa, J.L., Angulo Brown, F., Guzmán Vargas, L. (2008). El grado de multifractalidad y la simetría del espectro de singularidades para el análisis de series de interlatido cardiaco. En "La Física Biológica en México: Temas selectos 2". *Editado por Leopoldo García-Colín Scherer et al., El Colegio Nacional,* México 321-356.

Muñoz-Diosdado, A. (2009). Analysis of the relation between complexity and multifractality in cardiac interbeat intervals time series, *IFMBE Proceedings* 25(9): 1506-1509.

Muñoz-Diosdado, A., Almanza Veloz, V.H., Del Río Correa J.L. (2004). Multifractal analysis of aging and complexity in heartbeat time series, *AIP Conference Proceedings* 724: 186-191.

Muñoz-Diosdado, A., Gálvez-Coyt, G., Pérez Uribe, B.M., Arellanes González, J. (2009). Analysis of RR intervals time series of congestive heart failure patients with Higuchi's fractal dimension, *Proceedings IEEE EMBS Annual International Conference* 31: 3453-3456.

Muñoz-Diosdado, A., Jiménez-Flores, J.R., Méndez-Cruz, A.R. (2010). Analysis of Heartbeat Time Series from Young Subjects with Metabolic Syndrome. *Proceedings of the Cairo International Biomedical Engineering Conference* 5, 5-8.

Peng, C.K., Buldyrev, S.V., Goldberger, A.L., Havlin, S., Simons, S., Stanley, H.E. (1993). Finite-size effects on long-range correlations: implications for analyzing DNA sequences. *Phys. Rev.* E 47(5): 3730-3733.

Peng, C.- K., Buldyrev, S.V., Havlin, S., Simons, M., Stanley, H.E., Goldberger A.L. (1994). Mosaic organization of DNA nucleotides. *Phys Rev E* 49: 1685-1689.

Rodríguez-Morán, M., Salazar-Vázquez, B., Violante, R., Guerrero-Romero, F. (2004). Metabolic Syndrome among children and adolescents aged 10-18 years. *Diabetes Care* 27(10): 2516-2517.

Service, R.F. (1999). Complex systems: exploring the systems of life, *Science* 284, 80–81.

Stanley, H.E., Amaral, L.A.N., Goldberger, A.L., Havlin, S. , Ivanov, P.Ch., Peng, C.K. (1999). Statistical Physics and Physiology: Monofractal and multifractal approaches. *Physica A* 270: 309-324.

Stanley, H.G. (2000). Exotic statistical physics: Applications to biology, medicine and economics, *Physica A*, 285: 1-17.

Tapia, R. (2003). Prevalence and interrelation of chronic non transmisibles diseases and cardiovascular risk factors in Mexico: final results of the national survey of health (ENSA). 2000, *Arch. Cardiol.* Mexico 73: 62-77 (In Spanish).

Telesca, L., Colangelo, G., Lapenna, V., Heinicke, J., Koch, U. (2003). Quantitative dynamics in geophysical parameters simultaneously recorded in the Soos Nature Park (Czech Republic*). Fluctuation and Noise Letters* 3(1): L73-L82.

Turcotte, D.L. (1997). *Fractals and chaos in geology and geophysics. 2nd ed.,* Cambridge: Cambridge University Press.

Varotsos, P.A., Sarlis, N.V., Skordas, E.S. (2002). Long-range correlations in the electric signals that precede rupture. *Phys Rev E* 66: 011902.

WHO consultation. *Definition, diagnosis and classification of diabetes mellitus and its complications.* Part 1: Diagnosis and classification of diabetic mellitus. World Health Organization: Geneva, Switzerland.

In: Biotechnology
Editors: L. G. Torres Bustillos and I. Garcia Peña
ISBN 978-1-62081-071-2

Chapter 13

Optimization of the Ultrafiltration in Fermentation Broths

Carlos Orozco Álvarez**, Gerardo Albarrán Torres, Sergio García Salas and Leobardo Ordaz Contreras
Department of Bioengineering, UPIBI-IPN, Mexico

Abstract

Tangential flow filtration operations were studied to concentrate fermentation broths *Phaffia rhodozyma.* The first part discusses the methodology used to determine the best conditions as transmembrane pressure, feed flow, temperature and pH to maximize the filtration flux. A comparison is made between the ultrafiltration and microfiltration based on the flux.

In the second part, we discuss the scaling of the microfiltration and ultrafiltration using the mass transfer coefficient of polarizing gel model as a criterion for scaling. The final equation is deduced and also shows the sequence of calculation to determine the flow of feed which must operate pilot cartridges. Fluxes obtained at laboratory and pilot levels are compared and analyzes the importance of this criterion.

The third part is focused on obtaining a model of resistance. With this model, it is possible to estimate the value of the flux at both laboratory and pilot levels under any combination of operating conditions. It explains the methodology used for the determination of the components of the model and empirical correlations are developed based on operating conditions.

The final part deals with the optimization of energy consumption in tangential flow filtration operations. It explains the importance of energy consumption rate for the economy of the operation. Correlations are obtained for density, viscosity and rheological properties of fermentation broths in terms of cell concentration. These correlations are necessary to determine the pumping energy consumption per unit volume filtered. Operating conditions that minimize energy consumption are analyzed.

---

* tepoztlan61@yahoo.com.mx

Introduction

The production of salmon and rainbow trout raised in aquaculture farms has been increasing. In 2001, 45% of the salmon consumed came from these farms. Astaxanthin is the pigment responsible for the reddish color of these fish flesh, which being unable to synthesize it, have to consume it from commercial products (Ordaz et al., 2002). 35 to 75 mg of astaxanthin are employed per kilogram of fish food, and it represents from 10% to 15% of the cost of food.

Virtually, all of the astaxanthin consumed worldwide is produced by chemical synthesis. The price per kilo of commercial dye, in 2010 was 2500 U.S. dollars. The FDA has banned a series of synthetic dyes; on the other hand, natural dyes are widely recognized, that is why studies have been conducted to find a way of producing natural astaxanthin (Snodderly, 1995).

There are several natural sources of astaxanthin: phytoplankton, certain crustaceans, some algae and microorganisms. The microbial source is a viable alternative, especially when fermentation technology is developed, in which case it only remains to find the best producing organism, fermentation conditions and the recovery and purification process of astaxanthin.

A microorganism that has shown potential to be used for the production of astaxanthin is the yeast *Phaffia rhodozyma.* It is known that the specific content of astaxanthin in wild strains is usually about 400 ppm; while the mutant strains can reach up to 3000 ppm (Fleno et al., 1994).

It has been reported that when we use a ruptured yeast in feed formulation, the pigment is deposited more efficiently than when the cell is full (Fleno et al., 1994). Among the reported methods for cell disruption, we have the mechanical breakdown and the chemical treatment, having a better performance. The packed cell, ruptured or not, is subjected to a drying process to obtain the product in solid form.

Thus, to obtain the cell concentration, once the fermentation process is finished, we recommend operations such as the tangential microfiltration and the ultrafiltration. These have been displacing traditional operations such as centrifugation and conventional filtration because they reduce operating time, and also, equipment investment cost is lower (Orozco et al., 2003).

Based on this, this section studies the tangential microfiltration and ultrafiltration to concentrate the fermentation broths of *Phaffia rhodozyma*.

Methodology

Maximization of Filtrate Flux

The fermentation broths of *Phaffia rhodozyma* were obtained from its cultivation in a jacketed fermenter, type stirred tank, with a volume operation capacity of 0.015 m^3, and under the operating conditions reported in the literature (Bowen et al., 1996). It produces a type of cell concentration curve vs. absorbance used to calculate the level of concentration in the fermentation broth after each tangential flow filtration.

The first experiments were performed in a tangential microfiltration cartridge laboratory (pore diameter of 0.2 microns and membrane area of 0.042 m^2), where variables are investigated as feed flow: 5 to 30 $x10^{-6}$ m^3 / s, transmembrane pressure : 0 to 210 kPa, pH: 4.0 to 8.0, and temperature from 303 to 323 K. It evaluates the effect of these variables in flux as well as the effect of concentration of the fermentation performance of the membranes (Yeh et al., 2003).

The same experiment is repeated but now using a laboratory ultrafiltration cartridge with a molecular cutoff of 100 kDa and a membrane area of 0.042 m^2, to compare the results of both membranes and select only one (Orozco et al., 2003).

Then, like in a glossary, these terms are defined to clarify the experimental procedure (Figure 1):

- The cartridge is the module where the membrane is arranged in the form of hollow fibers.
- The feed flow is the flow of the fermentation broth supplied to the microfiltration and ultrafiltration cartridges.
- The filtrate flux is the filtratation flow obtained from the cartridge and divided by the area of the membrane.
- Transmembrane pressure is calculated as the average of the inlet and outlet pressures of the cartridge, and to this average, filtration pressure is subtracted.
- The term molecular weight cut-off (MWCO) of 100 kDa indicates that the ultrafiltration membrane can retain molecules or particles having a size equal to or greater than this value.

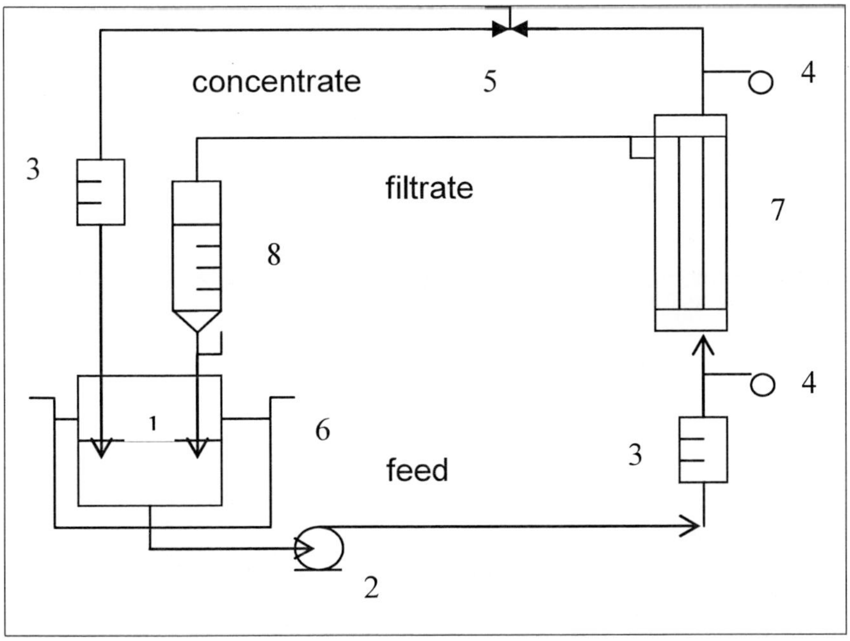

Figure 1. Micro and ultrafiltration laboratory equipment. (1) feed tank, (2) peristaltic pump, (3) rotameter, (4) manometer, (5) back pressure valve, (6) thermostat, (7) hollow fiber module (8) test tube.

Scaling

Scaling is performed tangential filtration processes using the mass transfer coefficient (k) of the polarizing gel model as a criterion for scaling (Russotti et al., 1995). Pilot cartridges are used microfiltration and ultrafiltration hollow fiber type membrane with areas of 0.12 m^2 and 0.46 m^2, respectively. Fermentation broths are concentrated from 50 to 5 L under the operating conditions found in the previous section.

On the other hand, the polarizing gel model allows calculation of the flux (J) through the following equation:

$$J = k \ln \frac{C_G}{C_b} \tag{1}$$

To achieve an acceptable scaling range, it must reach the same flux values at pilot and laboratory levels. This can be written as follows:

$$J_1 = J_2 \tag{2}$$

$$\left(k \ln \frac{C_G}{C_b} \right)_1 = \left(k \ln \frac{C_G}{C_b} \right)_2 \tag{3}$$

1: laboratory level
2: pilot level

The value of (C_b) vary in the same range for both levels, and the (C_G) is constant, therefore:

$$k_1 = k_2 \tag{4}$$

where (k) can be determined through correlations obtained by applying dimensional analysis, and generally in the membrane modules, the conditions for using the Leveque equation are met (Charcosset and Choplin, 1996; Lozano, et al, 2000):

$$k \frac{d_h}{D} = 1.86 \left(d_h v \frac{\rho}{\mu} \right)^{\frac{1}{3}} \left(\frac{\mu}{\rho D} \right)^{\frac{1}{3}} \left(\frac{d_h}{L} \right)^{\frac{1}{3}} \tag{5}$$

where (d_h) is the diameter of the hollow fiber, meanwhile (L) is the length of the hollow fiber, and (D) corresponds to the diffusion coefficient of solute retention, then arranging for (k):

$$k = 1.86 D^{\frac{2}{3}} \left(\frac{v}{d_h L} \right)^{\frac{1}{3}} \tag{6}$$

Taking (k) as scaling criterion and considering the constant diffusion coefficient D (same operating conditions at both levels), then the value of speed in the cartridge at a pilot level (v_2) is calculated using the following expression:

$$v_2 = \left(\frac{v_1}{d_{h1} L_1} \right) d_{h2} L_2 \tag{7}$$

Finally, using (v_2) and the number of hollow fibers in the pilot cartridge, it is already possible to calculate the feeding flow for the operation.

Resistance Model

Cartridges are characterized using deionized water to determine the membrane resistance (R_m) of the model of resistance (Graves et al., 2006). The resistance values because of the fouling (R_f) and the polarization of the concentration (φ) become zero (equation 8); this way, by plotting the inverse of the J vs. the inverse of transmembrane pressure (PTM), a straight line whose slope is the value of (R_m) is obtained.

$$J = \frac{PTM}{R_m + R_f + \phi PTM} \tag{8}$$

To determine R_f, φ, experiments were performed in steady-state conditions. This is achieved by recirculating the filtering and retaining it to the feed tank to maintain a constant yeast concentration, and measuring the filtration flux every 5 minutes until the value does not change (Orozco et al., 2003). To perform this, cell concentrations (C_L) from 10 to 60 kg/m^3, feed rates (v_{al}) of 0.1 to 1.0 m/s, and PTM from 0 to 207 kPa were used.

When plotting ($1/J$) vs. ($1/PTM$), at a certain speed and concentration, a straight line is obtained whose intersection is (ϕ), and the slope gives us ($R_m + R_f$). With these latest results, it is possible to get empirical correlations to express the dependence of (ϕ) and (R_f) according to the yeast concentration in the fermentation broths:

$$\phi = a(v_{al})^b (C_L)^c \tag{9}$$

$$Rf = d(v_{al})^e (C_L)^f \tag{10}$$

Optimization of Energy Consumption by Pumping

The density (ρ), viscosity (μ), consistency index (K) and the flow behavior index in power law (n) at different cell concentrations (C_L) are determined. The rheological properties (*n* and *K*) were determined at 303 K with a Haake viscometer Model RV20. To determine these properties the power law model is used:

$$\mu = K \cdot \gamma^{n-1} \tag{11}$$

Applying logarithms:

$$\ln \mu = \ln K + (n-1) \cdot \ln \gamma \tag{12}$$

Thus, by plotting ln (μ) vs. ln (γ), the slope is (*n-1*) and the intercept is ln (*K*).

On the other hand, the highest energy consumption in the process of tangential filtration is attributed to pumping (Yang et al., 1997). For fluids flow in a horizontal tube, the potential energy is considered as negligible as the same as changes in kinetic energy, then the net work required W_s is equal to –□P_t/□, so the power consumption is calculated as follows:

$$P = (\rho Q) W_s = Q(-\Delta P_t) \tag{13}$$

Substituting the expression for the pressure drop, the equation of power consumption for pumping is obtained (Geankoplis, 1998):

$$P = 2 \cdot K \cdot L \cdot \left[\left(\frac{3 \cdot n + 1}{m \cdot n \cdot \pi} \right) \cdot r^{-\left(\frac{3 \cdot n + 1}{n}\right)} \right]^{n} \cdot Q^{1+n} \tag{14}$$

To calculate the rheological properties (n, K), the feed flow (Q) and physical characteristics of the membrane module (L, length of the hollow fiber, m, number of hollow fibers) are needed.

The index of energy consumption per unit of filtration volume (*P/F*) in the ultrafiltration process of fermentation broths, trying different pressures, determines feed flows and yeast concentrations.

Results and Discussion

Maximization of Filtrate Flux

First it worked with cartridges of laboratory level using water as the fluid to know which ones were the maximum transmembrane pressure and feed flow conditions that the equipment would allow. The results are shown in Figure 2.

It is noticed that the equipment can work up to 210 kPa of transmembrane pressure, although the cartridge specifications recommend to work up to 140 kPa for a longer life span. It is also seen that the values of flux in the microfiltration cartridge are approximately nine times higher than those obtained with the ultrafiltration cartridge to any working pressure.

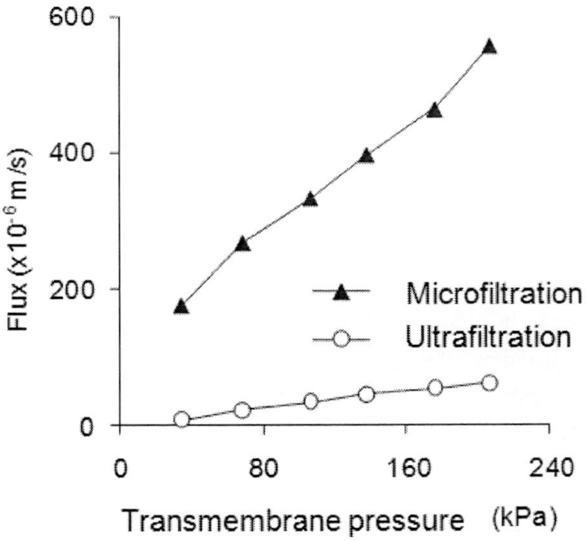

Figure 2. Characterization of microfiltration and ultrafiltration cartridges at laboratory level.

This was expected because the pore size of microfiltration is superior to the ultrafiltration one in several orders. With these results, it is possible to obtain the hydraulic resistance of the membrane which is simply the inverse of the slope of the previous graph. For microfiltration, the resistance is of 0.45 $x10^6$ kPa.s/m, while for the ultrafiltration resistance is up to eight times higher than that obtained for microfiltration. These results are consistent with those reported by other authors (Yeh et al., 2003).

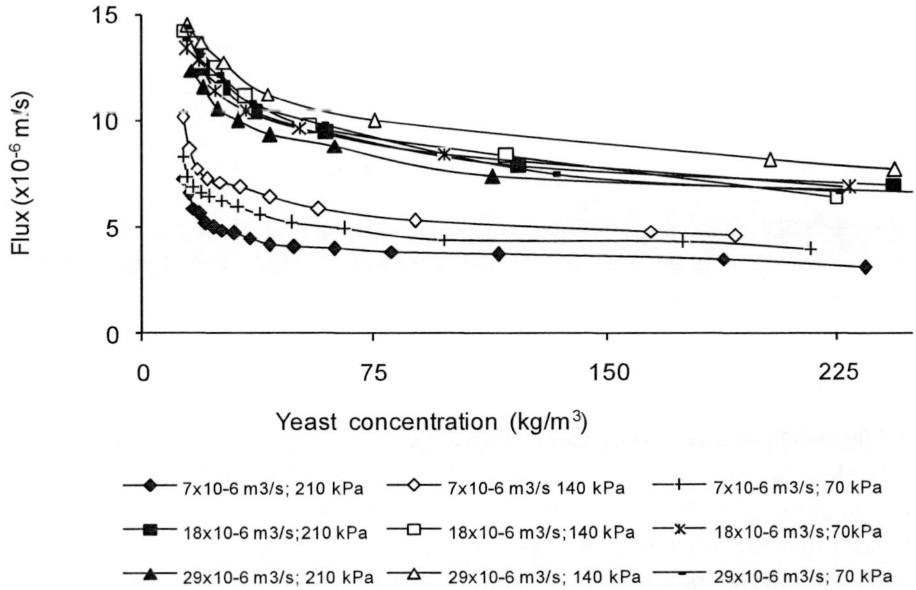

Figure 3. Effect of feed flow and transmembrane pressure in microfiltration: pH 6 and 303 K.

For the concentration of fermentation broths by tangential microfiltration, a series of tests were performed to find the best conditions of transmembrane pressure and feed flow, in which the highest values flux were obtained. In Figure 3, results are shown.

Flux values are much lower than those obtained when working only with water; decreasing from 420 to 8 $x10^{-6}$ m/s, on average, to 140 kPa. The second observation is that from a yeast concentration of 50 kg/m^3, flux remains practically constant although increasing the concentration (Naja et al., 2006). It is also noted that when working with feeding flows of only 7 $x10^{-6}$ m^3/s, the lowest values of flux is reached, but the pressure is increased from 30 to 210 kPa (Frenander and Jonsson, 1996).

When the feed flow is increased to 18 $x10^{-6}$ m^3/s, the highest flux is obtained, and in turn, the highest value is obtained at 70 kPa, and not as 140 kPa, as it was expected (Meireles et al., 2002). That is, the increment of the pressure causes a higher "crowding" of the yeast on the surface of the membrane, which can not be countered by the feed flow, causing a decrement in flux (Graves et al., 2006).

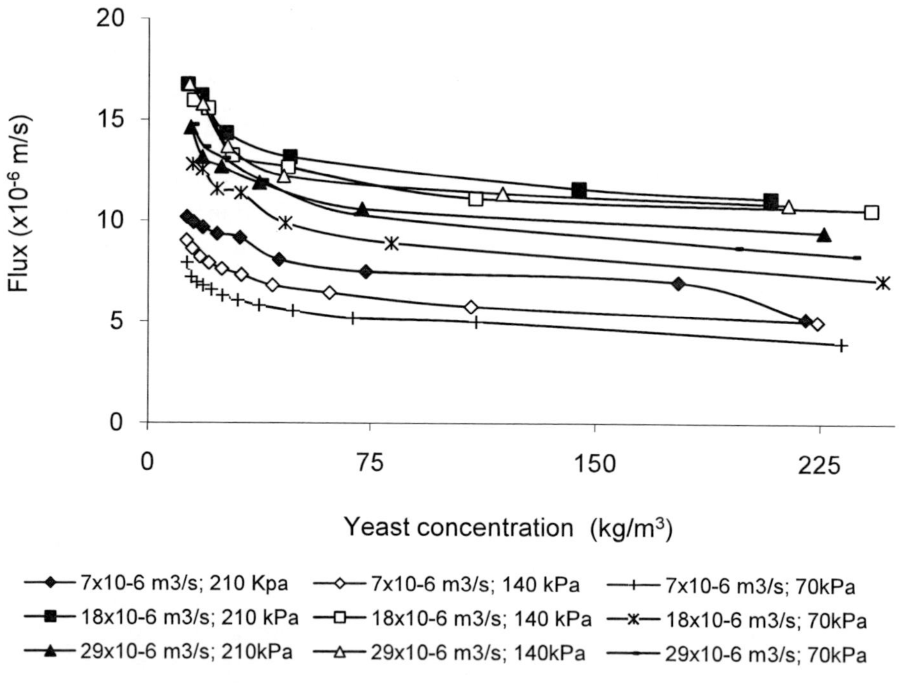

Figure 4. Effect of feed flow and transmembrane pressure in ultrafiltration, pH 6 and 303 K.

Even increasing the feed flow from 18 to 29 $x10^{-6}$ m^3/s, it is not possible to increase the flux (at 210 kPa) and the equipment does not let us work with more feed flows. This means that the feed flow, as well as the transmembrane pressure, has a significant impact on the performance of the membrane (Liew et al., 1995).

In Figure 4, the results from the concentration of the broths by ultrafiltration are shown. It notes that the average flux is much lower than that found with water: it reduces from 70 to 10 $x10^{-6}$ m/s operating at 140 kPa transmembrane pressure.

Flux is practically a horizontal line in the concentration range of 75 to 200 kg/m^3. The trend is that, at the same pressure, the flux increases if is increased the feed flow (Yeh and Tsai, 1998). The conditions that maximized the flux were: transmembrane pressure of 210 kPa, and feed flow of 18 x10^{-6} m^3/s.

Once transmembrane pressure conditions and the feed flow for the concentration of microfiltration (MF) and ultrafiltration (UF) broths were found, we proceeded to investigate the effect of pH. The results are summarized in Figure 5, in which we analyzed data on the tangentail microfiltration and ultrafiltration.

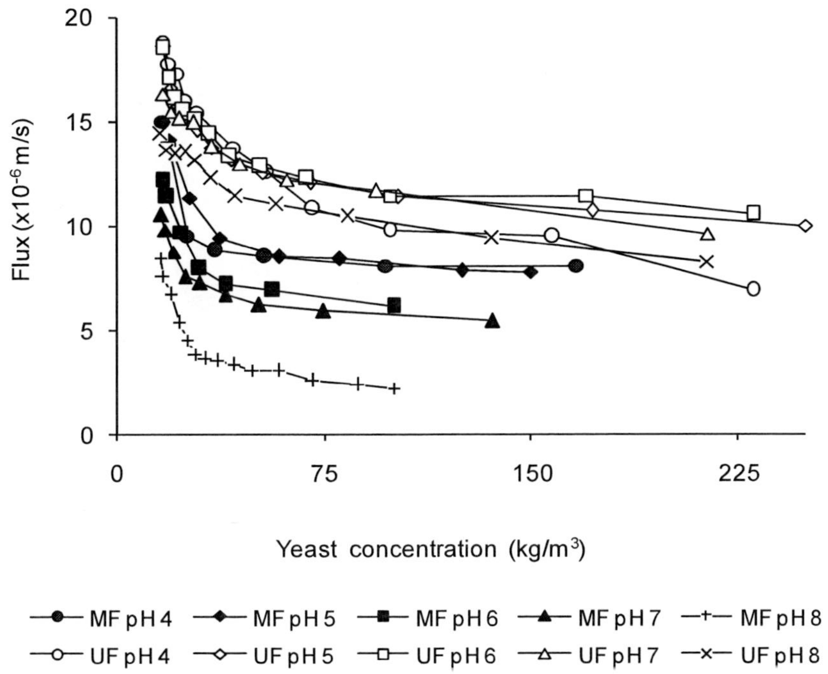

Figure 5. Effect of pH on microfiltration and ultrafiltration: 140 kPa, 303 K and 18 x10^{-6} m^3/s.

It is observed that the flux obtained with the microfiltration cartridge decreases as the pH increases (Graves et al., 2006) and is lower than those obtained with ultrafiltration. Thus, the highest flux is achieved using the ultrafiltration cartridge working at a pH of 5 or 6.

Some authors explain that the flux in ultrafiltration is higher than the one in microfiltration because yeast cannot penetrate into the holes of the membrane while it can do it in the microfiltration membrane, causing an obstruction in this one that causes the decrement in the flux.

The explanation that can be given about why the flux is higher at pH 5 or 6 is that the surface of yeast possesses a net charge to cause some repulsion between them, and this is reflected in a lower tendency to agglomerate in the surface of the membrane, leaving larger spaces between adjacent cells which ultimately results in higher flux values (Persson et al., 2003).

Finally, we proceeded to investigate the effect of temperature in the concentration of yeast broths obtaining the results summarized in Figure 6. Temperature has a direct effect on the flux, and a temperature of 323 K is obtained the highest values of flux. It can also be observed that the flux obtained in ultrafiltration is higher than that in the tangential microfiltration. The literature explains that when increasing the temperature, the transfer of materials increases back to the bulk of the feed stream, it means that the accumulation of yeast on the surface of the membrane decreases causing an increased flux (Yeh et al., 2003).

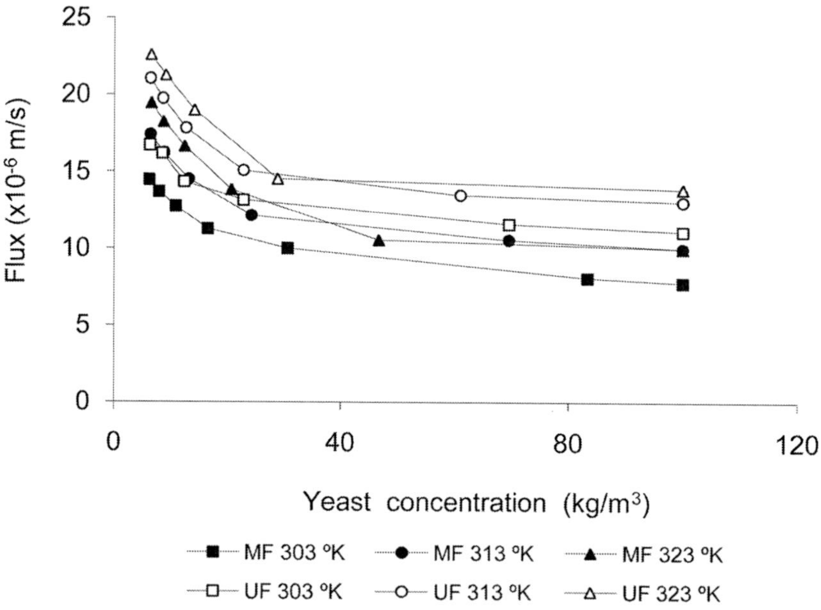

Figure 6. Effect of temperature on microfiltration and ultrafiltration: 140 kPa, pH 6 and 18 x10^{-6} m^3/s.

Scaling

Table 1 shows the calculations to scale microfiltration and ultrafiltration operations (Orozco et al., 2010). The shear rate of laboratory cartridges obtaining a value of 5857 s^{-1} for microfiltration and 3735 s^{-1} in ultrafiltration was calculated. Being the mass transfer coefficient *k* the scaling criteria, proceeded to a series of calculations to determine, finally, the feed flow of the pilot cartridges: 8.2 x10^{-5} m^3/s (4.9 Lpm) for microfiltration cartridge; 8 x10^{-4} m^3/s (48 Lpm) for the ultrafiltration cartridge.

When the feed flow at which the pilot cartridges would work is already known, proceeded to the concentration of fermentation broths by microfiltration and ultrafiltration. The results obtained at this pilot level and their comparison with those obtained at laboratory level can be observed in Figure 7.

As can be seen, ultrafiltration results are practically the same at both laboratory and pilot levels, indicating this similarity that scaling criteria was adequate.

Something similar happens with the results of tangential microfiltration; pilot level results are just inferior to the laboratory ones, but in a practical sense, they can be considered equal. This shows that the criteria used for scaling was also suitable (Meireles et al, 2002).

And as it had been pointed out above, the values of ultrafiltration flux were higher than those obtained by tangential microfiltration.

Table 1. Scaling procedure of tangential filtration operations.

Microfiltration laboratory cartridge type: CFP-2-E-4A		Ultrafiltration laboratory cartridge type: UFP-100-E-4	
diameter of the hollow fiber, d_{h1} (m)	0,001	diameter of the hollow fiber, d_{h1} (m)	0,001
length of the hollow fiber, L_1(m)	0,267	length of the hollow fiber, L_1(m)	0,267
cartridge area, a_c (m^2)	0,042	cartridge area, a_c (m^2)	0,042
area of hollow fiber, a_f (m^2)	8,4E-04	area of hollow fiber, a_f (m^2)	8,4E-04
number of hollow fibers	50	number of hollow fibers	50
Calculation of shear rate:		Calculation of shear rate:	
cartridge feed flow (Lpm)	1,725	cartridge feed flow (Lpm)	1,1
cartridge feed flow, F_a (m^3/s)	2,9E-05	cartridge feed flow, F_a (m^3/s)	1,8E-05
feed flow in hollow fiber (m^3/s)	5,8E-07	feed flow in hollow fiber (m^3/s)	3,7E-07
area of feed flow per hollow fiber (m^2)	7,9E-07	area of feed flow per hollow fiber (m^2)	7,9E-07
average rate in hollow fiber, v_1 (m/s)	0,732	average rate in hollow fiber, v_1 (m/s)	0,467
shear rate in hollow fiber, v_c (s^{-1})	5857	shear rate in hollow fiber, v_c (s^{-1})	3735
Pilot scaling pilot cartridge type: CFP-2-E-5A		Pilot scaling pilot cartridge type: UFP-100-H-9A	
diameter of hollow fiber, d_{h2} (m)	0,001	diameter of hollow fiber, d_{h2} (m)	0,002
length of hollow fiber, L_2 (m)	0,225	length of hollow fiber, L_2 (m)	0,488
cartridge area, a_{c2} (m^2)	0,120	cartridge area, a_{c2} (m^2)	0,460
area of hollow fiber, a_{f2} (m^2)	7,1E-04	area of hollow fiber, a_{f2} (m^2)	0,003
number of hollow fibers	170	number of hollow fibers	150
Calculation of feed flow:		Calculation of feed flow:	
average rate in hollow fiber, v_2 (m/s)	0,615	average rate in hollow fiber, v_2 (m/s)	1,704
area of feed flow per hollow fiber, (m^2)	7,9E-07	area of feed flow per hollow fiber, (m^2)	3,1E-06
feed flow in hollow fiber (m^3/s)	4,8E-07	feed flow in hollow fiber (m^3/s)	5,4E-06
cartridge feed flow, F_{a2} (m^3/s)	8,2E-05	cartridge feed flow, F_{a2} (m^3/s)	8,0E-04
cartridge feed flow (Lpm)	4,9	cartridge feed flow (Lpm)	48
shear rate, v_c (s^{-1})	4922	shear rate, v_c (s^{-1})	6818

Resistance Model

To develop the model of resistance expressed in Equation 8, first proceeded to obtain the term R_m (Liew et al, 1995). The results are shown in Figure 8. The inverse of the slope is the desired value, thus $R_m = 1.319 \times 10^6\ kPa \cdot s / m$

To estimate the values of R_f and ϕ in the resistance model, proceeded to the experimental determination of flux (J) at different transmembrane pressures (PTM), and we repeated this for different feed flows (v_{al})((Yeh et al., 2003). The results are concentrated in Table 2.

The results in Table 2 are plotted in Figures 9, 10 and 11 for further analysis.

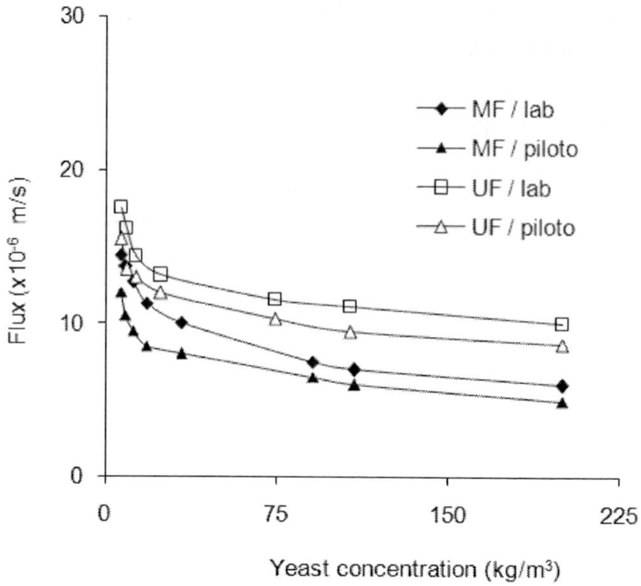

Figure 7. Comparison of microfiltration and ultrafiltration operations at laboratory and pilot scale.

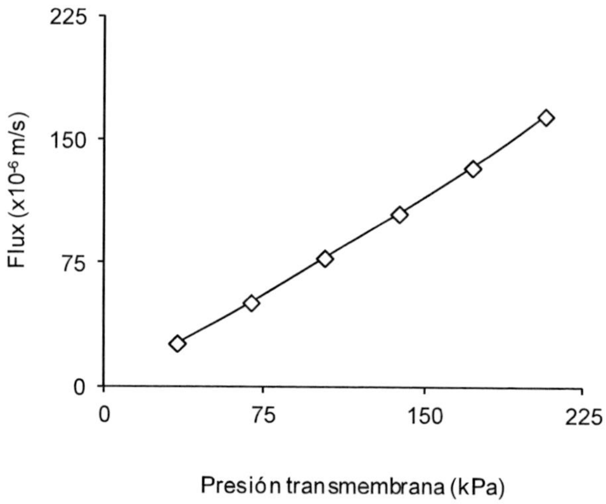

Figure 8. Determination of Rm in the laboratory ultrafiltration cartridge (MWCO100 kDa) using distilled water as the fluid.

The flux has a hyperbolic behavior when it is in function of the transmembrane pressure, working to the same feed rate and yeast concentration. This is due to the formation of a gel polarizing layer; so that, even though the pressure increases, the filtration flux does not. This way, the higher the feed rate is, higher flux values are obtained. If the yeast concentration is increased, the flux during this experimentation is reduced.

Table 2. Experimental data of the flux in the cartridge of MWCO100 kDa using the fermentation broth of *Phaffia rhodozyma* for the determination of ($R_m + R_f$) and ϕ

C_L (kg/m^3)	*PTM* (kPa)	v_{al} = 0.158 m/s J (x10^{-6} m/s)	v_{al} = 0.322 m/s J (x10^{-6} m/s)	v_{al} = 0.519 m/s J (x10^{-6} m/s)	v_{al} = 1.058 m/s J (x10^{-6} m/s)
10	34	8.25	10.77	12.49	15.28
	69	9.42	13.43	16.17	18.92
	103	10.33	14.78	17.71	20.64
	138	10.98	15.65	18.86	22.09
	172	11.57	15.94	19.49	23.04
	207	12.69	17.15	20.41	23.72
32	34	6.59	9.88	12.38	14.16
	69	6.33	10.25	13.17	17.51
	103	5.99	10.83	13.85	18.94
	138	6.08	10.79	14.32	19.53
	172	6.32	10.69	14.91	20.12
	207	6.78	11.27	16.54	21.69
52	34	4.63	7.17	10.30	13.57
	69	5.36	8.35	12.46	16.39
	103	5.88	9.42	12.98	17.23
	138	6.40	9.74	13.39	18.08
	172	6.59	10.22	13.62	18.37
	207	6.83	10.50	13.81	18.64

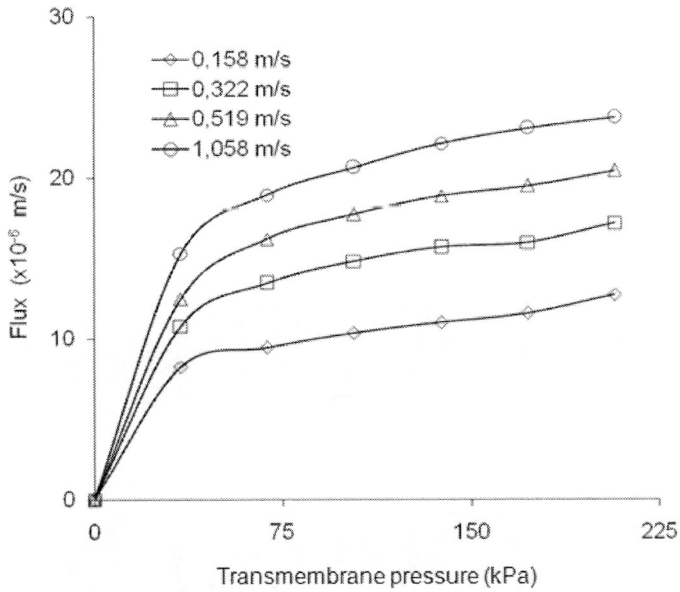

Figure 9. Effect of feed rate in ultrafiltration of fermentation broths of *Phaffia rhodozyma*, C_L = 10 kg/m^3.

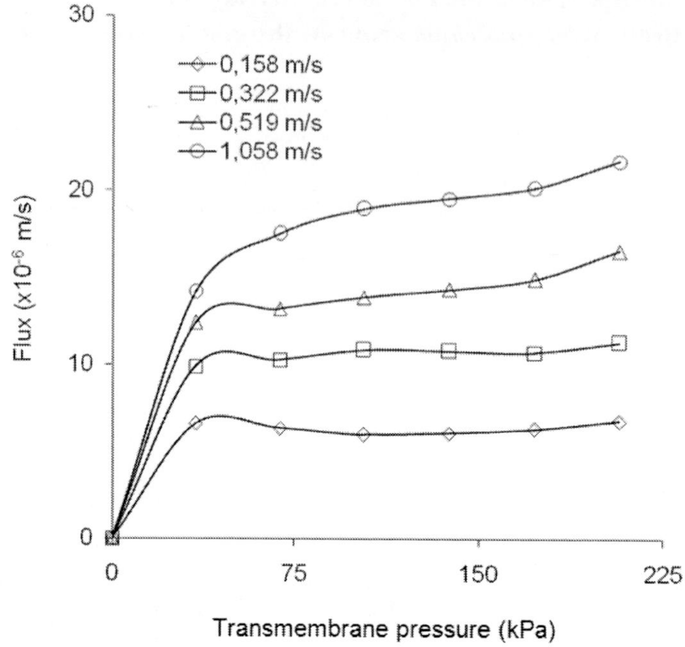

Figure 10. Effect of feed rate in ultrafiltration of fermentation broths of *Phaffia rhodozyma*, C_L = 32 kg/m^3.

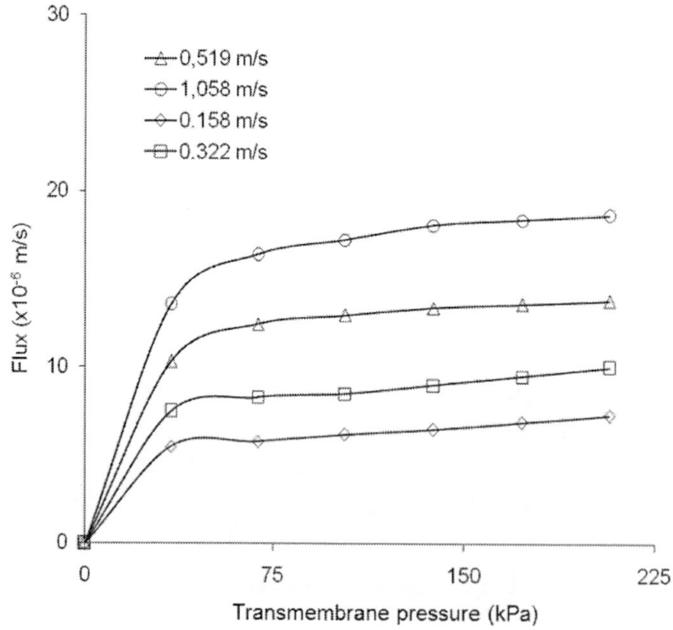

Figure 11. Effect of feed rate in ultrafiltration of fermentation broths of *Phaffia rhodozyma*, C_L = 52 kg/m^3.

From this experimental data, and using the Equation (8), results in Table 3 were obtained. By plotting (1/*J*) vs. (1/*PTM*), at a certain speed and concentration, is a straight line whose intersection is (ϕ) and the slope gives us ($R_m + R_f$).

Table 3. Results obtained from ϕ and R_f; $R_m = 1.319 \times 10^6$ kPa.s/m

C_L (kg/m^3)	v_{al} (m/s)	R_m+R_f (x10^6 kPas/m)	ϕ (x10^5 s/m)	R_f (x10^6 kPas/m)
10	0.158	3.480	0.643	2.161
	0.322	1.744	0.511	0.425
	0.519	1.485	0.422	0.165
	1.058	1.301	0.359	0.018
32	0.158	6.850	1.159	5.530
	0.322	3.722	0.828	2.403
	0.519	2.018	0.436	0.699
	1.058	1.478	0.370	0.159
52	0.158	4.842	1.274	3.523
	0.322	3.106	0.801	1.787
	0.519	2.302	0.619	0.983
	1.058	1.785	0.377	0.465

With these results, empirical correlations to express the dependence of (ϕ) and (R_f) can be obtained, according to the concentration of yeast in the fermentation broths:

$$\phi = 3.4678 \times 10^2 \cdot v_{al}^{-0.6455} \cdot C_L^{0.0290} \tag{24}$$

$$Rf = 2.4091 \times 10^4 \cdot v_{al}^{-1.9108} \cdot C_L^{1.9064} \tag{25}$$

Where:

ϕ: factor of concentration polarization (s/m)

R_f: resistance fouling (kPa.s/m)

v_{al}: feed rate (m/s)

C_L: yeast concentration (kg/m^3)

Optimizing Power Consumption for Pumping

To calculate the power consumption for pumping (*P*), through the use of Equation 14, the correlations of density, viscosity and rheological properties must be obtained depending on the concentration of yeast (Geankoplis, 1998).

Figure 12 shows the results of density (ρ), which increases proportionally as the cell concentration increases (C_L). Meanwhile, the viscosity (μ) of the broths increases exponentially from a yeast concentration of 40 kg/m^3 (Figure 13).

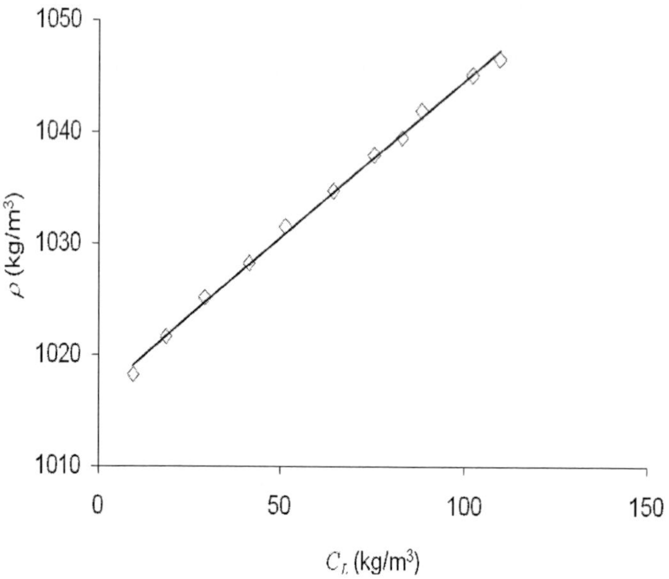

Figure 12. Effect of yeast concentration in the fermentation broth density of *Phaffia rhodozyma.*

Correlations that represent the dependence of the density and viscosity in function of the yeast concentration are presented below:

$$\rho = 0.2822 \cdot C_L + 1016.4 \tag{26}$$

$$\mu = 0.9265 \cdot e^{0.03 \cdot C_L} \tag{27}$$

Where:

ρ: density in kg/m^3

η: viscosity in Pa.s

C_L: yeast concentration in kg/m^3

The experimental results for the determination of the rheological properties of the broths are concentrated in Table 4. The shear stress was determined at each shear rate, and this was repeated for each yeast concentration (Charcosset and Choplin, 1996).

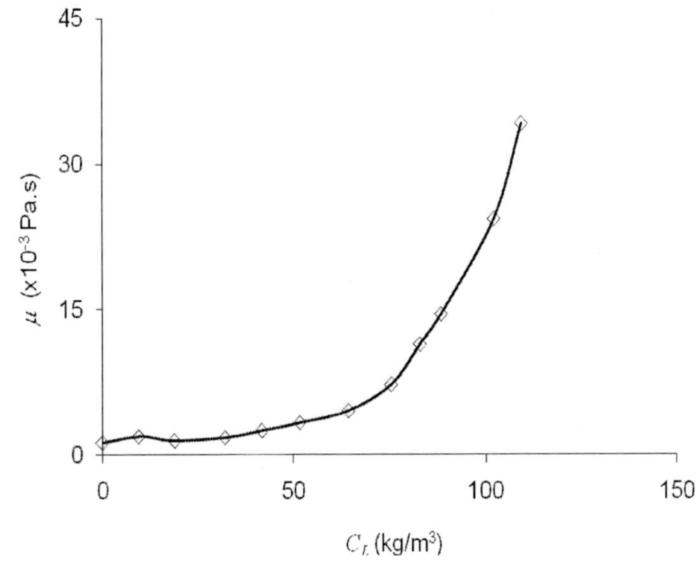

Figure 13. Effect of yeast concentration on the viscosity of the fermentation broth of *Phaffia rhodozyma.*

Table 4. Shear stress (τ) at different shear rates (γ) for fermentation broths of *Phaffia rhodozyma.*

	C_L (kg/m^3)											
γ (s-1)	0	10	19	32	42	52	65	76	83	88	102	109
	τ (Pa)	τ (Pa)	τ (Pa)	τ (Pa)	τ (Pa)	τ (Pa)	τ (Pa)	τ (Pa)	τ (Pa)	τ (Pa)	τ (Pa)	τ (Pa)
135		0.28	0.20	0.30	0.35	0.56	0.73	1.18	1.69	3.01	5.79	8.90
270		0.54	0.40	0.53	0.83	1.01	1.51	2.40	4.08	5.27	9.61	15.06
540	0.57	1.07	0.94	1.08	1.63	2.08	2.90	4.49	7.30	9.33	16.16	25.45
810	0.89	1.75	1.38	1.62	2.32	3.06	4.02	6.30	10.20	13.19	22.73	34.53
1080	1.16	2.19	1.74	2.14	2.86	3.93	5.19	8.03	13.12	17.23	29.19	43.43
1350	1.41	2.71	2.15	2.69	3.51	4.92	6.32	9.75	16.09	21.18	35.42	52.33
1620	1.76	3.01	2.26	3.02	4.27	5.76	7.57	11.57	18.87	25.10	42.36	61.77
1890	2.12	3.72	2.84	3.52	4.96	6.53	8.86	13.51	21.54	29.01	48.59	70.49
2160	2.56	4.25	3.25	3.92	5.57	7.41	10.09	15.54	24.74	32.75	55.18	79.03
2430	2.92	4.86	3.63	4.51	6.14	8.37	11.39	18.12	28.59	36.49	61.41	88.29
2700	3.40	5.18	4.05	5.13	6.88	9.30	12.89	20.63	32.57	40.41	69.42	98.61

As it was mentioned in the methodology, when plotting ln (μ) vs. ln (γ), the slope is (n-1) and the intercept is ln (K), which in turn $\mu = \tau/\gamma$.

Thus, Table 5 shows the values obtained for n and K. The increment of yeast concentration makes the n value decreases gradually from 1.0 to 0.8, indicating that the fermentation broth has a pseudo plastic behavior. Contrary to this, K starts rising exponentially to a value of 171 $\times10^{-3}$ Pa.s^n, indicating that the viscosity of the fermentation broth also increases exponentially (Yang-Ming et al., 1997). In Figures 14 and 15, the rheological behavior of n and K is shown for a better appreciation.

Correlations representing n and K dependence in the concentration of yeast are presented below:

$$n = 1.0434 \cdot e^{-0.002 \cdot C_L} \tag{28}$$

$$K = 0.6733 \cdot e^{0.045 \cdot C_L} \tag{29}$$

Where:

n: flow behavior index

K: consistency index in $Pa.s^n$

C_L: yeast concentration in kg/m^3

Once the correlations of the density, viscosity and rheological properties in function of the yeast concentration are determined, it is possible to calculate the energy consumption for pumping (P).

In Figure 16, we can see the effect of cell concentration in the filtration flow (F), the power consumption (P) and the index of energy consumption per volume of the obtained filtrate (P/F). It is evident that the power consumption increases with the increasing of yeast concentration (because the power is increased at the same rate as the viscosity and consistency index K does), while the filtrate flow decreases, and therefore, more energy is required to remove the same amount of filtering in a increasingly concentrated broth, this means that the index of energy consumption increases exponentially (Yang-Ming et al., 1997).

Table 5. Rheological properties of *Phaffia rhodozyma* fermentation broth at different concentrations of yeast.

C_L (kg/m^3)	n	K ($x10^{-3}$ $Pa*s^n$)
0	1.095	0.56
9.61	0.977	2.35
18.76	0.985	1.72
31.9	0.948	2.80
41.64	0.959	3.57
51.63	0.942	5.44
64.53	0.934	7.78
75.58	0.925	12.91
83.04	0.939	18.83
88.4	0.872	40.04
102.12	0.831	91.71
109.46	0.798	171.59

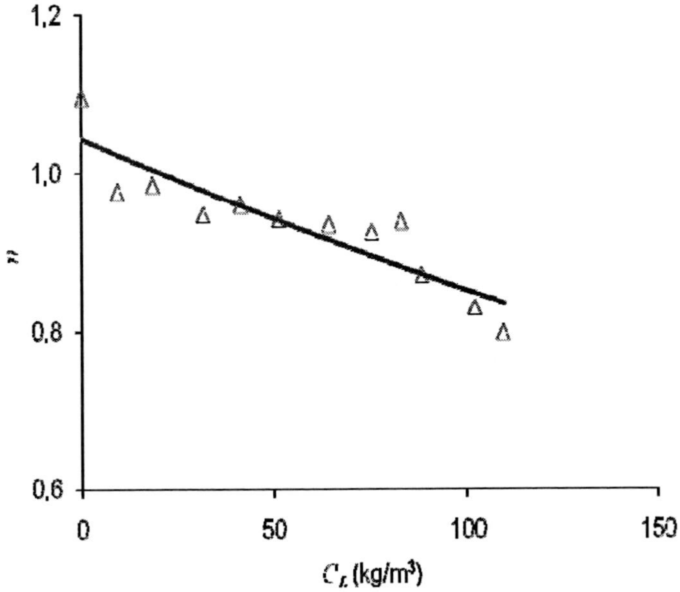

Figure 14. Effect of yeast concentration on the flow behavior index (n).

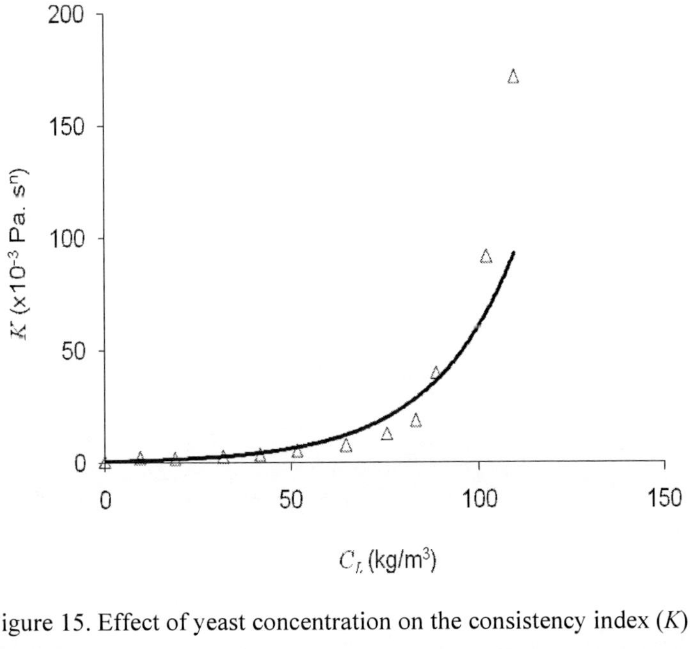

Figure 15. Effect of yeast concentration on the consistency index (K).

The same analysis can be done now testing a transmembrane pressure increase from 138 to 207 kPa, and maintaining the same feed flow. The results are shown in Figure 17. The behavior is the same and the results are very similar, so it can be said that variation in transmembrane pressure does not produce a significant change in the values P/F .

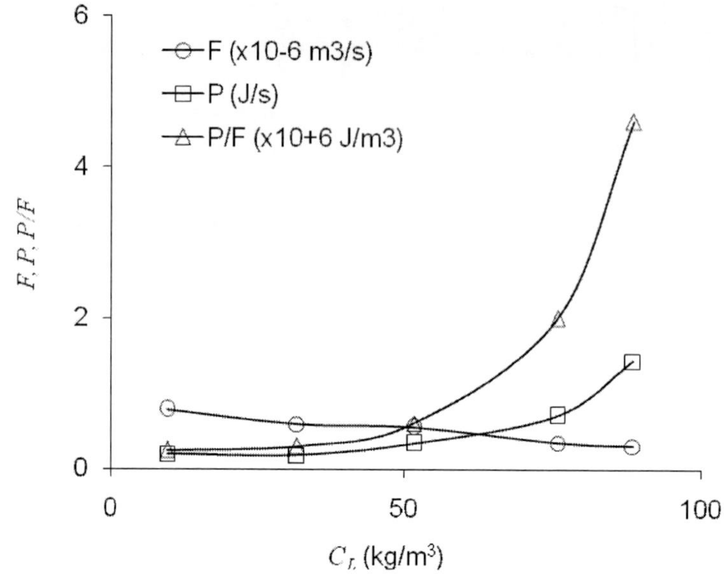

Figure 16. Effect of yeast concentration in the filtrate flow (*F*), the power consumption (*P*) and the index of energy consumption (*P/F*) in ultrafiltration. Transmembrane pressure: 138 kPa, feed flow rate: 20 $\times 10^{-6}$ m^3/s.

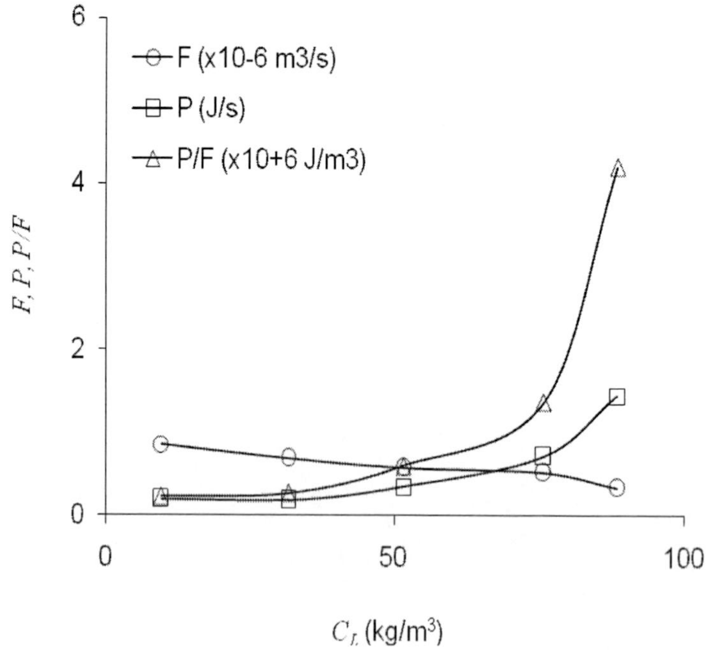

Figure 17. Effect of yeast concentration in the filtrate flow (*F*), power consumption (*P*) and index of energy consumption (*P/F*) in ultrafiltration. Transmembrane pressure: 207 kPa, feed flow rate: 20 $\times 10^{-6}$ m^3/s.

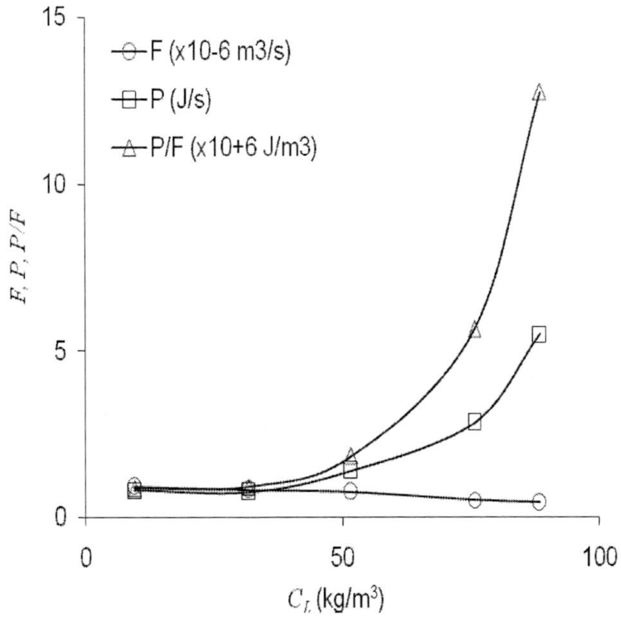

Figure 18. Effect of yeast concentration in the filtration flux (*F*), the power consumption (*P*) and the index of energy consumption (*P/F*) in ultrafiltration. Transmembrane pressure: 138 kPa, feed flow rate: 40 $x10^{-6}$ m^3/s.

Now the increment of feed flow rate is analyzed. When it is increased from 20 to 40 $x10^{-6}$ m^3/s, and maintaining the same pressure of 138 kPa, results are presented in Figure 18. The same trend was observed as previously discussed, however, doubling the feed flow will cause the energy consumption rate *P/F* to rise three times.

This means that energy consumption is tripled if the feed is duplicated and the same amount of filtering is achieved. The latter is due to "consolidation" of the cell layer formed on the surface of the membrane that does no longer permit an increment in the flux despite an increase in pressure and/or feed flow.

CONCLUSIONS

The concentration of fermentation broths of *Phaffia rhodozyma* is more effective when using the ultrafiltration process because they reach higher fluxes than when using the tangential microfiltration.

The filtration flux values obtained using both laboratory and pilot scales were very similar, indicating that the scaling approach (*k*) was appropriate.

Resistance model correlations of ϕ and R_f were determined; so that, the flux value at laboratory and pilot levels can be determined proposing the values of pressure, feed flow and yeast concentration.

The viscosity of the fermentation broths increases exponentially with the yeast concentration, while the density makes it lineally. On the other hand, the flow behavior index (*n*) of the fermentation broths is smaller than the unity; reflecting a pseudo plastic behavior, while the consistency index (*K*) rises exponentially with the increment of yeast concentration.

The index of power consumption per volume of the obtained filtrate (*P/F*) increases exponentially with the increasing of yeast concentration. The increment of the transmembrane pressure does not increase significantly this index, neither improves the flux. Any increment in the feed flow will improve slightly the flux, but it will also raise the power consumption index in the same proportion, and by a factor of three.

References

Bowen Fontana, J. D., M. F. Guimaraes, N. T. Martins, C. A. Fontana and M. Baron, (1996) Culture of the astaxanthinogenic yeast Phaffia rhodozyma in low-cost media, *Appl. Biochem. Biotec*hnol., 57-58: 413-422

Charcosset C., Choplin L., (1996). Ultrafiltration of non-newtonian fluids. *Journal of Membrane Science*, 115:147-160.

Fleno, B., Christensen, Y., Larsen, R., Johansen, S.R.,Johnson, E.A. (1994). *Astaxanthin producing yeast cells, methods for their preparation and their use,* US 5.356.810

Frenander, U., Jonsson, A.S. (1996), Cell harvesting by cross-flow microfiltration using a shear-enhanced module, *Biotechnology and Bioengineering*, 52:397-403.

Geankoplis, C.J. *Procesos de Transporte y Operaciones Unitarias, tercera edición,* México 1998.

Graves, K., Rozeboom, G., Heng, M., Glatz, Ch. (2006), Broth Conditions Determining Specific Cake Resistance During Microfiltration of *Bacillus subtilis. Biotechnology and Bioengineering*, 94:346-352.

Liew, M.K., Fane, A.G., Rogers, P.L. (1995), Hydraulic resistance and fouling of microfilter by Candida utilis in fermentation broth, *Biotechnology and Bioengineering,* 48, 108-117.

Meireles, M., Clifton, M., Aimar, P. (2002), Filtration yeast suspensions: experimental observations and modelling of dead-end filtration with a compressible cake. *Desalination*, 147: 19-23.

Naja, G., Volesky, B., Schnell, A. (2006). Performance analisis of an integrated tangencial microfilter-fermenter, *Chem Technol Biotechnol*, 81:648-658 (2006).

Ordaz, C.L., Ponce, N.T., García, S.S., Orozco, A.C. (2002). La astaxantina en el sector acuícola: usos, producción y perspectivas, *Tecnología de alimentos,* 37 (4), 16-20.

Orozco, A.C., Vidal, R.D., García, S.S., Ordaz, C.L. (2003). Concentración de suspensiones de levadura por filtración tangencial, *Tecnología de alimentos,* 38 (2), 7-17.

Orozco Álvarez Carlos, García Salas Sergio, Moreno Rivera Lourdes. (2010). Ultrafiltración de esporas en suspensión. *En VI Congreso Internacional de Ingeniería Bioquímica.* Guerrero-México.

Persson, A., Jonsson, A., Zacchi, G. (2003), Transmission of BSA during cross-flow microfiltration: Influence of pH and salt concentration, *Journal of Membrane Science*, 95:125-134.

Russotti G., Osawa A.E., Sitrin R.D., Buckland B.C., Adams W.R., Lee, S.S. (1995), Pilot-scale harvest of recombinant yeast employing microfiltration: a case of study. *J Biotechnol*. 42:235-246.

Snodderly, D.M. (1995). Evidence for protection against age-related macular degeneration by carotenoids and antioxidant vitamins. *Am. J. Clin. Nutr.,* 62(suppl):1448S-1461S.

Yang-Ming, Lo., Yang, S., Min, D.B. (1997). Ultrafiltration of Xanthan Gum fermentation broth: process and economic analyses. *Journal of Food Engineering,* 31:219-236.

Yeh, H.M., Tsai, J.W. (1998). Membrane ultrafiltration in multipass hollow-fiber modules. *Journal of Membrane Science*, 142:61-73.

Yeh, H.M., Wu, H.P., DongJ.F. (2003). Effects of design and operating parameters on the declination of permeate *flux* for membrane ultrafiltration along hollow-fiber modules. *Journal of Membrane Science*, 213:33-44.

In: Biotechnology
Editors: L. G. Torres Bustillos and I. Garcia Peña
ISBN 978-1-62081-071-2

Chapter 14

NEW APPROACHES TO MICROBIAL ENHANCED OIL RECOVERY (MEOR)

***Norma G. Rojas-Avelizapa*[*,1], *Regina Hernandez-Gamaz*[1] *and Luis G. Torres*[2]**
CICATA Queretaro
[1]IPN Mexico
[2]UPIBI, IPN, Mexico

INTRODUCTION

Biotechnology and Their Applications in the MEOR Processes

The operation of an oil reservoir basically occurs in three stages. In the first, oil is drained naturally into the wells under the influence of pressure gradient between the bottoms of the wells within the reservoir. When the pressure of oil reservoir is inadequate, or when wells are producing significant amounts of fluids such as water and gas, then starts the second phase, which involves the injecting into the reservoir of a less expensive fluid than oil to maintain the pressure gradient. In these first two stages, about 25% to 30% of original oil is recovered in place (OOIP), leaving the rest of the oil trapped in the pores of the reservoir structure, besides the presence of natural fractures or high permeability regions causing the injected water flows through channels of least resistance, leaving substantial amounts of oil trapped in the formation. After primary and secondary recoveries, the reservoir still contains great amounts of oil, which it is estimated about 60-80%. The recovery of this remaining and entrapped oil requires advanced alternatives such as enhanced oil recovery (EOR) technologies.

The global market for enhanced oil recovery technologies was $4.7 billion in 2009, but is expected to increase to $16.3 billion in 2014, for a 5-year compound annual growth rate (CAGR) of 28.4% (www.bccresearch.com). Enhanced oil recovery (EOR) technologies are becoming increasingly significant to global oil supply by increasing or reviving oil

[*] Email: ngrojas2003@yahoo.com.mx

production at reserves or oil fields depleted of more easily recoverable oil through primary or secondary oil-recovery methods and technologies. Then, the objective of those technologies is to increase the amount of crude oil that can be extracted from an oil field. Using EOR technologies, about 30-60% of the original oil in reservoirs can be extracted.

Due to no (missing text) all EOR technologies are applicable for most reservoirs; the potential use of microorganisms or their by-products has gained attention. Biotechnologically-enhanced oil recovery processes, known as microbial enhanced oil recovery (MEOR), refers to the use of microorganisms and their metabolic products (biosurfactants, biopolymers, biomass, acids, solvents, gases and also enzymes) to increase oil recovery from exhausted and negligible reservoirs, thus extending the life of the oil wells (Bryant and Lockhart, 2000). MEOR also could involve flooding with *ex situ*-produced agents by industrial or pilot scale fermentation. Microbial events that take place in the reservoir facilitate the movement of oil out of a well. Other applications of biotechnology in MEOR include genetic engineering techniques and recombinant DNA technology which are used to develop specific bacterial strains according with the MEOR project.

Most of the MEOR processes, particularly the early methods, involved the injection of microorganisms into the reservoir. Unfortunately, some operators have had bad experiences during normal waterflooding operations because microorganisms caused plugging of the wells or they contributed to corrosion problems by the production of hydrogen sulfide. It has been suggested that not only bacteria would cause plugging, but also the by-products of their metabolism, such as ferric hydroxide (Updegraff, 1983).

To eliminate this problem, some alternatives have been proposed using spores instead of vegetative cells because of their smaller size (Hitzman, 1962). Lappin-Scott et al., (1988) proposed to use ultra microbacteria with a diameter less than 0.3 μm. Microorganisms in whatever MEOR application must be smaller than the pore size in the formation. Davis and Updegraff, reported that the pore diameter should be at least twice the diameter of the microbial cells (David and Updegraff, 1954). Chang and Yen (1984) suggested the use of a lysogenic bacteria strain. rRegardless of the controversial results of MEOR projects, in 2003, more than 400 field tests had been conducted in the United States (Van Hamme et al., 2003).

In all reported studies, the microorganisms selected for use in MEOR had a maximum growth temperatures below 80°C until it was discovered that some microorganisms grow at temperatures near 121°C. There is a patent about how to produce modified microorganisms suitable for use in MEOR which are able to live at extreme temperatures, high pressures, extreme pHs and high salinities (Premuzic, 1996).

Most of the available literature of MEOR studies is based on laboratory data and it has been difficult to extrapolate these results to predict that will happen in the field based from the results of another field mainly because of the reservoir characteristics. In most cases, the interpretation of field data also has problematic results because of diverse variables that change during the recovery process (Ollivier, 2005). Some authors have reported the results of MEOR field trials (Davis, 1967; Lazar 1991; Maudgalya, 2007). Of the 407 examples they mentioned, 333 were concerned with repairing wellbore or formation damage but not specifically MEOR. Only a few reports explain the mechanisms of oil recovery. Previous situations could explain why MEOR has not gain enough attention in the oil industry. Other possible explanations on why MEOR has not been completely accepted refers to: a) microbial performance in the laboratory is not the same as in the field and as a consequence, laboratory experiments cannot predict the results in the field (Zhang and Xiang, 2010), b) well stimulation is not a MEOR process (Bryant and Lockhart, 2002), c) the monitoring time of a MEOR process is too short to see its effect (Moses, 1991).

A successful and reported project is that of the North Blowhorn Creek Unit field situated in Lamar Co., AL. which had 20 injector wells and 32 producing wells. The MEOR process involved the addition of KNO_3 and $NaHPO_4$ to the waterflood. In 2001, US Department of Energy (DOE) reported an increase of oil reserves of 400,000 to 600,000 bbl, decreasing the decline rate of oil from 18.9% to 7-12% per year, and extending the economic life of the field by 5-11 years.

Now we can mention that a group of MEOR technologies has been successfully demonstrated in areas such as improved oil mobilization, single well stimulation, profile modification and reduced souring, however, more welldocumented studies are needed.

As it has been described in several reports, MEOR methods offer distinct advantages over traditional EOR methods. First, microbial technologies are environmentally friendly, typically consisting of bacterial stimulation which can be found in the environment and "food-type" nutrients. Secondly, microbial systems are considered as low-cost solutions (consistently in the range of $2/bbl). Finally, microbial technologies require only minimal modifications to existing facilities (Zhang and Xiang, 2002). MEOR technologies have demonstrated to be able to extend the production life of the well and sustain its economic viability, which in turn reduces premature abandonment of oil fields. Additionally, microbial solutions in many cases avoid the use of toxic compounds or biocides reducing the environmental impact of oil extraction (Lazar and Yen, 2007).

Microbial Diversity in Oil Reservoirs

Prokaryotes are a very important group in the biosphere due to its abundance and diversity, most of these microorganisms have not been adequately studied mainly due to problems associated with culture media and because some are even considered non-cultivable microorganisms (Torsvik et al., 2002). Bacteria and archaea have demonstrated to have diverse and spectacular metabolic and adaptation capabilities to extreme environments. They can grow in a wide range of temperatures, pH, salinity and oxygen concentrations. They are also responsible for promoting the functioning of biogeochemical cycles and maintaining the process of decomposition, detoxification and primary production in the biosphere (Torsvik et al., 2002; Øvreås, 2000). Because of its abundance, functional diversity and the proportion of the total biomass, the limited knowledge of their diversity is surprising (Whitman et al., 1998).

In oil reservoirs, and specifically in the subsurface rock, there is a unique atmosphere where living conditions are very restrictive to most microorganisms due to considerable depth, high concentration of hydrocarbons, oxygen-limited conditions, low nutrient content, high salinity, high pressures and extreme temperature. Consequently, this environment provides ecological niches that can be occupied exclusively by prokaryotes. These extreme conditions coupled with the abundance of not easily assimilative carbon sources or deficient in energy intake suggest that the majority of microorganisms are inactive or adapted to have a low metabolic activity (D'Hondt et al., 2002).

In past decades, different microorganisms have been isolated from oil wells, and have been described by their metabolism such as fermentative, methanogens, iron reducers, sulphate reducers, nitrate reducers, hydrocarbon oxidizing and other aerobic organisms (Magot et al., 2000; Voordouw et al., 1996).

Apparently, anaerobic organotrophic organisms are widely distributed in oil fields and could be considered as indigenous organisms (Bonch-Osmolovskaya et al., 2003). The genera *Thermoanaerobacter, Thermotoga, Thermococcus*, which are isolated very often in oil reservoirs, can be indicators of native organisms (Dahle et al., 2008; Magot, 2000). Mesophilic organisms are common in wells with high temperatures, although it is estimated that many of them could reach the reservoir by drilling, pumping processes or by injection of seawater (Dahle et al., 2008). Expanding knowledge of the microbial communities in oil field environments could help to infer the dynamics of the community, and the recognition of specific metabolism for biotechnological applications including other than MEOR (Van Hamme et al., 2003).

With regard to MEOR technologies, it can be said that important progress has been made at lab and field applications, sometimes in close collaboration with the oil industry. However, it requires a greater understanding of the microbial diversity, the microbial mechanisms in a complex environment, which could lead to improve oil recovery techniques (Li et al., 2006). Earlier studies of microbial diversity of oil fields have revealed the presence of diverse procariotic groups in different proportions, classifying them according with their metabolism, rather than taxonomic clades (Bonch-Osmolovskaya et al., 2003; Yamane et al., 2008). Since 1926, microbiologists had described the cultivable microorganisms, however, a detailed description of biodiversity was considered until recently. Particularly molecular techniques, which are more sensitive, have been used to describe the microbial diversity of reservoirs from 1996. In a Canadian reservoir with a temperature of 25°C and brine injection, microorganisms were identified as members of the family *Desulfobacteriaceae, Desulfovibrionaceae*, other members were also found such as *Clostridium, Eubacterium, Synergistes, Thiomicrospira, Arcobacter, Campylobacter* and *Oceanospirillum.* The dominant microbial group corresponded to sulphate-reducing bacteria (SRB) (Voordouw et al., 1996).

Hyperthermophiles are an important group in petroleum microbiology; these microorganisms are represented by different taxonomic groups, allowing them a wide range of metabolic activities (Stetter and Huber, 1999). Studies of an oil reservoir from Kubiki Japan at 50-58°C, using both molecular and cultivation techniques revealed the presence of *Thermococcus* and *Thermotog*a (Takahata et al., 2000). Making a thermophilic anaerobic enrichment, Orphan et al., (2000) found the genera *Thermoanaerobacter, Thermococcus, Methanobacterium, Methanococcus* and *Methanoculleus.* Also, by means of ribosomal libraries, the representative clones belonged to the genera *Thermoanaerobacter, Thermococcus, Desulfothiovibrio, Aminobacterium, Acidaminococcus, Pseudomonas, Halomonas, Acinetobacter, Sphingomonas, Methylobacterium* and *Desulfomicrobium* were detected (Orphan et al., 2000).

In oil field environments, there are a significant number of Archaea whichrarely gets the cultivation of such microbial groups, but the use of DNA-based methods are more sensitive and specific, getting its identification. Watanabe et al, in 2002 found by a ribosomal library construction: *Methanosaeta*, *Methanomethylovorans* and *Methanomicrobiaceae*. The study also showed that archaeal cells represented more than 10% of the total microbial count (Watanabe et al., 2002). Potential applications of Archaea in MEOR are promising because they contain unique membrane lipids, which are an excellent source for the formation of liposomes with remarkable thermostability that could allow the stabilization of emulsions of oil in brine (Schiraldi et al., 2002).

In a study conducted in Nizhnevartovsk (Siberia, Russia), cultivable microorganisms belonging to the genera *Petrotoga, Thermococcus, Thermotoga, Thermosipho, Thermoanaerobacter* and *Geobacillus* were identified. Also, by probes, the microbial groups: *Thermodesulfobacterium, Geotoga, Desulfurococcus, Aquificales/Desulfurobacterium* and *Thermovibrio* were identified. This study involved the use of probes for the detection of microbial groups of interest to industry, so organisms like sulphate-reducing bacteria can be detected in a sensitive and specific form (Bonch-Osmolovskaya et al., 2003).

Other studies have reported the finding of common inhabitants of soil in a variety of oil field studies. Some reasons could include the use of molecular techniques, which sensitivity could detect contaminants in a reservoir or organisms that could reach samples by operational situations. A study performed on Hejime, Japan, reported the finding of species such as *Ochrobactrum anthropi, Burkholderia cepacia, Stenotrophomonas maltophilia, Propionibacterium acnes,* and *Brevundimonas diminuta* (Yoshida et al., 2005). Microbial diversity is highly dependent of oil reservoir conditions. Grabowski et al, (2005) described the diversity of a low temperature reservoir where a variety of organisms were found in horizontal drilling wells and that operated using drilling muds (see Table 1). Despite the high temperature of many oil reservoirs, a study of a 75ºC reservoir located in Hebei, China reported the finding of the following genera: *Wolinella, Burkholderia, Serratia, Pseudomonas, Mycobacterium* and *Sulfurospirillum* and also some hyperthermophiles as *Thermotoga Thermoanaerobacter* and *Thermodesulfovibrio* (Li et al., 2006). From these results, the presence of mesophilic organisms is evident, although it is unknown if they are relevant to the ecology of the reservoir (Von Der Weid et al., 2008). The same case was observed in samples of oil from China and Japan in which *Acinetobacter, Propionicabterium, Sphingobium, Bacillus, Clostridium, Petrotoga* and *Synergistales* were found. Microbial groups, identified in most cases, are considered transient microbes although the genera of *Petrotoga* are characteristic of such an environment (Yamane et al., 2008). There are descriptions of complex microbial communitics within the reservoir, Dahle et al, (2008) analyzed the microbial diversity in the water of a high temperature well. Microbial groups from the most abundant to least abundant were: *Firmicutes> Bacteroidetes> Proteobacteria> Spirochaetes> Thermotogales*. While the Archaea found were: *Methanococcus, Thermococcus* and *Methanolobus*. Among the cultivable microorganisms, members of the genera: *Thermosipho, Thermotoga*, *Anaerophaga* and *Thermovirga* were identified (Dahle et al., 2008). This study defined a complex microbial community, however, the diversity description generated little information and then it is relevant to begin and infer the involvement of microorganisms in the ecological context of the environment. Table 2 shows, in detail, some of the recently published studies that describe the microbial communities found in oil fields. There are parameters that can help correlate the information obtained, such as the well temperature, the methodologies used for microbial detection, exploitation characteristics of the reservoir, among others. Studies of microbial diversity are commonly limited to the taxonomic description and only in some studies, the groups are evaluated metabolically (Bonch-Osmolovskaya et al., 2003; Yamane et al., 2008). It is possible that an approach based on molecules such as RNA and proteins provide further functional information of the microbial diversity of these environments (Ferrer et al., 2009).

Table 1. Diversity of microorganisms that can be used to develop new or improve current MEOR technologies

Order	Genera	Metabolism	Possible application in MEOR
Methanomicrobiales	*Methanocorpusculum, Methanoculleus, Methanospirillum, Methanocalculus,*	Methanogens	CH_4 production for well pressurization
Methanosarcinales	*Methanosarcina, Methanolobus, Methanosaeta, Methanomethylovorans*		
Methanobacteriales	*Methanobacterium*		
Methanococcales	*Methanococcus*		
Desulfuromonadales	*Desulfuromonas*	Sulfate-reducing and sulfur-reducing bacteria, anaerobic, oxidize acetate and other simple carbon sources	Decrease or eliminate souring
Desulfovibrionales	*Desulfovibrio, Desulfomicrobium*		
Synergistales	*Desulfothiovibrio,*		
Desulfurococcales	*Desulfurococcus*		
Desulfobacterales	*Desulfobacterium*		
Nitrospirales	*Thermodesulfovibrio*		
Thermotogales	*Thermotoga, Thermosipho, Petrotoga, Geotoga*	Anaerobe, fermentative bacteria, hypertermophile, NaCl resistant	CO_2 and ethanol production to well pressurization and oil viscosity reduction
Order	Genera	Metabolism	Possible application in MEOR
Thermoanaerobacterales	*Thermoanaerobacter, Caldanaerobacter*	Anaerobe, fermentative bacteria, hipertermophile, NaCl resistant	CO_2 and solvent production
Thermococcales	*Thermococcus*	Hyperthermophilic, anaerobic, mixed heterotrophic, and carboxydotrophiceuryarchaeon	Growth under high pressure conditions
Pseudomonadales	*Pseudomonas, Acinetobacter*	Extraordinary wide metabolism	Production of byproducts such as biosurfactants

Order	Genera	Metabolism	Possible application in MEOR
Clostridiales	*Clostridium, Eubacterium, Acetobacterium*	Anaerobic, spore and non-spore-forming, Fermentative and temperature resistant	Spore-forming application and fermentative products as acid and gas
Thiotrichales	*Thiomicrospira*	Isolated from deep-sea hydrothermal vents. Obligatelychemolithoautotrophic sulfur-oxidizing bacteria	Sulfur-oxidation and contribution to sulfur mobilization inside oil reservoir
Oceanospirillales	*Oceanospirillum*	Halophilic and chemoorganotrophic bacteria, with tolerance to NaCl and grow at different temperatures	Use of biomass to plugging
Bacillales	*Bacillus, Geobacillus*	Aerobic or facultative, chemoorganotrophic and mesophilic bacteria	Biosurfactant production
Aquificales	*Thermovibrio*	Extremely thermophilic, chemolithoautotrophic nitrate-reducing bacterium	Displacement of sulphate-reducing bacteriafor souring prevention
Campylobacterales	*Wolinella,Sulfospirillum*	Heterotrophic, sulfur-reducing, microaerophilic, vibrioid from oxidized marine surface sediment.	Souring prevention
Rhodocyclales	*Azoarcus*	Nitrogen fixing proteobacteria	Nitrogen supply to oil reservoir
Rhodocyclales	*Thauera*	Anaerobic bacteria usingselenate as electron donor	Mineral mobilization in oil reservoir
Deferribacterales	*Geovibrio*	Gram negative reduces Fe(III), Mn(IV), U(VI), Cr(VI), nitrate, sulphate, sulphite, or thiosulphate with acetate as the electron donor.	Mineral mobilization and rock wettability changes
Cytophagales	*Cytophaga*	Aerobic cellulose-degrading bacteria also degrade agar and chitin	Degradation of biomacromolecules used in oil recovery

Table 2. Common microorganisms found in oil reservoir environments

Sampling reservoir	Bacteria	Archaea	Reference
Reservoir in west Canada, 25°C, producing by water injection	*Desulfovibrionaceae, Desulfobacteriaceae, Clostridium sp., Eubacterium sp., Synergistes sp., Thiomicrospira, Arcobacter, Campylobacter, Oceanospirillum spp.*	-	Voordouw et al., 1996
Oil reserve in Niigata, Japan, 50-58°C	*Thermotoga*	*Thermococcus*	Takahata et al., 1999
High temperature wells in Santa Clara, USA., 50-105°C	*Thermoanaerobacter, Methanobacterium, Methanococcus, Methanoculleus, Thermoanaerobacter, Desulfothiovibrio, Aminobacterium, Acidaminococcus, Pseudomonas, Halomonas, Acinetobacter, Sphingomonas, Methylobacterium, Desulfomicrobium*	*Thermococcus*	Orphan et al., 2000
Sampling reservoir	Bacteria	Archaea	Reference
Oil in Kuji, Iwate, Japan	-	*Methanosaeta concilii, Methanomethylovorans hollandica,* member of the family *Methanomicrobiaceae*	Watanabe et al., 2002
Reservoir in Samotlor, Siberia, Rusia 4°C-40°C	*Petrotoga, Thermotoga, Thermosipho, Thermoanaerobacter, Geobacillus, Thermodesulfobacterium, Geotoga, Thermus, Aquificales/Desulfurobacterium -Thermovibrio.*	*Desulfurococcus, Thermococcus*	Bonch-Osmolovskaya et al, 2003
Storage pile in Hejime, Japan.	*Ochrobactrum anthropi, Burkholderia cepacia, Stenotrophomonas maltophilia, Propionibacterium acnes* and *Brevundimonas diminuta*	-	Yoshida et al., 2005
Reservoir in Pelican, Canada. 18°–20°C. Horizontal wells using drilling muds.	*Lactosphaera, Thauera, Tannerella, Pseudomonas, Propionicimonas, Desulfomicrobium, Desulfovibrio, Acetobacterium,Azoarcus, Rhodoferax, Diaphorobacter, Pandorea, Pseudomonas, Desulfomicrobium, Desulfovibrio, Desulfuromonas, Campylobacter, Arcobacter, Sulfurospirillum, Cytophaga, Rikenella, Bacteroides, Tannerella, Acetobacterium, Lactosphaera, Clostridium, Sedimentibacter, Fusibacter, Holdemania, Desulfurosporosinus, Ruminococcus, Geovibrio, Spirochaeta.*	*Methanocorpusculum, Methanospirillum, Methanocalculus, Methanosarcina, Methanosaeta, Methanolobus*	Grabowski et al., 2005
Formation water of reservoir in Heibei, China, 75°C	*Wolinella, Burkholderia, Serratia, Pseudomonas, Sulfurospirillum, Mycobacterium, Thermotoga, Thermodesulfovibrio, Thermoanaerobacter*	-	Li et al, 2006
Platform Troll C in Norway, 70°C	*Arcobacter, Anaerophaga, Geotoga, Thermotoga, Thermosipho, Thermovirga, Halanaerobium, Clostridium, Clostridium, Thermovirga, Caminicella, Thermacetogenium, Desulfovibrio, Pelobacter, Agarivorans, Thermosipho, Anaerophaga*	*Methanococcus, Methanolobus, Thermococcus.*	Dahle et al, 2008

Sampling reservoir	Bacteria	Archaea	Reference
Virgen field, Brasil, 60°C Core washing	*Pelomonas, Pseudomonas, Variovorax, Rubrivivax, Comamonas, Azoarcus, Bradyrhizobiaceae, Rhodopseudomonas, Methylobacterium, Ochrobactrum, Rhodobacter, Defluvibacter, Xanthomonas, Stenotrophomonas, Bacillus, Oceanobacillus, Brevibacillus, Actinobacteria, Arthrobacter*	-	Von Der Weid et al., 2008
Oil from different reservoirs in China and Japan	*Acinetobacter, Propionibacterium, Sphingobium, Bacillus, Petrotoga, Clostridium* and *Synergistales*	-	Yamane et al., 2008

(-) no reported.

Previous studies carried out by our research group (Hernandez-Gama et al., 2011) identified the dominant bacteria from three different wells in a Mexican reservoir located in Chicontepec Veracruz. Members of *Thermotoga neapolitana, Thermoanaerobacter pseudethanolicus* and *Caldanaerobacter subterraneus* were detected in the oil reservoir (Figure 1 and 2). Also, it was demonstrated that these microorganisms produce CO_2 after 10 days of incubation.

Important efforts have been done for alternatives to oil recovery. Hernandez-Gama (2011) reported the evaluation of hydrophobicity of microbial cells by a microbial adherence to hydrocarbons assay (MATH). The test was performed considering two aliphatic hydrocarbons n-heptane and n-hexane and diesel, microbial cultures tested were both anaerobic and fermentative from 3 different wells from Chicontepec and Samaria reservoirs.

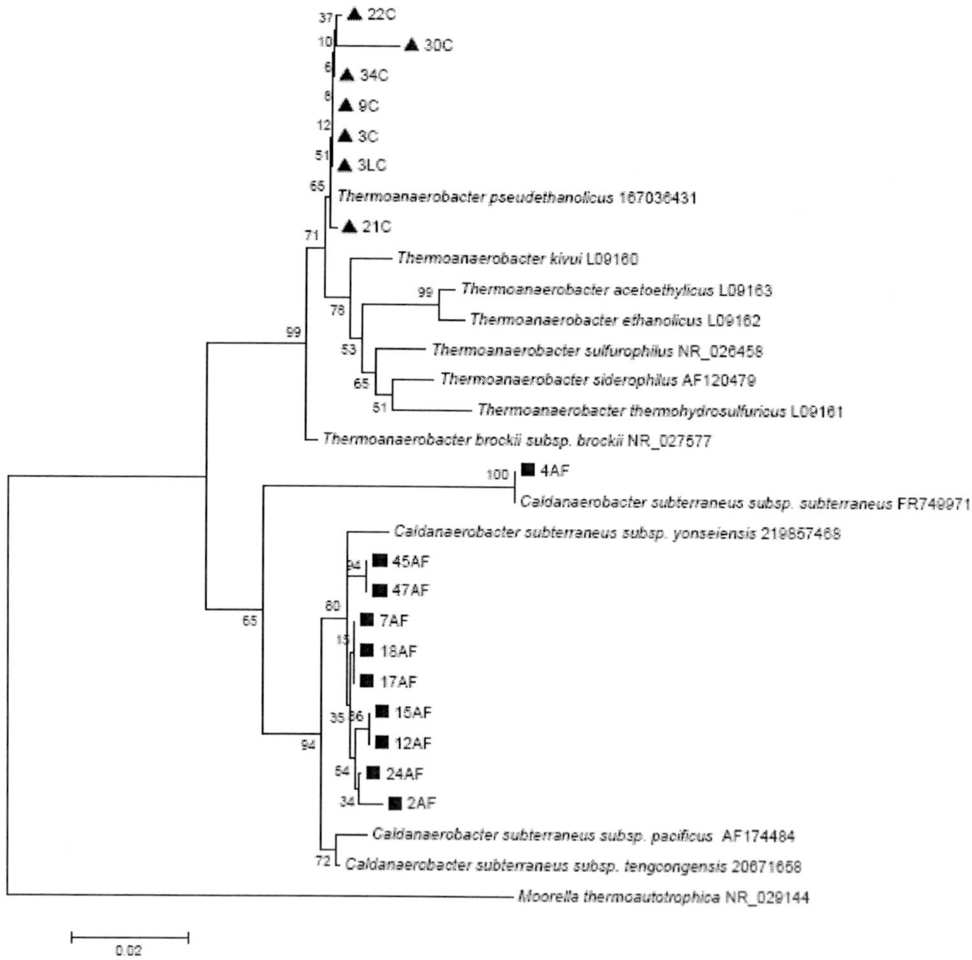

Figure 1. Neighbor-joining phylogenetic tree of 16S rRNA gene sequences cloned directly from consortia from Coyotes oil reservoir (black triangle) and consortia from Agua Fria oil reservoir (back squares). Sequences were aligned with ClustalX, and distances were calculated with the Jukes-Cantor substitution model. The scale bar indicates the nucleotide substitutions per site. Number at the branches indicates the bootstrap values of 1,000 resampling. *Moorella thermoautotrophica* served as the out-group.

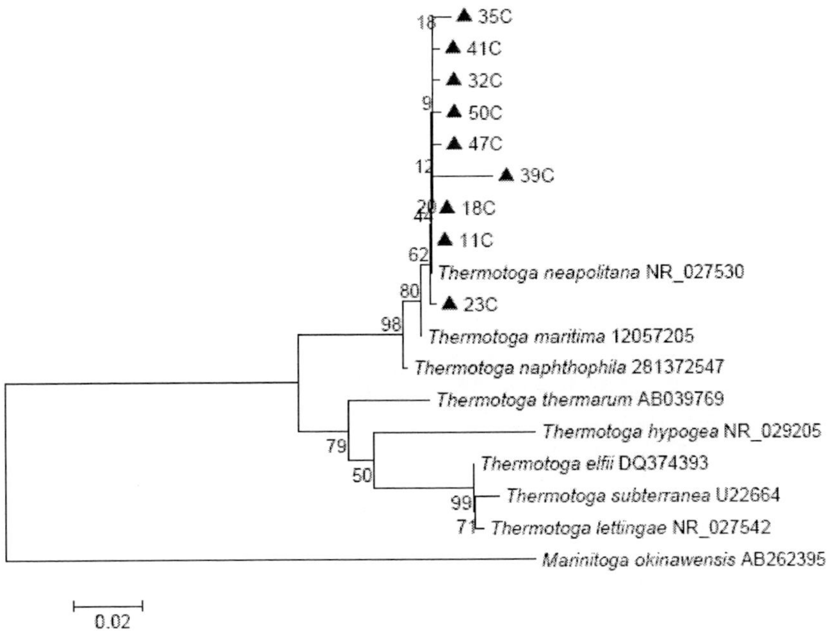

Figure 2. Neighbor-joining phylogenetic tree of 16S rRNA gene sequences cloned directly from consortia from Coyotes oil reservoir (black triangle). Distances were calculated with the Kimura 2-parameter substitution model. The scale bar indicates the nucleotide substitutions per site. Number at the branches indicates the bootstrap values of 1,000 resamplings. *Marinitoga okinawensis* served as outgroup.

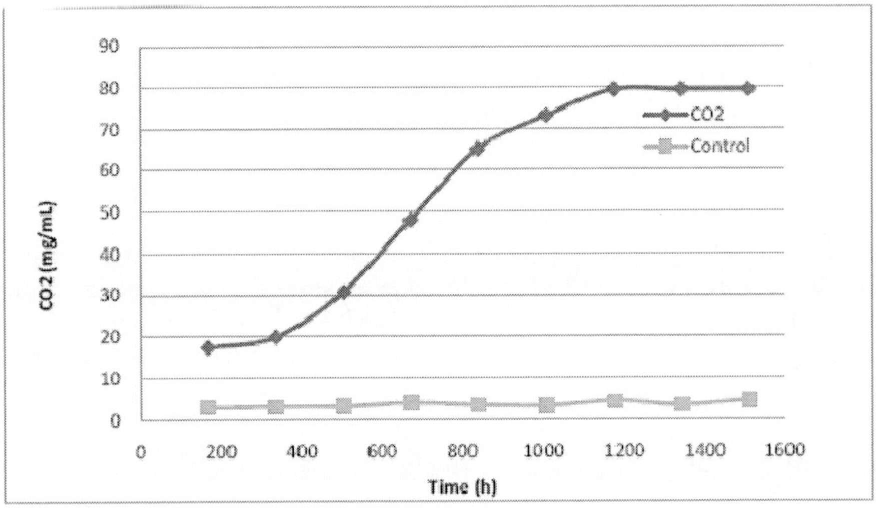

Figure 3. Carbon dioxide production by microbial cultures from Chicontepec oil field.

Torres et al., (2007a) have developed a technique for the preparation of oil/water emulsions, stabilized by solids. These emulsions can be prepared using the right amounts of oil, water, solid (sand, bentonite, kaolinite, etc) and a small amount of surfactant, for changing the hydrofobicity of the particles. Once good emulsions were obtained, these were characterized by rheometry. Plots of shear rate versus shear stress, as well as oscilatory rhelogy (G´, G´´ and complex viscosity as a function of amplitude ω), are employed to define if emulsions are highly stable, stable or unstable.

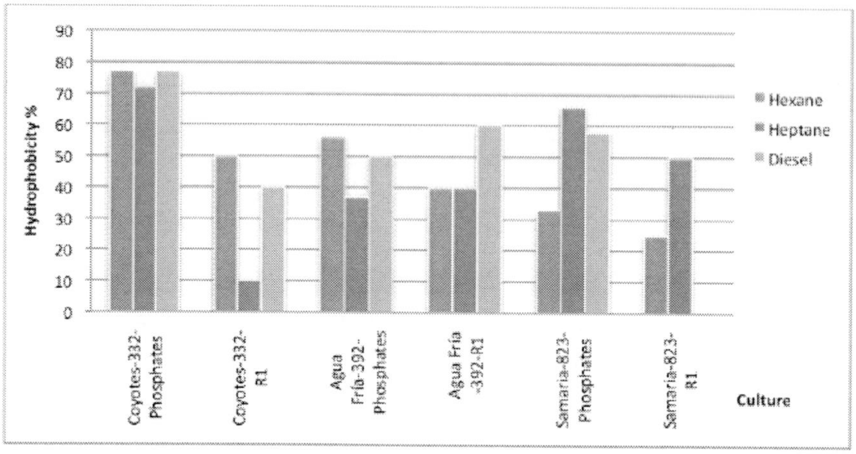

Figure 4. MATH assay of fermentative microbial culture from Chicontepec and Samaria oil reservoirs.

The same technique has been applied to creosote (Torres et al, 2007b) as a basis for a cleaning technique for contaminated aquifers. The aim of our studies is to use microbial cells as solids and find the physiological characteristics that allow the cells to stabilize an oil emulsion. It has been published that formation of emulsions inside the oil well is an aid to enhance the oil recovery.

The first step for our work is to study the cells as solids. The knowledgement of the surface properties of the cells is very necessary. If particles are a sort of hydrophobic, that will determine the suitability to produce stable oil/water emulsions. As an example, three model compounds (hexane, heptane and diesel) have been used to evaluate the hydrophobic character of three different consortia (i.e., Coyotes, Agua Fria and Samaria microorganisms), using the very simple method identified as MATH.

As shown in Figure 4, in general, microbial cultures had an intermediate to high hydrophobicity (30-40% intermediate hydrophobicity and ≥40% high hydrophobicity), except for the Coyotes culture in R1 media with heptane as solvent. The Samaria culture in the media R1 added with diesel, in which the hydrophobicity was inexistent. It can also be seen that there were differences between the various solvents tested, however, any pattern can be associated with the solvent type. In order to select microorganisms able to stabilize emulsions, the best candidates are the organisms with intermediate hydrophobicity, a characteristic that will allow to stabilize oil droplets in water at the interface and enhance oil recovery.

BIOTECHNOLOGY AND THEIR APPLICATIONS IN THE EOR PROCESSES

The historical use of microbial products and their effectiveness in a wide range of oil field applications have demonstrated the importance of biotechnology in the oil industry. Over the past 15 years, their use has grown as an alternative to conventional technologies. The cumulative genetic information present in a diverse collection of microorganisms indicates an enormous biochemical potential for new products and technologies. Biotechnology for the oil field is only in its infancy. It is clear that the number of applications will only continue to increase. Widely reported successes in the application of microbial products have caused their use to gain widespread acceptance. It is expected that in the next 100 years to see a gradual replacement of conventional technologies with new and improved biotechnological applications.

Some problems that have been solved by biotechnological or microbial products include paraffin control, scale and corrosion control, waterflood treatments, well stimulation and frac damage repair avoiding the use of expensive alternatives such as thermal and chemical treatment (paraffin deposition), toxic compounds that control scale through either chelation or dispersant mechanisms and polymers or gels that cause decreasing permeability and blocking flow from producing zones.

Our research group has been proposed different approaches to the EOR and MEOR processes. Very recently, we have reported the characterization of crudes arising from the zone of Chicontepec, Veracruz, Mexico (see Figure 6).

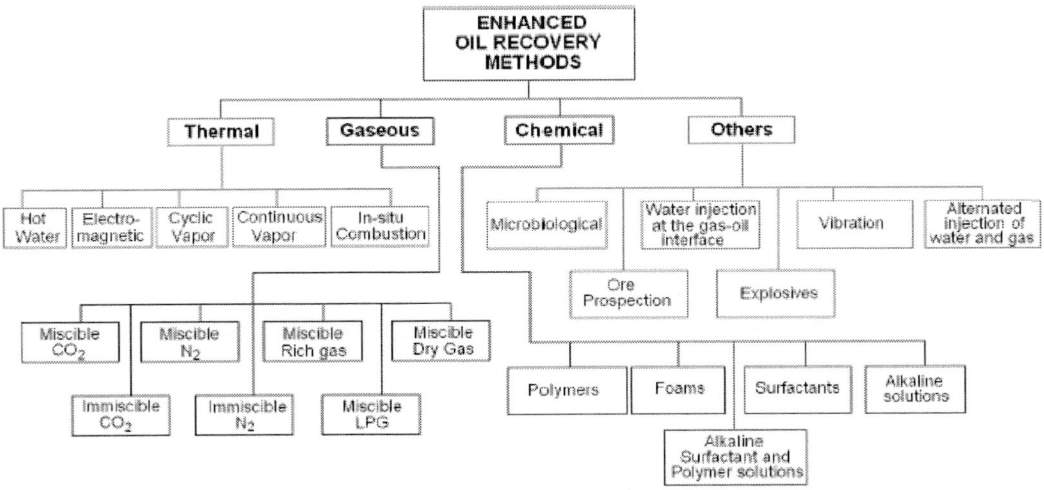

Figure 5. Resume of the main EOR process developed and employed at laboratory or field level.

Exploitation of petroleum in the Chicontepec Basin is an especially important project for Mexico, since it represents an oil production equivalent to 39% of the national hydrocarbons reserve, which amounts to about 17.7 billion equivalent oil barrels. The *Paleocanal de Chicontepec* is located in the geological basin of Tampico-Misantla, east of the Tuxpan Platform (*Faja de Oro*). This oil field covers an area of about 3,800 km^2. The zone covers twelve municipalities, nine from Veracruz State and three from Puebla State. The PEMEX objective is to develop a basin with the capability of producing between 550,000 to 700,000 barrels/day by 2017. For that, it is necessary to develop specialized technologies which substantially enhance well productivity while ensuring minimum cost (Torres et al., 2011a).

The main conclusions of the work are that the viscosity and surface tension of these samples is correlated with the fraction of n-pentane insoluble components (asphaltenes) and inversely correlated with the fraction of aromatic compounds. The acid components in the crude oils are correlated with surface tension while the basic components are negatively correlated with it. Finally, the base number is correlated with viscosity.

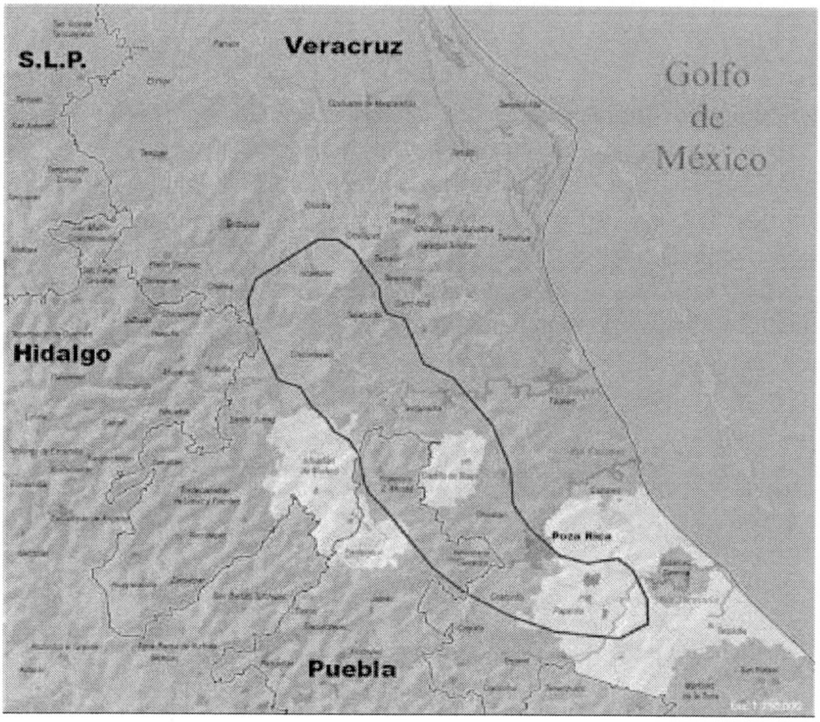

Figure 6. Localization of the Chicontepec Paleocanal (Veracruz, Puebla, and Hidalgo States in Mexico): Adapted from PEMEX (2010).

On the other hand, our research group has also been investigating the suitability of using biosurfactants in the EOR process, instead of the traditional organic products, synthesized by organic chemistry processes. Why this proposal? As stated by Chen et al., (2010) compared with conventional surfactants, their lower toxicity, higher biodegradability, higher tolerance to pH, temperature, and salinity, and their ability to be synthesized from a variety of non-petrochemical sources have increased their potential attraction. However, this has to be set against a greater difficulty in large-scale production and purification.

Torres et al., (2011b) recently reported a paper where the performance of three natural surfactants (one produced by a bacteria and two of vegetal origin) in comparison to synthetics surfactants (cationic, anionic, nonionic and zwitterionic) for its potential use in enhanced oil recovery EOR application was compared. The basis of comparison was basically the surfactant's solutions surface tension against air at room temperature, in the presence and absence of brine (salinity of about 4.3%).

Biosurfactants behaved quite well in the presence of brine, regarding the surface tension (ST), the effect of the heat treatment (HT) and the foamability assessments (FA). Figure 7 shows the ST of biosurfactants in water and brine, in comparison to that showed by synthetic products. Figure 7 also shows the behavior of the rhamnolipid in water and brine. Note the low values (less or equal to 32 than mN/m) reached for the rhamnolipid at any assessed concentration, at concentrations between 0.01 and 01.1% w/w.

Conclusions of the work were that biosurfactants have the potential for the EOR process. The rhamnolipid produced by *Pseudomonas* and natural surfactants developed characteristics as good or even better than the synthetic surfactants normally employed in this process in terms of surface tensions, foaming capabilities, behavior at high salinity (46,800 mg/L) and resistance to high temperatures (even 70°C).

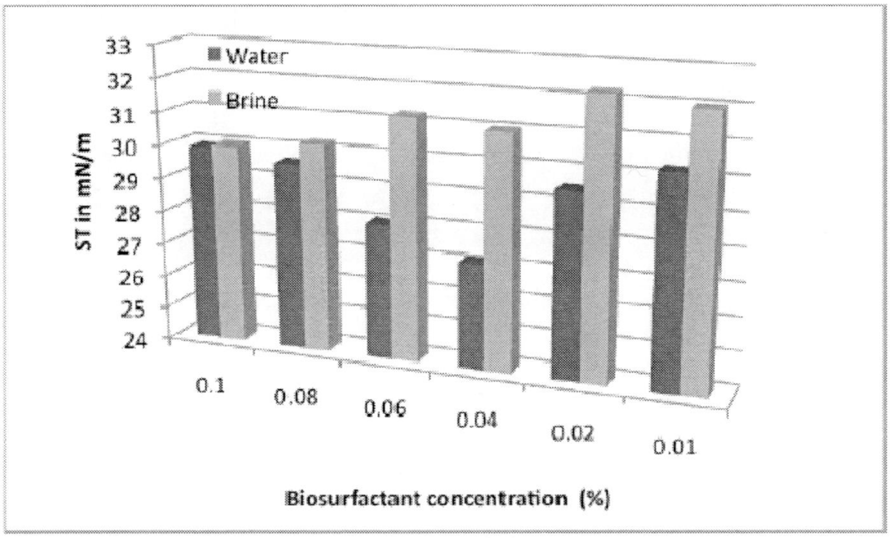

Figure 7. Surface tension ST for the rhamnolipid solutions in presence or not of brine. Adapted from Torres et al., (2011b).

It is well known that well oil fermentative bacteria produce some specific metabolites which can modify the surface tension of both oil and brine in the oil well. With the purpose of identifying those metabolites and their effect over the oil and brine, our research group carried out an in-vitro study (Torres et al., 2011c) where fermentative bacteria arising from a production well (from Chicontepec zone) were grown on different culture media and conditions. Biomass growth, the sugar consumption, the production of CO_2 and defined metabolites (such as acetone, total solvents and total acids) as well as the surface tensions of both the aqueous phase (brine) and organic phase (oil) were monitored along the fermentation process. Among the many results of the paper, the production of the metabolites of interest was confirmed, as well as the changes in the ST during the process. Figure 8 shows the ST for these experiments under different conditions.

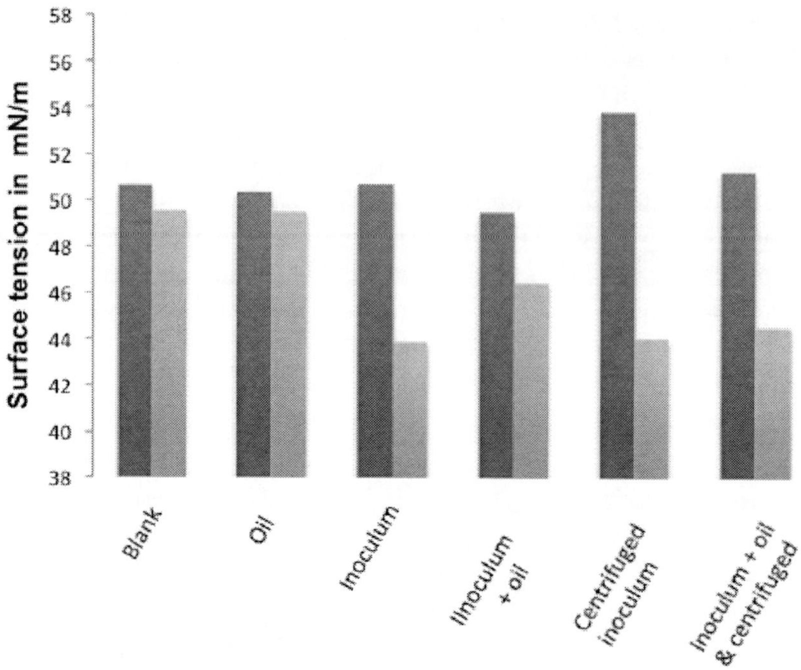

Figure 8. Surface tension for experiments with real brine. Initial (dark) and final (pale) values. Adapted from Torres et al., (2011c).

In other studies, our research group has worked on the development of EOR techniques which employ bioproducts. That is the case of humic acids. Humic acids are the product of vegetal material degradation. It has been reported that humic acids were first discovered in the sea and natural basins, but later on, there has been a great interest in the humic acids found on soils (Ramirez et al., 2011). Humic acids are a rather complex mixture of monomers that interact at certain pH and ionic strengths and form giant macromolecules. A proposed scheme of a humic acid is shown in Figure 9.

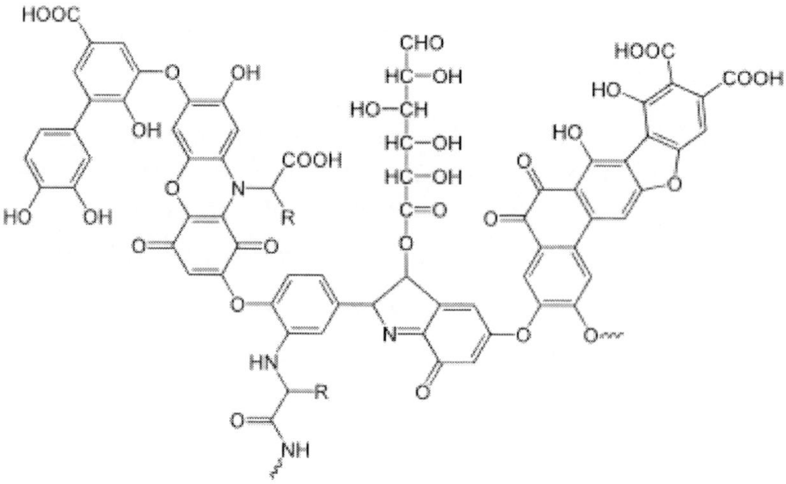

Figure 9. Schematic representation of a portion of humic acids. Adapted from Ramirez et al.,(2011).

Nowadays, humic acids (HA) have been produced by the biodegradation of grass, tree cuttings, etc. Ramirez et al., (2011) have reported the two different procedures to obtain HA by composting of vegetal residues, sludges and *tezontle.* They also have developed an extraction method to obtain the humic acids as dry material, susceptible of commercialization.

In the same paper, they have reported some chemical characteristics of the two HA, regarding their contents of the main functional groups reported for HA (i.e., the content of acidic and phenolic groups), as well as some surface characteristics such as the critical micellar concentrations (CMC), the minimum surface tensions (MST) and the effect of pH and come divalent cations (Ca^{++} and Mg^{++}) over the HA surface activity.

Using one of this HA's, we have studied the interfacial tension (IFT) between the HA solutions and three Mexican crudes, arising from Chicontepec. These studies (Martinez et al, 2011) are pioneers in the field, since as far as we know; nobody has proposed the use of HA for EOR processes. In a first work, the IFT among HA solutions prepared in water or in brine and the crude oils, were evaluated using a simple Du Nouy ring-tensiometer. The three crudes were also characterized in terms of important physical-chemical characteristics such as acid and basic indexes, content of n-pentane insolubles, aromatics, polar fractions, among others.

Figure 10 shows the IFT for one of the crudes (Coyotes zone) against the HA solutions, prepare in both water and brine. As observed, IFT were as low as 13 mN/m, for an HA concentration of 1 g/L (0.1% w/w). Besides, brine (with a salt content of approximately 46 g/L) did not affect the IFT values to a large extent.

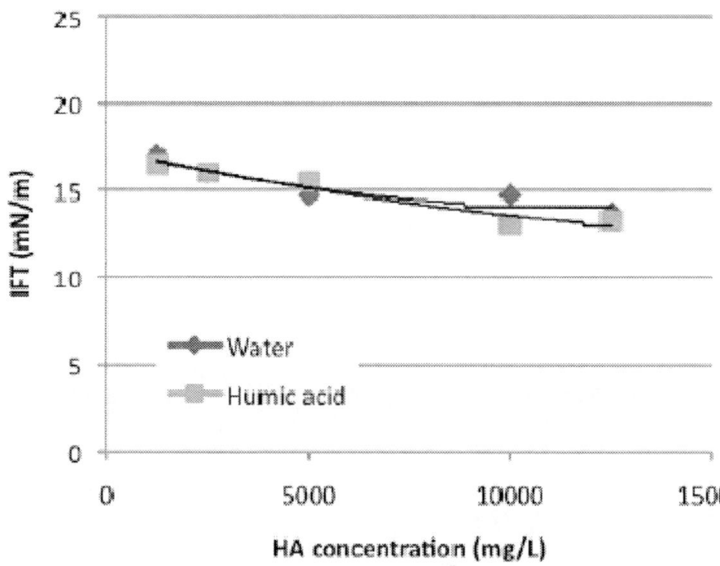

Figure 10. IFT for the Coyotes crude, against HAs solutions (water and brine). From Martinez (2011).

Interaction between polymers and surfactants has interesting applications in the field of cosmetics, pharmaceutics, printing industry and EOR. In EOR operation, there are two fundamental necessities. One is to inject surfactants that can diminish the surface tension (and this the IFT values between rock, oil and brine) and the other is to obtain viscosities higher than that of water, with the purpose of displacing the oil inside the capillary of rocks (Levaro, 2011). The combination of both effects has been studied, mixing one polymer which promotes an augmentation in viscosity and one surfactant that lower the surface tension. The polysaccharide produced by *Xanthomonas campestris* has been chosen as a model to work with xanthan gum. This polymer has been characterized widely in terms of its production, recovery, as well as its chemical and rheological characteristics. Years ago, some papers related to those issues had been published (Hannote et al., 1991, Torres et al., 1991, Albiter et al, 1992, Sanchez et al, 1992, Torres et al., 1992, Flores et al, 1994, Brito et al., 1995, John et al., 1995, Torres et al., 1995). On the other hand, it has been basicallyworked with the rhamnolipid produced by *Pseudomonas aeruginosa*, which has tremendous potential to be used in EOR or MEOR practice (Chen et al, 2010). Combinations of xanthan-cationic and xanthan-anionic surfactants have also been tried (See figure 11).

The mixtures of xanthan with the biosurfactant or organic surfactants using two techniques have been characterized. One is the use of the Du Noy ring tensiometer to measure the ST of the solutions prepared in water or brine. The other is the use of a rheometer to measure the relationship between the shear rate and shear stress of the same solutions. Raw data can be modeled by the power law or any other rheological equation.

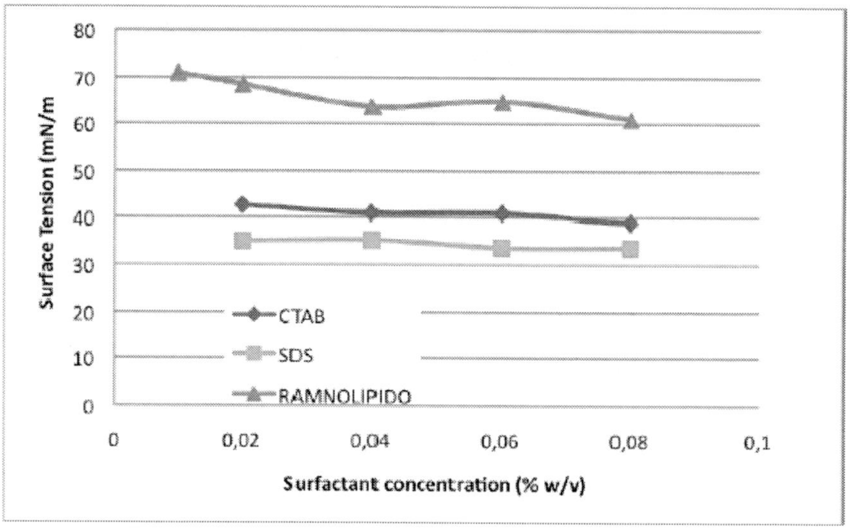

Figure 11. Surface tension for the mixtures xanthan gum (0.1% w/v)-surfactant. Measurements at 25°C. From Levaro (2011).

General Conclusions

A wide spectrum of research projects has been developed by our research group in order to promote new insights in MEOR process. The more interesting lines that we are working are those enlisted below:

1. Studies on the microbial diversity in oil wells (including molecular biology methods).
2. Generation of microbial metabolic products such as biosurfactants, biopolymers, biomass, acids, solvents and gases.

3. Use of biomass (microbial cells) as stabilizers for oil in water emulsions, as a basis for the development of MEOR technologies.
4. Xanthan gum production and characterization of novel products.
5. Mexican crudes characterization.
6. Application of surfactants (synthetic, microbial products and vegetal gums, as well as humic acids arising from composting processes) in EOR.
7. Study of the surface tensions of mixtures of oil, culture media and cells in presence and absence of brine.
8. Interactions between a biopolymer (such as xanthan gum) and surfactants (including the rhamnolipid produced by *Pseudomonas aeruginosa*).

Development of these lines will state, in the next future, basis for new and suitable technologies to improve the petroleum industry in Mexico, particularly in the field of oil production. Every day, new challenges are established due to the diminution of the oil reserves, the availability of more complex oils (more heavy, more viscous and with higher levels of sulfur and nitrogen concentrations) and the necessity of new oil products.

ACKNOWLEDGEMENTS

Part of this work has been developed in conjunction with M in C. Ana. Muñoz (IMP-Mexico). The participation of other colleagues and students is recognized and appreciated.

REFERENCES

Albiter, V., Torres, L.G., Galindo, E. (1992). Recovery of xanthan gum from fermentation broths by precipitation in a stirred tank. *Proc. Biochem*. 29:187-196.

Bonch-Osmolovskaya, E., Miroshnichenko, M., Lebedinsky, A., Cherny, N., Nazina, T., Ivoilov, V. (2003). Radioisotopic, culture-based, and oligonucleotide microchip analyses of thermophilic microbial communities in a continental high-temperature petroleum reservoir. *Appl. Environ. Microbiol*, 69:6143-6151.

Brito, E., Torrres, L., Galindo, E. (1995). Diffusion behavior of ammonium in xanthan gum solutions. *Biotechnol. Progr.* 11:221-223.

Bryant, S., Lockhart, T. (2000). Reservoir engineering analysis of microbial enhanced oil recovery. *In*: Proceedings of the SPE Annual Technical Conference and Exhibition, Dallas, TX.

Chang, P., Yen, T. (1984). Interaction of *Escherichia coli* B and B/4 and bacteriophage T4D with berea sandstone rock in relationship to enhanced oil recovery. *Appl. Environ. Microbiol.* 47:544-550.

Chen, M.L., Penfold, J., Thomas, R.K., Smyth, T.J.P., Perfumo, A., Marchant, R., Banat, I.M., Stevenson, P., Parry, A., Tucker, I., Grillo, I. (2010). Solution self-assembly and adsorption at the air-water interface of the monorhamnose and dirhamnose rhamnolipids and their mixtures. *Langmuir* 26:18281–18292.

Dahle, H., Garsho, F., Madsen, M., Birkeland, N. (2008). Microbial community structure analysis of produced water from a high-temperature North Sea oil-field. *Antonie van Leeuwenhoek,* 93:37-49.

Davis, J. (1967). Petroleum Microbiology. Amsterdam New York NY. Elsevier Publishing Company.

Davis, J., Updegraff D. (1954). Microbiology in the petroleum industry. *Bacteriol. Revs.* 18:215-238.

D'Hondt, S., Rutherford, S., Spivack, A. (2002). Metabolic activity of subsurface life in deep-sea sediments. *Science,* 295:2067-2070.

Ferrer, M., Beloqui, A., Timmis, K., Golyshin, P. (2009). Metagenomics for mining new genetic resources of microbial communities. *J. Mol. Microbiol. Biotechnol.* 16:109-112.

Flores, F., Torres, L., Galindo, E. (1994). Effect of the dissolved oxygen tension during fermentation on the production and quality of xanthan gum. *J. Biotechnol.* 34: 165-173.

Grabowski, A., Nercessian, O., Fayolle, F., Blanchet, D., Jeanthon, C. (2005). Microbial diversity in production waters of a low-temperature biodegraded oil reservoir. *FEMS Microbiol. Ecol.* 54:427-443.

Hall, C., Tharakan, P., Hallock, J., Cleveland, C., Jefferson, M. (2003). Hydrocarbons and the evolution of human culture. *Nature*, 426:318-322.

Hannote, M., Flores, F., Torres, L., Galindo, E. (1991). Apparent yield stress estimation in xanthan gum solutions and fermentation broths by a low-cost viscometer. *Chem. Eng. J.* 45:B49-B56.

Herbert, R. (1986). The ecology and physiology of psychrophilic microorganisms. *In*: Herbert, R. and Codd, G., (Eds), Microbes in Extreme Environments, Academic Press Inc., New York, pp. 1–23.

Hitzman, D. (1962). Microbiological secondary recovery US Patent 3,032.472.

Jenney, F., Adams, M. (2008). Hydrogenases of the model hyperthermophiles. *Ann. N.Y. Acad. Sci.* 1125:252-266.

John, A.H., Bujalski, W., Nienow, A.W., Sanchez, A., Torres, L., Galindo, E. (1995). Studies of an independently-driven, dual impeller protofermenter with and without a draft tube: power and hold-up. Transactions of the Institution of Chemical Engineers (Part A). 73:535-541.

Lappin-Scott, H., Cusack, F., Costerton J.W. (1988). Nutrient resuscitation and growth of starved cells in sandstone cores: a novel approach to enhanced oil recovery. *Appl. Environ. Microbiol.* 54:1373-1382.

Lazar, I. (1991). MEOR field trials carried out over the world during the last 35 years. *In*: E.C. Donaldson (Ed), Microbial enhancement of oil recovery-recent advances. Elsevier Science Publisher, pp. 485-530.

Lazar, I., Petrisor, I.G., Yen, T.F. (2007). Microbial enhanced oil recovery (MEOR). *Petroleum Sci. Technol.* 25:1353-1366.

Levaro, D. (2011). Interaction between a biopolymer and surfactants, including the rhamnolipid produced by *Pseudomonas aeruginosa.* Bachelor´s thesis in Biotechnology. UPIBI-IPN. Mexico (In Spanish).

Li, H., Yang, S.Z., Mu, B.Z., Rong, Z.F., Zhang, J. (2006). Molecular analysis of the bacterial community in a continental high-temperature and water-flooded petroleum reservoir. *FEMS Microbiol. Lett.* 257:92-98.

Magot, M., Ollivier, B., Patel B. (2000). Microbiology of petroleum reservoirs. *Antonie van Leeuwenhoek*, 77:103–116.

Martinez, I.K. (2011). Interfacial tension between three Mexican crudes and humic acid solutions. Bachelor´s thesis in Biotechnology. UPIBI-IPN. Mexico (In Spanish).

Maudgalya, S., Knapp, R, McInerney, M. (2007). Microbial enhanced oil recovery technologies a review of the past, present and future SPE Production and Operations Symposium Oklahoma City OK.

Moses, V. (1991). "MEOR In The Field: Why So Little?", Microbial Enhancement of Oil Recovery-Recent Advances, Elsevier Science Publishing Company, Inc., New York City, pp. 21-28.

Ollivier, B., Magot, M. (2005). Petroleum Microbiology. Washington DC. ASM Press.

Orphan, V., Taylor, L., Hafenbrandl, D., Delong, E. (2000). Culture-dependent and culture-independent characterization of microbial assemblages associated with high-temperature petroleum reservoirs. *Appl. Environ. Microbiol.* 66:700-711.

Øvreås, L. (2000). Population and community level approaches for analyzing microbial diversity in natural environments. *Ecol. Lett.* 3:236-251.

Premuzic, E., Lin, M. (1996). Process for producing modified microorganisms for oil treatment at high temperatures, pressures and salinity US Patent, 492,828.

PEMEX (2010) Internet resources. Consulted 2010. www.pemex.gob.mx.

Ramirez, V., Rodriguez-Valadez, F., Torres, L.G. (2011). Do humic acid´s chemical characteristics determine their surface tension behavior and pseudo-CMC values in solution? *J. Colloid Surf.* Submitted.

Sanchez, A., Martinez, A., Torres, L., Galindo, E. (1992). Power consumption of three impeller combinations in mixing xanthan fermentation broths. *Process Biochem.* 27:351-365.

Schiraldi, C., Giuliano, M., DeRosa, M. (2002). Perspectives on biotechnological applications of archaea. *Archaea*, 1:75–86.

Stetter, K., Huber, R. (1999). The role of hyperthermophilic prokaryotes in oil fields. In: Proceedings of the 8th International Symposium on Microbial Ecology Bell CR, Brylinsky M, Johnson-Green P (Ed) Atlantic Canada Society for Microbial Ecology, Halifax, Canada.

Takahata, Y., Nishijima, M., Hoaki, T., Maruyama, T. (2000). Distribution and physiological characteristics of hyperthermophiles in the Kubiki oil reservoir in Niigata, Japan. *Appl. Environ. Microbiol.* 66:73-79.

Torres, L., Galindo, E., Brito, E., Choplin, L. (1991). Viscous behavior of xanthan aqueous solutions from a variant strain of *Xanthomonas campestris. J. Ferment. Bioeng.* 75:58-64.

Torres, L.G., Nienow, A.W., Sánchez, A., Galindo E. (1992). The characterization of a viscoelasticity parameter and other rheological properties of various xanthan gum fermentation broths and solutions. *Bioprocess Eng.* 9:231-237.

Torres, L.G., Flores, F., Galindo, E. (1995). Apparent yield stress of xanthan solutions and broths. *Bioeng. Process*, 12 41-46.

Torres, L.G., Iturbe, R., Snowden, M.J., Leharne, S.A. (2007a). Preparation of o/w emulsions stabilized with solid particles and their characterization through oscillatory rheology and other measurements. *Colloids Surf. A. Physicochemical and Engineering Aspects.* 302:439-448.

Torres, L., Iturbe, R., Choudhry, B., Leharne, S. (2007b). Can Pickering emulsion formation aid the removal of creosote DNAPL from porous media? *Chemosphere*, 71:123-132.

Torres, L.G, Muñoz, A., Avendaño, J.R, Leharne, S.A. (2011a). Viscosity, surface tension and other physical-chemical characteristics of three crude oils arising from Chicontepec, Mexico. *J. Petroleum Sci. Technol.* In press.

Torres, L.G., Moctezuma, A., Avendaño, J.R., Muñoz, A., Gracida, J. (2011b). Comparison of bio- and synthetic surfactants for EOR. *J. Petroleum Sci. Eng.* 76: 6-11.

Torres, L.G., Ramirez, O., Rojas-Avelizapa, N., Muñoz A. (2011c). Effect of microbial metabolites production over the surface tension of Mexican brine/crude after a fermentation process. *J. Petroleum Sci. Technol.* Submitted.

Torsvik, V., Øvreås, L., Thingstad, T. (2002). Prokaryotic diversity magnitude, dynamics and controlling factors. *Science*, 296:1064-1066.

Updegraff, D. (1983). Plugging and penetration of petroleum reservoir by microorganisms In: Proceedings of the International Conference on Microbial Enhancement of Oil Recovery, 80/85

Van Hamme, J., Singh, A., Ward, O. (2003). Recent Advances in Petroleum Microbiology. *Microbiol. Mol. Biol. Revs*. 67:503-549.

Von Der Weid, I., Korenblum, E., Jurelevicius, D., Rosado, A., Dino, R. (2008). Molecular diversity of bacterial communities from subseafloor rock. *J. Microbiol. Biotechnol*. 18:5–14.

Voordouw, G., Armstrong, S.M., Reimer, M.F., Fouts, B., Telang, A.J., Shen, Y., Gevertz, D. (1996). Characterization of 16S rRNA genes from oil field microbial communities indicates the presence of a variety of sulfate-reducing, fermentative, and sulfide-oxidizing bacteria. *Appl. Environ. Microbiol*. 62:1623-1629.

Watanabe, K., Kodama, Y., Hamamura, N., Kaku, N. (2002). Diversity, abundance and activity of archeal populations in oil-contaminated groundwater accumulated at the bottom of a underground crude oil storage cavity. *Appl. Environ. Microbiol.* 68:3899-3907.

Whitman, W.B., Bowen, T.L., Boone, D.R. (2006). The methanogenic bacteria. *In*: Dworkin, M., Falkow, S., Rosenberg, E., Schleifer, K.H. and Stackebrandt, E. (Eds) The Prokaryotes. A Handbook on the Biology of Bacteria: Ecophysiology and Biochemistry, vol. 3, pp. 165–207. New York: Springer.

Yamane, K., Maki, H., Nakayama, T., Nakajima, T., Nomura, N., Uchiyama, H., (2008). Diversity and similarity of microbial communities in petroleum crude oils produced in Asia. *Biosci. Biotechnol. Biochem. 72:*2831-2839.

Yoshida, N., Yagi, K., Sato, D., Watanabe, N., Kuroishi, T., Nishimoto, K. (2005). Bacterial communities in petroleum oil in stockpiles. *J. Biosci. Bioeng.* 99:143-149.

Zhang, X., Xiang, T. (2010). Review on Microbial Enhanced Oil Recovery Technology and Development in China. *Int. J. Petroleum Sci. Technol*. 4: 61-80.

In: Biotechnology
Editors: L.G. Torres Bustillos and I. Garcia Peña
ISBN 978-1-62081-071-2

Chapter 15

MICROBIAL GROWTH MODELING AND ANALYSIS

Edgar Salgado-Manjarrez*, Juan S. Aranda-Barradas and Inés García-Peña

Bioprocess and Bioengineering departments, UPIBI IPN, Mexico

INTRODUCTION

Human beings have exploited the catalytic abilities of microorganisms for millennia, although it was not consciously until the discovery of microorganisms and enzymes. Since then, the number of products and services using microorganisms has multiplied geometrically. It is quite clear to us now that cells possess tremendous catalytic activities, as life can thrive in almost every corner of our planet, often forming *consortia* for degrading substances that hardly could be considered as substrates for life. These amazing capabilities make microorganisms and other cultured cells powerful catalyst that can be exploited just as they are found in nature or even improved by modifying their physiology. At the middle of the 20th century, changes in the physiology of cells were produced by mutation, a cluttered chaotic process that required lots of experimentation; latter on, by the 1970's, recombinant DNA technologies gave birth to the so called "genetic engineering", by which specific genes are introduced in a cell, modifying more specifically its physiology. However, the last 30 years have taught us that, in many cases, a "cut", "mix" and "paste" logic is not very reliable for modifying the catalytic properties of cells: there exist several control systems that often make the results unpredictable. The genomics, proteomics, metabonomics and most –omics are advancing our understanding of cells physiology at different degrees and levels. From an engineering perspective, being able to truly reengineer cell systems, predicting the results of genetic manipulations, is almost the "Holy Grail" of biotechnology.

Cell growth is an extremely complex process: thousands of reactions among thousands of substances happen simultaneously, catalyzed by thousands of enzymes, everything orchestrated by thousands of genes and other genetic control systems. Nothing yet invented by humanity reaches such complexity and yet, every year we get closer to be able to truly

* Email: esalgado@ipn.mx

reengineer cells or design novel cell systems. The complexity of the task involves the conjunction of several disciplines and the manipulation of huge amounts of data. Additionally to this biological knowledge, important concepts from reaction engineering, computer science and chemical kinetics are needed to develop a quantitative approach to cell metabolism, in order to exploit biological systems in rational biomolecule production processes. Systems Biology is considered as the part of Bioinformatics that using a systems approach, intends to unveil the interactions in biochemical systems; to some extent, it is hardly differentiated from Computational Biology, an area that uses specifically mathematical modeling and simulation to understand complex biological systems. As a part of this areas, Metabolic Engineering uses a computational, systems biology approach for manipulating biochemical and metabolic systems in a rational way through an estimation of fluxes inside the cells, accomplished by means of modeling and, more specifically, Metabolic Flux Analysis also known as Metabolic Flux Balancing. It is through this formalism that, presumably, we will in the future be able to reengineer cells or create novel cell systems.

CELL KINETICS AND DYNAMICS OF CELL CULTURES

Modeling of reacting biological systems requires a translation of basically two aspects of the system characteristics to a mathematical structure:

a) The system's changes in time considering external factors, such as homogeneity of the reaction medium, transport phenomena, pH and temperature. This behavior is depicted by the dynamical description of cell cultures. In a mathematical way, cellular culture dynamics is expressed by time-dependent differential equations that consider variations in the main features of the system (e.g. chemical species concentrations, temperature, pH) as produced by inputs, outputs, chemical reactions or any external factor to the reacting system.
b) The changes in the chemical or biochemical species that occur in the reacting system which are represented by a kinetics theory applicable to the involved chemical or biochemical reactions. In biological systems such as cell cultures aiming at producing some economically valuable metabolite, cell kinetics should comprise the suitable biological knowledge for the production process. So, kinetics of biological reactions can include only macroscopic concentrations and lumped parameters or, in an increasingly detailed depiction, data from intracellular compartments, metabolomics, reaction thermodynamics, genomics or proteomics.

The roughest form of cellular growth and production kinetics description consists in the so-called unstructured models that express simplified relations between reactive concentrations and kinetic rates. Introducing comprehensive knowledge on the main biochemical pathways allows for the analysis of metabolic fluxes that results in improved assessments of process yields. In general, the mathematical complexity of kinetic models is proportional to the description level they propose.

CELL GROWTH STOICHIOMETRY

Stoichiometry deals with the balancing of chemical reactions. Cell reactions are not different from any other reaction in this aspect, and then the methods used are equally applicable. The cell complexity arises from the number of reactions. In order to facilitate comprehension, it is both, useful and illustrating to treat the cell as the product and catalyzer of a very simple reaction, namely,

Substrates → Cell mass + Excreted metabolic products + Heat

Taking into account that C, H, O and N represent most of cell mass, cell requires that these atoms be present in the substrates. The reaction can be written considering abstract carbon and nitrogen sources and a metabolic product, as

$$\eta_S\, C_{a_S} H_{b_S} O_{c_S} N_{d_S} + \eta_O\, O_2 + \eta_N\, C_{a_N} H_{b_N} O_{c_N} N_{d_N} \rightarrow$$
$$\eta_X\, CH_{b_X} O_{c_X} N_{d_X} + \eta_P\, C_{a_P} H_{b_P} O_{c_P} N_{d_P} + \eta_C\, CO_2 + \eta_A\, H_2O \quad (1)$$

where it has been assumed that growth is aerobic, hence the O_2, and CO_2 is produced, as well as H_2O as part of cell metabolism. In this equation η_j represent the stoichiometric coefficients, the number of moles of each substance.

All elements in a reaction must balance, for the carbon for instance,

$$\eta S \cdot aS + \eta N \cdot aN = \eta X \cdot 1 + \eta P \cdot aP + C \cdot 1$$

or as a summation,

$$\sum_j (\eta_j \cdot a_j) = 0 \quad . \quad (2)$$

which can be written in matrix form as:

$$\sum_j (\eta_j \cdot a_j) = \begin{bmatrix} a_S & a_N & 1 & a_P & 1 \end{bmatrix} \begin{bmatrix} -\eta_S \\ -\eta_N \\ \eta_X \\ \eta_P \\ \eta_C \end{bmatrix} = 0 \quad . \quad (3)$$

and considering every element in the reaction,

$$\begin{bmatrix} a_S & 0 & a_N & 1 & a_P & 1 & 0 \\ b_S & 0 & b_N & b_X & b_P & 0 & 2 \\ c_S & 2 & c_N & c_X & c_P & 2 & 1 \\ d_S & 0 & d_N & d_X & d_P & 0 & 0 \end{bmatrix} \cdot \begin{bmatrix} -\eta_S \\ -\eta_O \\ -\eta_N \\ \eta_X \\ \eta_P \\ \eta_C \\ \eta_A \end{bmatrix} = \begin{bmatrix} 0 \\ 0 \\ 0 \\ 0 \\ 0 \\ 0 \\ 0 \end{bmatrix} . \quad (4)$$

which can be written in compact form as

$$\mathbf{E}\cdot\boldsymbol{\eta}=\mathbf{0}\,. \tag{5}$$

where **E** is the elemental matrix with *k* rows and *n* columns, for *k* elements and *n* substances. $\boldsymbol{\eta}$ is the stoichiometric coefficients vector of size *n*. Notice that the reactants' or substrates' stoichiometric coefficients have a minus sign, since these substances are consumed in the chemical reaction.

In order to get a solution for the system of equations, such as equation (5) row number in **E** must be equal to the vector size $\boldsymbol{\eta}$. Otherwise, if there are more rows than columns and the system is underdetermined, or overdetermined, if there are more rows than columns. In terms of elements and substances, the degrees of freedom are $F = n - k$. For calculation purposes, the system can be re-written in terms of measured and calculated sub-matrices,

$$\mathbf{E}\cdot\boldsymbol{\eta}=\begin{bmatrix}\mathbf{E}^{c} & \mathbf{E}^{m}\end{bmatrix}\cdot\begin{bmatrix}\boldsymbol{\eta}^{c}\\ \boldsymbol{\eta}^{m}\end{bmatrix}=\mathbf{0} \tag{6}$$

$$\mathbf{E}^{c}\cdot\boldsymbol{\eta}^{c}+\mathbf{E}^{m}\cdot\boldsymbol{\eta}^{m}=\mathbf{0}\,. \tag{7}$$

then the system can be solved as:

$$\boldsymbol{\eta}^{c}=-\left(\mathbf{E}^{c}\right)^{-1}\cdot\mathbf{E}^{m}\cdot\boldsymbol{\eta}^{m}\,. \tag{8}$$

It must be noted that $\mathbf{E}^{c}$ has to be invertible, so it must be square, its determinant must be different from zero and must be of full rank. If these conditions are fulfilled the system is said to be *observable*. In this way unknown coefficients in the reaction can be calculated from some known ones.

This previous stoichiometric analysis can be extended to a kinetic approach in the form:

$$\mathbf{r}^{c}=-\left(\mathbf{E}^{c}\right)^{-1}\cdot\mathbf{E}^{m}\cdot\mathbf{r}^{m}\,. \tag{9}$$

where the $\boldsymbol{\eta}$ vectors of stoichiometric coefficients are replaced by volumetric rate **r** vectors. This comprehensive approach allows for the estimation of valuable metabolites in production systems through the modeling of kinetic behavior by a stoichiometric analysis. Barrera-Martínez et al. (2011) settled a stoichiometric black box model for the assessment of bioethanol production in *Saccharomyces cerevisiae* fed-batch cultures, as illustrated in Figure 1.

By measuring the biomass accumulation, the glucose consumption and the oxygen supply volumetric rates, the estimation of produced ethanol was possible from a simplified stoichiometric description of the production system. As it can be observed, however, that the estimation error has some tendency and, in general doesn't follow quite well the experimental values. A better estimation can be obtained with a metabolic model, as explained latter.

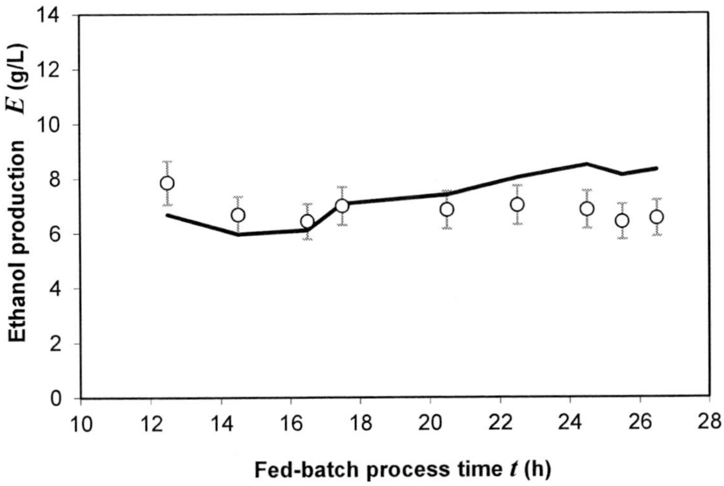

Figure 1. Ethanol production by *Saccharomyses cerevisiae* in a fed-batch system: experimental data (markers) and estimated concentrations by means of a stoichiometric black box analysis of cell growth (continuous line).

NON-STRUCTURED MODELING

Stoichiometry does not give information about the rate at which a reaction takes place. In fact, there is no way to predict this information; it has to be determined experimentally. An apparatus commonly use for both, kinetic studies and production scale is the tank reactor as depicted in Figure 2.

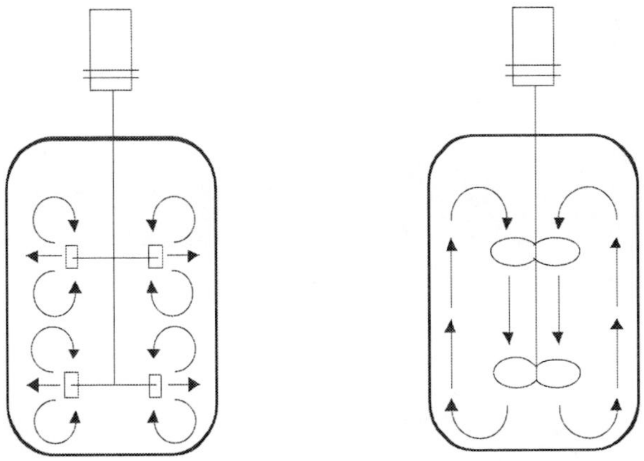

Figure 2. Classical stirred tank bioreactors.

A balance carried out over the solutes in mass or moles, $\mathbf{y}$, in such a system reads,

$$\frac{d(V\mathbf{y})}{dt} = \mathbf{R} + F_0\mathbf{y}_0 - F\mathbf{y} \tag{10}$$

here **R** stands for the rate of change at which the substance is consumed or produced by the reaction. A similar balance can be written for the *mass* in the system, from which it can be concluded that as long as the density changes are negligible,

$$\frac{dV}{dt} \approx F_0 - F \tag{11}$$

as no mass is produced nor consumed in non-nuclear reactions. Combining the previous equations the final equation is as follows,

$$\frac{d\mathbf{y}}{dt} = \mathbf{r} + D(\mathbf{y}_0 - \mathbf{y}) \tag{12}$$

where **r** is the volumetric rate of change and $D = F_0/V$ is the dilution rate. This rather simple equation expresses the dynamics of solutes in the system: how the concentration of substances in a reactor changes depending on their input and output concentrations, the feed rate and the reaction rate. As such, the balance equation can be used for determining reaction rates from experimental dynamic information but it can also be used for the simulation of the behavior of the system under different conditions given some kinetic information or models.

Some simple models have been proposed that fit rather well most of the observed behaviors. For instance, the classic model of Monod,

$$\mu = \frac{r_X}{X} = \mu_{max} \frac{S}{S + K_S}. \tag{13}$$

expresses the *specific growth rate* in function of a substrate concentration. Substrate consumption can be accounted by a model like the one of Pirt,

$$\sigma = \frac{r_S}{X} = \frac{\mu}{Y_{XS}} + m. \tag{14}$$

where σ stands for the *specific consumption rate*, Y_{XS} is the biomass yield on the substrate and m is a coefficient for accounting of all the substrate that is not used for maintenance. A similar model can also be used for the products,

$$\pi = \frac{r_P}{X} = \alpha\mu + \beta \tag{15}$$

originally proposed by Luedeking and Piret. In this equation α can be interpreted as product yield or proportionality between production and growth; and β stands for the non-growth associated production.

As these kind of models assume that biomass is a "complex" substance, they are considered non-structured. In spite of this extreme simplification, non-structured models have been used successfully for describing a number of metabolite production systems (Noorman et al. 1993). Production of biomolecules issued from the energetic or primary metabolism, such as ethanol (Aranda and Salgado, 2002; Barrera-Martínez et al., 2011), polyols (Delia et al., 1999; Aranda et al., 2000), simple organic acids or amino acids are classically depicted through unstructured models. Processes aiming at producing secondary metabolites, such as enzymes (Martínez-Trujillo et al., 2008; Martínez-Trujillo et al., 2009) or antibiotics, can be equally described with unstructured models. Wastewater treatment processes are also successfully modeled by this unstructured approach (Rivera-Salvador et al., 2009). As a particular example, Aranda et al., (2000) used a non-structured model, based on the Monod kinetic model considering oxygen as the limiting substrate and couplet to a simple mass transfer model to demonstrate how *Candida parapsilosis* always produces xylitol, but this alcohol only accumulates under oxygen limitation. What is more, the model suggests that there exist an optimal oxygen transfer rate for xylitol production, as depicted in Figure 3.

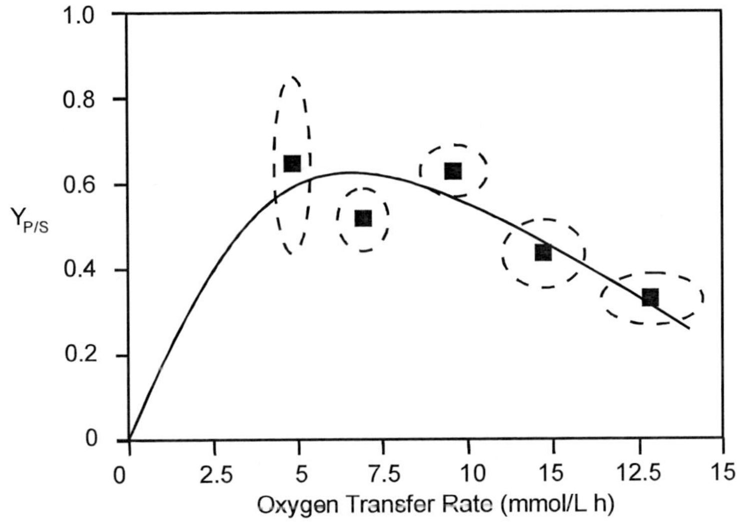

Figure 3. Xylitol production yields by *Candida parapsilosis* as a function of the oxygen transfer rate in the culture.

Indeed, non-structured models have been quite useful and its usefulness will surely continue for some years to come. However, since the 1960's a better, more realistic, approach started to develop, one that through reactions and/or substances, introduced some structure to the biomass.

Metabolic Modeling

In bioreactors with cell cultures, catalytic reactions take place mostly inside the cells or in their surfaces. In the external medium, reactions are not usually catalyzed, what is observed in the medium is the net flow of substances from and into the cells, so, instead of considering the cells as one complex substance, it is more realistic to consider it as a *biotic* phase suspended in an *abiotic* phase or medium.

A mass balance on the abiotic phase reads,

$$\frac{d(V_A \mathbf{y}_A)}{dt} = F_{A,0}\,\mathbf{y}_{A,0} - F_A\,\mathbf{y}_A + \mathbf{R}_I \quad . \tag{16}$$

notice that concentrations are relative to the abiotic phase which volume relates to the total volume by a biotic phase volume fraction, φ, so $V_A = (1\text{-}\varphi)\cdot V$, and:

$$\frac{d([1-\varphi]\,\mathbf{y}_A)}{dt} = D([1-\varphi_0]\,\mathbf{y}_{A,0} - [1-\varphi]\,\mathbf{y}_A) + \mathbf{r}_I \quad . \tag{17}$$

as $[1-\varphi]\,\mathbf{y}_A = \mathbf{y}$ so the mass balance reduces to:

$$\frac{d\mathbf{y}}{dt} = D(\mathbf{y}_0 - \mathbf{y}) + \mathbf{r}_I \,. \tag{18}$$

that is identical to equation (12) except that here the kinetic term has become a transport rate, not a reaction rate. Clearly, this truly matches to reality. It should be also noted that in the previous equation volumetric exchange rates are referred to total volume, V, the sum of both volume phases.

The mass balance for the biotic phase is:

$$\frac{d(V_B \mathbf{y}_B)}{dt} = F_{B,0}\,\mathbf{y}_{B,0} - F_B\,\mathbf{y}_B + (\mathbf{R}_1 + \mathbf{R}_2 + \ldots + \mathbf{R}_m) - \mathbf{R}_I \,. \tag{19}$$

where the rate terms in parenthesis indicate the metabolic reactions that contribute to the production or consumption of a particular metabolite. Taking into account the biotic phase volume:

$$\frac{d(\varphi\,\mathbf{y}_B)}{dt} = D(\varphi_0\,\mathbf{y}_{B,0} - \varphi\,\mathbf{y}_B) + \sum_j \mathbf{r}_j - \mathbf{r}_I \quad . \tag{20}$$

as in the previous case, concentrations can be referred to the total volume,

$$\frac{d\dot{\mathbf{y}}}{dt} = D(\dot{\mathbf{y}}_0 - \dot{\mathbf{y}}) + \sum_j \mathbf{r}_j - \mathbf{r}_I \,. \tag{21}$$

the upper dot is to distinguish them from abiotic phase concentrations. Notice, however, that volumetric exchange rates and volumetric reaction rates are all referred to total volume.

Production or consumption rates of a given substance in a reaction can be expressed as the product of a stoichiometric coefficient and the reaction rate, which is usually equal to the reaction rate of a reference substance in the reaction, which has a unity stoichiometric coefficient. According to this,

$$r_i = \sum_j \alpha_{ji} r_j \tag{22}$$

so if the whole set of metabolic reactions are, in matrix form, for every substrate, metabolic product and biomass constituent,

$$\mathbf{r}_{S,i} = -\sum_j \alpha_{ji}\mathbf{r}_j = \mathbf{A}^T \cdot \mathbf{r}_M \, . \tag{23}$$

$$\mathbf{r}_{P,i} = \sum_j \beta_{ji}\mathbf{r}_j = \mathbf{B}^T \cdot \mathbf{r}_M \, . \tag{24}$$

$$\mathbf{r}_{X,i} = -\sum_j \gamma_{ji}\mathbf{r}_j = \mathbf{\Gamma}^T \cdot \mathbf{r}_M \, . \tag{25}$$

Matrices **A**, **B**, and **Γ** have *m* rows and *N*, *M* and *L* columns, respectively. Also, in these matrices the rows correspond to the reactions and the columns to the substances, this is the reason to transpose when calculating the rates for substances. The whole set of balanced metabolic reactions can be written as,

$$\mathbf{A}\cdot\mathbf{S} + \mathbf{B}\cdot\mathbf{P} + \mathbf{\Gamma}\cdot\mathbf{X} = \mathbf{0} \, . \tag{26}$$

where **S**, **P**, and **X**, stand for the substrate, metabolic products and biomass constituents (not their rates). Taking this into account,

$$\sum_{j=1}^{m} \mathbf{r}_j = \begin{bmatrix} \mathbf{A}^T \\ \mathbf{B}^T \\ \mathbf{\Gamma}^T \end{bmatrix} \cdot \mathbf{r}_M = \mathbf{Y}\cdot\mathbf{r}_M \, . \tag{27}$$

where Y is the stoichiometric matrix, finally:

$$\frac{d\mathbf{y}}{dt} = D(\mathbf{y}_0 - \mathbf{y}) + \mathbf{Y}\cdot\mathbf{r}_M - \mathbf{r}_I \, . \tag{28}$$

the sum of all substances in this reaction is equal to the mass of the biotic phase, i.e, the biomass,

$$\frac{dX}{dt} = D(X_0 - X) + \sum_{i=1}^{n}\sum_{j=1}^{m} r_{ij} - \sum_{i=1}^{n} r_I \, . \tag{29}$$

or,

$$\frac{1}{X}\frac{dX}{dt} = D\left(\frac{X_0}{X} - 1\right) + \sum_{i=1}^{n}\sum_{j=1}^{m} q_{ij} - \sum_{i=1}^{n} q_I \,. \tag{30}$$

where rates q are *specific*. By comparison of equations 29, 30 and 12, it becomes clear that,

$$\mu = \sum_{i=1}^{n}\sum_{j=1}^{m} q_{ij} - \sum_{i=1}^{n} q_I \,. \tag{31}$$

i.e., the specific growth rate is the result of the sum of exchange rates, increasing the biomass when entering, decreasing it when exiting; plus the net conversion rates of each substance inside the cell due to every metabolic reaction.

As the reference to the total volume might be confusing, it is more convenient to rewrite this equation with concentrations and rates referred to cell mass, as follows: Biomass constituents are $\varphi \cdot \mathbf{y}_B = \varphi \cdot \rho_X \cdot \mathbf{x} = X \cdot \mathbf{x}$, where ρ_X is the biomass density and $\mathbf{x}$ are the concentrations referred to total biomass.

$$\frac{d(X\,\mathbf{x})}{dt} = D(X_0\,\mathbf{x}_0 - X\,\mathbf{x}) + \mathbf{Y} \cdot \mathbf{r}_M - \mathbf{r}_I \,. \tag{32}$$

which leads to:

$$\frac{d\mathbf{x}}{dt} = D\frac{X_0}{X}(\mathbf{x}_0 - \mathbf{x}) + \mathbf{Y} \cdot \mathbf{q}_M - \mathbf{q}_I - \mathbf{x}\left(\sum_{i=1}^{n}\sum_{j=1}^{m} q_{ij} - \sum_{i=1}^{n} q_I\right). \tag{33}$$

or,

$$\frac{d\mathbf{x}}{dt} = D\frac{X_0}{X}(\mathbf{x}_0 - \mathbf{x}) + \mathbf{Y} \cdot \mathbf{q}_M - \mathbf{q}_I - \mu\,\mathbf{x} \,. \tag{34}$$

This equation states that biotic phase composition changes in response to feed and discharges; metabolic reactions; exchange rates with the abiotic phase and concentrations are diluted proportionally to growth rate. If desired, a more succinct form of this equation can be written by grouping the metabolic reactions and exchange rates,

$$\mathbf{r} = \begin{bmatrix} \mathbf{r}_M \\ \mathbf{r}_I \end{bmatrix}. \tag{35}$$

and grouping all stoichiometric and exchange coefficients in a metabolic matrix (Noorman et al, 1993),

$$\mathbf{M} = \begin{bmatrix} \mathbf{Y} \begin{matrix} -\mathbf{I} \\ \mathbf{0} \end{matrix} \end{bmatrix}. \tag{36}$$

so,

$$\frac{d\mathbf{x}}{dt} = D\frac{X_0}{X}(\mathbf{x}_0 - \mathbf{x}) + \mathbf{M}\cdot\mathbf{q} - \mu\cdot\mathbf{x}. \tag{37}$$

or by considering the three sets for substrates, products and biomass constituents,

$$\frac{d\mathbf{x}}{dt} = D\frac{X_0}{X}(\mathbf{x}_0 - \mathbf{x}) + \begin{bmatrix} \mathbf{A}^T & -\mathbf{I} & \mathbf{0} \\ \mathbf{B}^T & \mathbf{0} & -\mathbf{I} \\ \mathbf{\Gamma}^T & \mathbf{0} & \mathbf{0} \end{bmatrix}\cdot\mathbf{q} - \mu\cdot\mathbf{x}. \tag{38}$$

This equation sets the basis for manipulating the huge amounts of data involved in cell metabolism.

Steady State Metabolic Flux Analysis

Equations (18) and (34), state how the biomass composition changes during cell culture, the only difference between them is how the concentration is expressed: one uses total culture volume as reference, while the other uses biomass weight. Assuming steady state and sterile feed,

$$0 = -D\dot{\mathbf{y}} + \mathbf{Y}\cdot\mathbf{r}_M - \mathbf{r}_I. \tag{39}$$

$$0 = \mathbf{Y}\cdot\mathbf{q}_M - \mathbf{q}_I - \mu\mathbf{x}. \tag{40}$$

and,

$$\mathbf{Y}\cdot\mathbf{r}_M = D\dot{\mathbf{y}} + \mathbf{r}_I \tag{41}$$

$$\mathbf{Y}\cdot\mathbf{q}_M = \mathbf{q}_I + \mu\mathbf{x}. \tag{42}$$

which give a convenient way for estimating metabolic reaction rates or fluxes in function of exchange (or "conversion") rates and a culture or biomass dilution rate *if steady state is achieved.*

As not necessarily all exchange rates are measured it is more convenient to take the concise forms,

$$\mathbf{M} \cdot \mathbf{r} = D \cdot \mathbf{x} . \tag{43}$$

$$\mathbf{M} \cdot \mathbf{q} = \mu \cdot \mathbf{x} . \tag{44}$$

Mass fractions of metabolic intermediaries are usually quite low, accounting for ca. 5% of biomass weight so for these substances (for the sake of brevity, in the following only the version with specific rates is considered, but the reasoning holds for the version with volumetric rates),

$$\mathbf{M} \cdot \mathbf{q} = \mathbf{0} \tag{45}$$

Now, this equation is identical in form to equation (5), so the same reasoning applies for calculating some rates or fluxes in function of others,

$$\mathbf{M}^c \cdot \mathbf{q}^c + \mathbf{M}^m \cdot \mathbf{q}^m = \mathbf{0} \tag{46}$$

$$\mathbf{q}^c = -\left(\mathbf{M}^c\right)^{-1} \cdot \mathbf{M}^m \cdot \mathbf{q}^m . \tag{47}$$

This simple equation is the basis for steady-state metabolic flux analysis: the estimation of fluxes in a proposed biochemical network in function of others, usually external, fluxes (Vallino and Stephanopoulos, 1993).

If convenient, the metabolic matrix can be expressed as follows,

$$\begin{bmatrix} \mathbf{A}^T & -\mathbf{I} & \mathbf{0} \\ \mathbf{B}^T & \mathbf{0} & -\mathbf{I} \\ \mathbf{\Gamma}^T & \mathbf{0} & \mathbf{0} \end{bmatrix} \cdot \begin{bmatrix} \mathbf{q}_M \\ \mathbf{q}_S \\ \mathbf{q}_P \end{bmatrix} = \begin{bmatrix} \mathbf{A}^T \\ \mathbf{B}^T \\ \mathbf{\Gamma}^T \end{bmatrix} \cdot \mathbf{q}_M + \begin{bmatrix} -\mathbf{I} & \mathbf{0} \\ \mathbf{0} & -\mathbf{I} \\ \mathbf{0} & \mathbf{0} \end{bmatrix} \cdot \begin{bmatrix} \mathbf{q}_S \\ \mathbf{q}_P \end{bmatrix} \tag{48}$$

or,

$$\mathbf{Y} \cdot \mathbf{q}_M = \mathbf{q}_I \tag{49}$$

$$\begin{bmatrix} \mathbf{A}^T \\ \mathbf{B}^T \\ \mathbf{\Gamma}^T \end{bmatrix} \cdot \mathbf{q}_M = \begin{bmatrix} \mathbf{q}_S \\ \mathbf{q}_P \\ \mathbf{0} \end{bmatrix} . \tag{50}$$

Which can be written as (Nielsen and Villadsen,1994):

$$\begin{bmatrix} \mathbf{q}_I^m \\ \mathbf{q}_I^c \\ \mathbf{0} \end{bmatrix} = \begin{bmatrix} \mathbf{Y}_1 & \mathbf{Y}_2 \\ \mathbf{Y}_3 & \mathbf{Y}_4 \\ \mathbf{Y}_5 & \mathbf{Y}_6 \end{bmatrix} \cdot \begin{bmatrix} \mathbf{q}_M^m \\ \mathbf{q}_M^c \end{bmatrix} \tag{51}$$

where some elements of $\mathbf{q}_S$ and $\mathbf{q}_P$ are in $\mathbf{q}_I^m$ and others in $\mathbf{q}_I^c$. In $\mathbf{q}_M^m$ are found a minimum set of rates that can determine the system while q_M^m and $\mathbf{q}_M^c$ only reflects the mentioned arrangement but does not imply that any of $\mathbf{q}_M$ is actually measured. Following this rearrangement, the solution to vectors $\mathbf{q}_I^m$, $\mathbf{q}_M^c$ and $\mathbf{q}_I^c$ are:

$$\begin{aligned} \mathbf{q}_M^m &= \left(\mathbf{Y}_1 - \mathbf{Y}_2\mathbf{Y}_6^{-1}\mathbf{Y}_5\right) \cdot \mathbf{q}_I^m \\ \mathbf{q}_M^c &= -\mathbf{Y}_6^{-1}\mathbf{Y}_5\left(\mathbf{Y}_1 - \mathbf{Y}_2\mathbf{Y}_6^{-1}\mathbf{Y}_5\right)^{-1} \cdot \mathbf{q}_I^m \\ \mathbf{q}_I^c &= \left(\mathbf{Y}_3 - \mathbf{Y}_4\mathbf{Y}_6^{-1}\mathbf{Y}_5\right) \cdot \left(\mathbf{Y}_1 - \mathbf{Y}_2\mathbf{Y}_6^{-1}\mathbf{Y}_5\right)^{-1} \cdot \mathbf{q}_I^m \end{aligned} \tag{52}$$

It is clear that,

$$\mathbf{q}^c = \begin{pmatrix} \mathbf{q}_M^m \\ \mathbf{q}_M^c \\ \mathbf{q}_I^c \end{pmatrix} \tag{53}$$

as in the equivalent equation (8).

Table 1. Metabolic reactions for a *Saccharomyces cerevisiae* simplified model

Flux	Condensed reaction
r_1	Glucose + $2H_3PO_4$ + 2ADP + $2NAD^+$ → 2Pyruvate + $2H_2O$ + 2ATP + 2NADH
r_2	Glucose + Pyruvate + Acetyl-CoA + NH_3 + 27.5ATP + $24.1H_2O$ → Biomass + CoA + 27.5ADP + $27.5H_3PO_4$
r_3	Pyruvate + CoA + NAD + → Acetyl-CoA + CO_2 + NADH
r_4	Acetyl-CoA + $2H_2O$ + H_3PO4 + ADP + $4NAD^+$ → $2CO_2$ + CoA + ATP + 4NADH
r_5	Acetyl-CoA + H_3PO_4 + ADP + 2NADH → Ethanol + CoA + H_2O + ATP + $2NAD^+$
r_6	3ADP + $3H_3PO4$ + NADH + 0.5O2 → 3ATP + NAD+ + $4H_2O$

It is important to mention that the previous analysis only provide us for a way of inferring some fluxes from others: the solution is mathematically correct, however, as everything comes from an assumed, albeit very possible, stoichiometry, the calculated fluxes can be considered very probable, but should not be assumed as certain. Any other plausible stoichiometry that makes the system observable, would give a set of fluxes that satisfies the constraints but the values can be different between two proposed stoichiometries even in some common paths. The analysis presented above was used by Barrera-Martínez et al.

(2011) for estimating the physiological state of yeast cultures with a simplified metabolic model (Table 1). They showed, for instance, that for increasing ethanol production rate, a high substrate inflow and a low oxygen one are required, whereas with a controlled glucose feed under aerobic conditions maximizes oxidative catabolism that is required for high biomass yields, as shown in Figure 4.

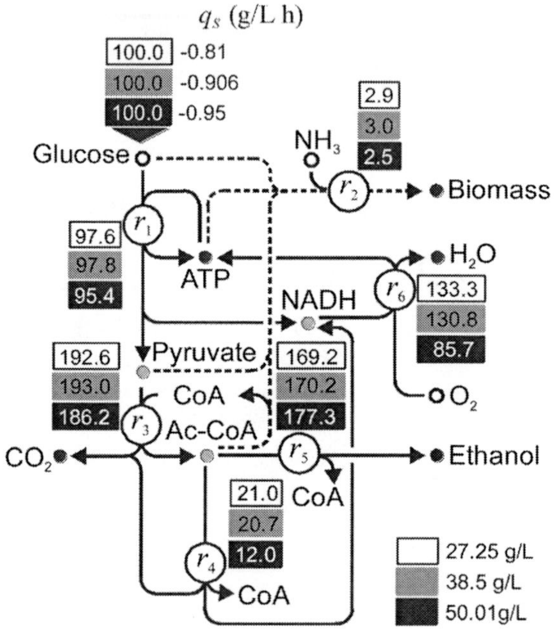

Figure 4. A minimal metabolic network for a yeast culture producing biomas and bioethanol. Metabolic fluxes for different glucose concentrations in the fed stream (27.2, 38.5 and 50.0 g/L) are referred to the glucose uptake (100%).

Overdetermined Systems

When estimations for fluxes are made with the least amount of information required to specify the system, the solution is as good as the input data. Unfortunately, measurements always have experimental error so the estimated fluxes reflect the propagation of experimental error. If more rates than necessary are measured, the redundant information can be used to obtain a "best solution" under a given policy. A least squares solution, for instance, can be obtained as follows:

$$\mathbf{q}^c = -\left(\mathbf{M}^T \cdot \mathbf{M}^c\right)^{-1} \cdot \mathbf{M}^T \cdot \mathbf{M}^m \cdot \mathbf{q}^m . \tag{54}$$

Even measurements can also be corrected from the least squares solution, by introducing the latter solution in Equation (47).

$$\left(\mathbf{M}_m - \mathbf{M}_c \cdot \left(\mathbf{M}_c^T \cdot \mathbf{M}_c\right)^{-1} \cdot \mathbf{M}_c^T \cdot \mathbf{M}_m\right) \cdot \mathbf{q}_m = 0 = \mathbf{R} \cdot \mathbf{q}_m \tag{55}$$

Here **R** is the *redundancy* matrix, and usually has not full rank. When dependent rows are eliminated a reduced, full rank set is obtained,

$$\mathbf{R}_R \cdot \mathbf{q}_m = 0 \tag{56}$$

Because of experimental errors, this equality is not satisfied in practice. If the error, $\boldsymbol{\delta}$, is substracted from these measured rates with noise, $\overline{\mathbf{q}}_m$, the true values of the measured rates should be obtained,

$$\mathbf{R}_R \cdot \left(\overline{\mathbf{q}}_m - \boldsymbol{\delta}\right) = 0\,. \tag{57}$$

from this equation the residuals, $\boldsymbol{\varepsilon}$, can be defined,

$$\mathbf{R} \cdot \overline{\mathbf{q}}_m = \mathbf{R} \cdot \boldsymbol{\delta} = \boldsymbol{\varepsilon} \neq \mathbf{0} \tag{58}$$

An estimation of the error, can be obtained as follows

$$\min_{\delta} \Phi = \boldsymbol{\delta}^T \cdot \mathbf{V}_m^{-1} \cdot \boldsymbol{\delta} \text{ subject to } \mathbf{R} \cdot \boldsymbol{\delta} = \boldsymbol{\varepsilon}\,. \tag{59}$$

here, the measurements variance is used as weight. The solution to this problem is,

$$\hat{\boldsymbol{\delta}} = \mathbf{V}_m \cdot \mathbf{R}_R^T \cdot \left(\mathbf{R}_R \cdot \mathbf{V}_m \cdot \mathbf{R}_R^T\right)^{-1} \cdot \boldsymbol{\varepsilon} \tag{60}$$

The assessment of $\hat{\delta}$ is useful for estimating the rate confidence intervals, as for the respiratory coefficient (RQ) in a biomass culture of *Saccharomyces cerevisiae* (Figure 5).

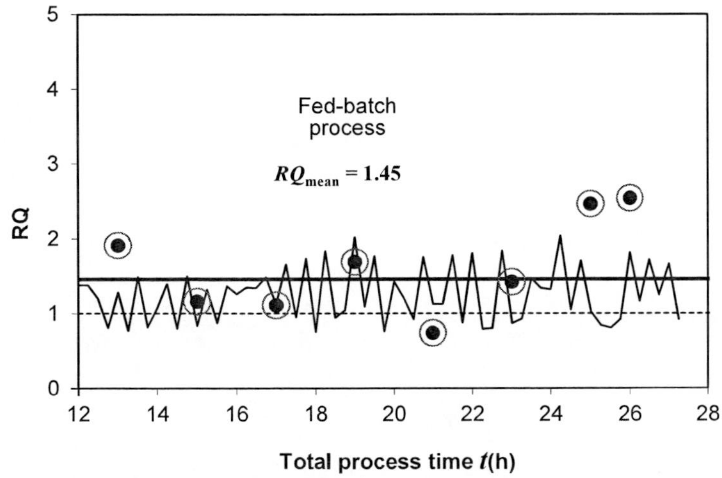

Figure 5. Estimated RQ for a yeast biomass fed-batch production process. Circles surrounding the calculated values (black markers) represent the estimated RQ confidence intervals.

UNDERDETERMINED SYSTEMS

The proposed stoichiometry is a set of constraints about what is thought the cell can do. When the measured variables are not enough to determine the system, it can yet be solved if more constraints or a suitable objective function is added. It is possible, for instance, to find the fluxes that would maximize cell or product yield.

This is carried out with linear programming methods for which it is necessary that all fluxes be positive in the equation, so the stoichiometry is rewritten splitting reversible reactions as a forward and a reverse reaction, this gives an extended system,

$$\mathbf{M}_{ex} \cdot \mathbf{q}_{ex} = \mathbf{0} \tag{61}$$

an objective function is specified and the solution found by a linear programming method, such as the simplex.

Through the use of this strategy, Aranda-Barradas et al. (2010) demonstrated that biomass yields could be estimated within an error of 12.8% of the experimental value.

Table 2. Transposed total stoichiometric matrix for the metabolic network for the xylose to xylitol bioconversion by *Candida parapsilosis*

Metabolites	Reactions													
	r_3	r_2	r_1	r_{10}	r_5	r_6	r_7	r_8	r_9	r_4	r_{11}	r_{12}	r_{13}	r_{14}
Xylose	0	−1	−1	0	0	0	0	0	0	0	0	0	0	0
O_2	0	−0.5	0	0	0	0	0	0	0	0	0	0	0	0
Xylitol	−1	0	1	0	0	0	0	0	0	0	0	0	0	0
Biomass	0	5	0	0	0	0	0	0	0	0	0	0	0	0
NH_3	0	−0.6	0	0	0	0	0	0	0	0	0	0	0	0
CO_2	0	0	0	3	0	0	0	0	0	0	0	1	0	0
H_2O	0	2	0	0	0	0	0	0	0	0	0	0	0	0
ADP	0	−3	0	0	0	0	0	1	−4	1	0	0	0	0
ATP	0	3	0	0	0	0	0	−1	4	−1	0	0	0	0
NAD^+	−1	1	0	−4	0	0	0	0	−2	0	0	0	0	0
NADH	1	−1	0	4	0	0	0	0	2	0	0	0	0	0
NADPH	0	0	−1	0	0	0	0	0	0	0	1	1	0	0
$NADP^+$	0	0	1	0	0	0	0	0	0	0	−1	−1	0	0
Erythrose-4-P	0	0	0	0	0	1	−1	0	0	0	0	0	0	0
Fructose-6-P	0	0	0	0	0	1	1	−1	0	0	−1	0	0	0
6-P-Gluconate	0	0	0	0	0	0	0	0	0	0	1	−1	0	0
Glyceraldehyde-3-P	0	0	0	0	1	−1	1	2	−1	0	0	0	0	0
Ribose-5-P	0	0	0	0	−1	0	0	0	0	0	0	0	1	0
Ribulose-5-P	0	0	0	0	0	0	0	0	0	0	0	1	−1	−1
Sedoheptulose-7-P	0	0	0	0	1	−1	0	0	0	0	0	0	0	0
Xylulose	1	0	0	0	0	0	0	0	0	−1	0	0	0	0
Xylulose-5-P	0	0	0	0	−1	0	−1	0	0	1	0	0	0	1
Pyruvate	0	0	0	−1	0	0	0	0	1	0	0	0	0	0

For instance, Aranda et al. (2010) proposed that minimizing growth would maximize xylitol production as xylose is split into these mean products. So the following objective function was proposed,

$$Min\, \mu = \sum_{i=1}^{n} \eta_i q_{M,i} - \sum_{i=1}^{n} q_I \tag{62}$$

subject to:

$$\mathbf{Y}_6 \cdot \mathbf{q}_M^c = \mathbf{0}$$

$$-q_{M,1} - q_{M,2} = q_S$$

$$-q_{M,3} - q_{M,1} = q_P$$

$$q_{M,1}, q_{M,2}, q_{M,3}, q_{M,4}, q_{M,10}, q_{M,12} > 10$$

The first constraint implies the pseudo-steady state assumption; while the 2nd and 3rd constraints are for the xylitol assimilation and production whilst the latter represents irreversibility constraints. The metabolic set is considered in Table 2, and the metabolic network is shown in Figure 6.

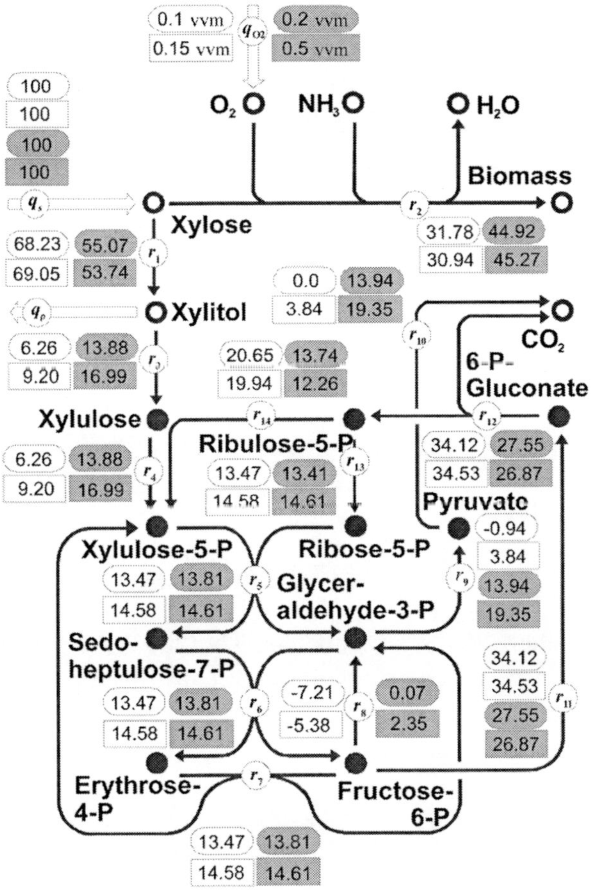

Figure 6. Metabolic fluxes r in xylose to xylitol bioconversion at four different aeration rates ((○) 0.1, (□) 0.15, (●) 0.2, (■) 0.5 vvm) expressed as a percentage of xylose intake flow q_S. The metabolic fluxes of the reactions for the replenishing of NAD+ and NADPH are also shown. Filled circles symbolize metabolic compounds in pseudo-steady state.

^{13}C Flux Analysis

When reaction fluxes are estimated by means of measured external conversion rates, they should only be considered as a feasible solution for there is no certainty that fluxes are actually as calculated. Measuring some of the internal fluxes would confirm the calculated fluxes if coincidental, or it would increase the chances of having the actual fluxes as the information increases. Unfortunately, it is still quite difficult to measure internal fluxes. Most common methods involves the use of at least one radioactively labeled substrate at a very specific position, for instance the first carbon, and the subsequent analysis of labeled metabolites distribution once steady-state is established (Antoniewicz et al., 2007; Shastri and Morgan, 2007). As the cell metabolizes the labeled substrate, the marked atoms end at particular positions and the proportions of this distribution reveals the rates of reactions.

The isotopic labeling technique works well in heterotrophic metabolism but does not work with autotrophic organisms. In this case, as only one-carbon molecule is used as substrate, at the end all carbons in the cell are labeled so there is no way of knowing the fluxes. A transient analysis must be carried out, the mathematical complexity and the computational methods involved are much more complicated. Young et al. (2008) have proposed a simplified method that significantly reduces the computational requirements.

Concluding Remarks

This work summarizes some examples where mathematical models successfully represent different biological process. Microbial growth modelling allows us to enhance the processes and predict some changes that can drive a better performance. The development of mathematical models to describe the process and improve the understanding of the metabolism is still limited but, since this is a powerful tool for the design of optimized engineering processes, every effort is valuable.

References

Antoniewicz MR, Kelleher JK, Stephanopoulos G. (2007) Elementary metabolite units (EMU): A novel framework for modeling isotopic distributions, *Metabolic Engineering*, 9: 68-86.

Aranda Barradas JS, Salgado Manjarrez E (2002). Producción de biomasa de *Saccharomyces cerevisiae* y usos en el sector alimentario, *Tecnología de Alimentos*, 37: 7-15.

Aranda-Barradas J, Delia ML, Riba JP. (2000). Kinetic study and modelling of the xylitol production using *Candida parapsilosis* in oxygen-limited culture conditions, *Bioprocess Engineering*, 22: 219-225.

Aranda-Barradas JS, Garibay-Orijel C, Badillo-Corona JA, Salgado-Manjarrez E. (2010). A stoichiometric analysis of biological xylitol production, *Biochemical Engineering Journal,* 50: 1-9.

Barrera-Martínez I, González García RA, Salgado-Manjarrez E, Aranda-Barradas JS. (2011). A simple metabolic flux balance analysis of biomass and bioethanol production in *Saccharomyces cerevisiae* fed-batch cultures, *Biotechnology and Bioprocess Engineering*, 16: 13-22.

Delia ML, Aranda-Barradas J, Riba JP. (1997). Production de xylitol par *Candida parapsilosis* : Métabolisme et oxygène, *Bulletin de la Société Française de Microbiologie*, 147-156.

Martínez-Trujillo A, Aranda JS, Aguilar-Osorio G. (2008). Kinetic study on inducibility of polygalacturonases from *Aspergillus flavipes* FP-500, *Electronic Journal of Biotechnology*, 11: 1-8.

Martínez-Trujillo A, Aranda JS, Gómez-Sánchez C, Trejo-Aguilar B, Aguilar-Osorio G. (2009). Constitutive and inducible pectinolytic enzymes from *Aspergillus favipes* FP-500 and their modulation by pH and carbon source, *Brazilian Journal of Microbiology*, 40: 1-8.

Nielsen J and Villadsen J. (1994). Analysis of reaction rates, in Bioreaction Engineering Principles. Plenum Press, U.S.A., 97-162.

Noorman HJ, Heijnen JJ, Luyben, ChAM. (1993). Linear relations in microbial reaction systems: A general overview of their origin, form and use. *Biotechnology and Bioengineering*, 38: 603-618.

Rivera-Salvador V, Aranda-Barradas JS, Espinosa-Solares T, Robles-Martínez F, Toledo JU. (2009). El modelo de digestión anaeróbica IWA-ADM1: Una revisión de su evolución, *Ingeniería Agrícola y Biosistemas*, 1: 109-117.

Shastri AA and Morgan JA. (2007). A transient isotopic labeling methodology for 13C metabolic flux analysis of photoautotrophic microorganisms, *Phytochemistry,* 68: 2302–2312.

Vallino JJ and Stephanopoulos G. (1993). Metabolic flux distributions in *Corynebacterium glutamicum* during growth and lysine overproduction, *Biotechnology and Bioengineering*, 41: 633-646.

Young JD, Walther JL, Antoniewicz MR, Yoo H, Stephanopoulos G. (2008). An Elementary Metabolite Unit (EMU) Based Method of Isotopically Nonstationary Flux Analysis, *Biotechnology and Bioengineering*, 99:686-699.

In: Biotechnology
Editors: L.G. Torres Bustillos and I. Garcia Peña
ISBN 978-1-62081-071-2

Chapter 16

PLANT POLYSACCHARIDES AS A NEW SOURCE FOR COAGULANT-FLOCCULANT AIDS AND SURFACTANTS

Luis G. Torres Bustillos*
Bioprocesses department UPIBI-IPN, Mexico City, Mexico

ABSTRACT

Polysaccharides are polymeric carbohydrate structures, formed of repeating units (either mono- or disaccharides) joined together by O-glycosidic bonds. Virtually, polysaccharides can be found in any vegetal, specially forming structural complexes. Non-starch water soluble polysaccharides are usually isolated by dissolving-purification-alcohol precipitation process. Lipids can be removed by solvent extraction of raw materials. Protein can be removed by enzyme digestion, while insoluble fiber can be removed by centrifugation of polysaccharide solutions. Finally, it can be said that most of purification cost comes from alcohol consumption.

Plant polysaccharides have been employed in a wide range of applications. Very traditional gums, such as Arabic, karaya and tragacanth gums, which in turn are exudates gums, are known from ancient times. Their recent applications range from food (beverages bases, chewing gums, confectionary and frostings, dairy products analogs, fats and oils, gelatins, puddings and fillings, hard candy, etc.) to non-food applications (pharmaceutical industry, suspending agents, etching and plating solutions, foam stabilizers, dispersant in paint and insecticidal emulsions, etc). More specifically, these polysaccharides are very interesting because of their emulsion properties such as surface tension, emulsion capacity and emulsion stability, as well as their rheological properties.

Besides these applications nowadays the use of these natural gums in the treatment of municipal and industrial wastewaters has been reported. These products act as coagulant-flocculants, substituting Fe and Al salts+ synthetic polymers in the treatment of contaminated waters. Sludges and waters produced in the treatment contain fewer metals and in the case of the sludges, they result more biodegradable.

On the other hand, some vegetal gums have been employed as surfactants-emulsifiers in the field of remediation of contaminated soils. Some of these products have been employed successfully in the washing of soils contaminated with petroleum

* Email: LTorresB@ipn.mx

hydrocarbons, different metals, pharmaceutical products or pesticides. They are used in smaller quantities than their counterpart (i.e, synthetic surfactants). An additional purpose of the use of vegetal products is that wastewater generated are biotreated more easily that the wastewaters generated when using synthetic products.

A second application in the treatment of contaminated soils, is its use in small quantities for the biodegradation of products present in contaminated soils, only as an aid, to enhance the desorption of organic products and thus, the microorganisms can use more efficiently the organic compounds as carbon or energy sources. The benefit of using natural products is the concern of leaving non-degraded products in soils which could be hardly biodegraded in long periods.

This chapter describes the polysaccharides, emphasizing in those of vegetal origin. Also, a description of some of the functional properties of the vegetal products is made. Finally, the main applications of natural products as coagulant-flocculants or surfactants are discussed.

INTRODUCTION

Polysaccharides are polymeric carbohydrate structures, formed of repeating units (either mono- or disaccharides) joined together by O-glycosidic bonds. These structures are often linear, but may contain various degrees of branching. Polysaccharides are often quite heterogeneous, containing slight modifications of the repeating unit. They are natural products obtained from a wide variety of plants and microorganisms (i.e. *leguminoseae*, bacteria, yeast, fungi and seaweeds). Examples of *leguminosae* are galactomanans guar, locust bean and mezquite seed gums. From the group of bacterial polysaccharides, it is possible to mention xanthan gum, microbial alginate, and pulullan. In the group of seaweed polysaccharides are comprehend agars, alginate, and carrageenans.

Polysaccharides have a general formula of $C_x(H_2O)_y$ where x is usually a large number between 200 and 2,500. Considering that the repeating units in the polymer backbone are often six-carbon monosaccharides, the general formula can also be represented as $(C_6H_{10}O_5)_n$, where $40 \leq n \leq 3000$.

A general classification of polysaccharides comprehends storage and structural polymers. The first group includes starches (plants) and glycogen (animals), while the second group includes cellulose, hemicelluloses, arabinoxylanes (plant) and chitin (animal).

Regarding plant polysaccharides, they are classified in accord to the specific vegetal origin as follows:

1. Higher plant cell walls insoluble: cellulose
2. Higher plant cell wall soluble: pectin
3. Higher plant seeds: cereal starch, guar gum, locust bean gum
4. Higher plant tuber and root: potato starch, tapioca starch and
5. Higher plant exudates: gum arabic, gum tragacanth.

See some structures of vegetal polysaccharides in figure 1.

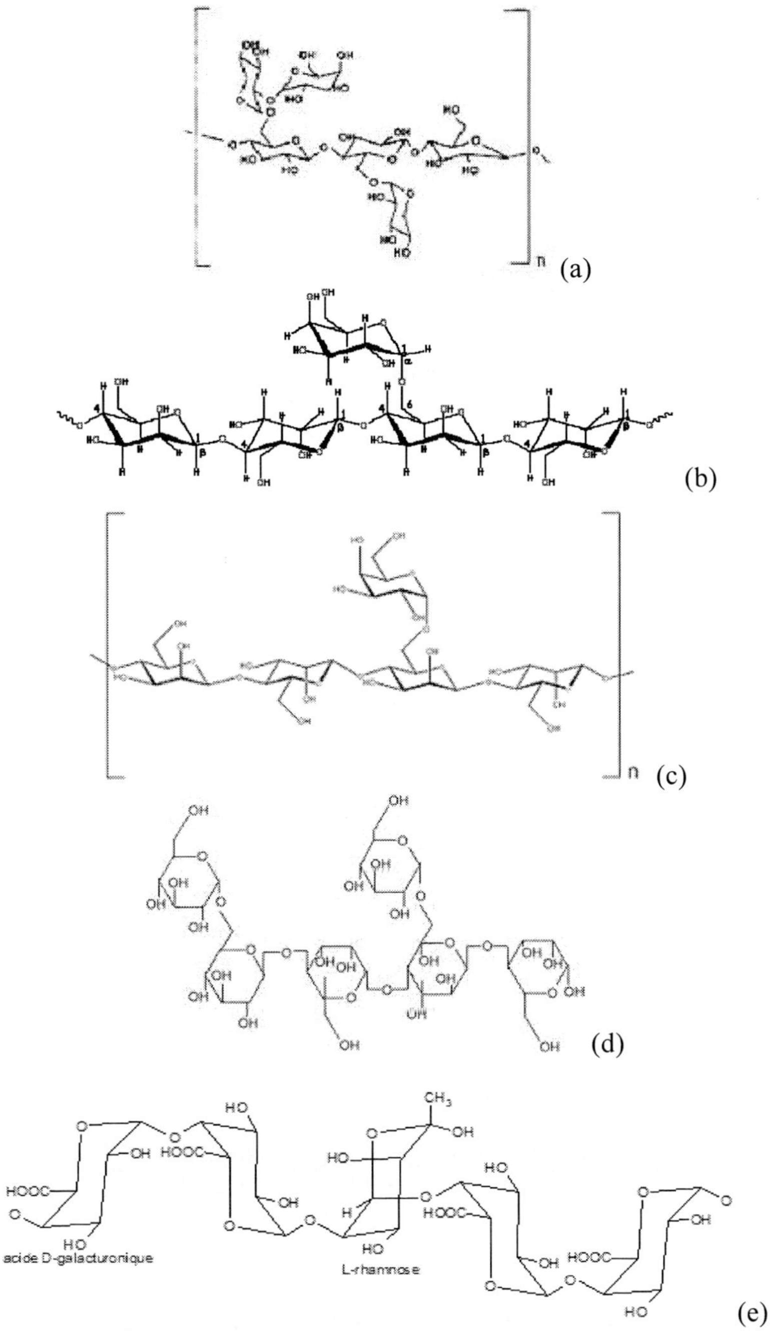

Figure 1. Structures of some plant polysaccharides. a) Tamarind seed gum b) Locust bean gum c) Tara gum. d) Guar gum and e) Pectin.

Regarding their structure, polysaccharides can be divided into:

1. Linear: amylose, cellulose, pectin, alginates
2. Short branched: guar gum, locust bean gum, xanthan gum and
3. Branch-on-branch: amylopectin, gum arabic, arabinoxylan.

Finally, regarding the monomers they contain, they can be named as:

1. Homoglycans: starch, cellulose
2. Diheteroglycans: agars, alginate, carrageenans and
3. Triheteroglycans: xanthan, gellan, arabinoxylan.

Polysaccharides, regarding their charges and distribution can be neutral (such as amylose, amylopectin, cellulose, guar gum, etc) or anionic (including alginates, carrageenans, gellan, gum arabic, xanthan, etc).

Polysaccharides have been used in many applications, but the use of these natural products in wastewater treatment (coagulant-flocculant aids) and as soil washing surfactants will be described in this chapter. Pictures of some polysaccharides producing-plants are shown at figure 2.

a) *Prosopis laevigata* tree

b) *Opuntia ficus-indica* and prickly pear

Figure 2. (Continued on next page).

c) *Prosopis laevigata* pod

d) *Ceratonia siliqua* tree

e) Guar gum pod

Figure 2. Trees and pods form some plants that produce galactomannans or arabinoxylanes.

PLANT POLYSACCHARIDES SOURCES

Virtually, polysaccharides can be found in any vegetal, specially forming structural complexes. Buckeridge et al. (1995) have reported a list of *leguminosae* that contain galactomannans in their seeds. Some of them are shown at table 5. In some cases, the yield of galactomannans and the Man:Gal ratios are included. It is important to take into account that seeds not always contain galactomannans, but other sugars such as xyloses, glucose, etc. besides; every seed contains carbohydrates, fats, ash, protein and crude fiber.

Buckeridge et al. (1995) have discussed the fact that the man: gal ratio together with the yield (%) are correlated with the taxonomic position (at the subfamily level), at least for the genera *leguminoseae.*

Table 1. Occurrence, yield and Man:Gal ratio of galactomannans from selected species

Species	Yield (%)	Ratio man:gal	Species	Yield (%)	Ratio man:gal
*Casia grandis**	37.5	1.7	*Inga uruguensis**	NR	NR
*Senna reticulate**	17.2	1.4	*Piptadenia gonocantha**	NR	NR
*Leucaena sp**	14.9	1.3	*Aeschinomene paniculata**	NR	NR
*Leucena pulverulenta**	15.4	1.2	*Centrolobium robustum**	NR	NR
*Mimosa sp**	26.4	0.9	*Macaerium sp.**	NR	NR
*Mimosa platphylla**	18.5	0.9	*Macuna sp.**	NR	NR
*Mimosa scabrella**	27.1	1.0	*Sophora tormentosa**	NR	NR
*Prosopis juliflora**	27.0	1.2	*Trigonella foenumgraecum***	NR	1.2
*Bowdichia virgiloides**	8.4	0.9	*Cyamopsis tetragonobulus***	NR	1.7
*Crotalaria micans**	16.6	2.1	*Ceratonia siliqua***	NR	3.7
*Crotalaria juncea**	25.6	2.5	*Solanum lycopersicum***	NR	NR
*Indigofera suffructicosa**	18.2	1.1	*Lupinus albus****	NR	NR
*Acasia farnesiana**	NR	NR	*Annona muricata*****	NR	NR
*Anadenanthera colubrine**	NR	NR	*Arenga saccharifera*****	NR	NR
*Anadenanthera falcate**	NR	NR	*Cocos nucifera*****	NR	NR
*Caliandra selloi**	NR	NR	*Sophora japonica*****	NR	NR
*Caliandra bracteosa**	NR	NR	*Convovolus tricolor*****	NR	NR
*Enterolibium sp.**	NR	NR	*Cassia angustifolia******	NR	NR
*Inga marginata**	NR	NR	*Caesalpina spinosa**	NR	3.0

Adapted from *Buckeridge et al. (1995). **Otegui (2007), ***Mohamed and Rayas-Duarte (1995), ****Kooiman (1971), *****Sanghi et al. (2002).

PLANT POLYSACCHARIDES EXTRACTION

Nishinari et al. (2011) reported characteristics and applications of three important groups of storage plant polysaccharides: xyloglucans, galactomannans and glucomannans. The first group includes seeds of *Tamarindus, Impatiens, Annona, Hymenaea* and *Detarium.*

The seeds from *Trigonella foenum-graecum (*fenugreek gum) *Cyamopsis tetragonoloba* (guar gum) *Ceratonia siliqua* (locust bean gum), *A. Konjac* plant c. Coch (KGM) and *Caesalpina spinosa* (tara gum) contains galactomannans.

Non-starch water soluble polysaccharides are usually isolated by dissolving-purification-alcohol precipitation process.

Lipids can be removed by solvent extraction of raw materials. Protein can be removed by enzyme digestion, while insoluble fiber can be removed by centrifugation of polysaccharide solutions. Finally, it can be said that most of purification cost comes from alcohol consumption.

Torres et al. (2011) have studied the polysaccharide contents of three species of *Annona* genera, very common in Mexico: *A. cherimola, A. muricata* and *A. diversifoila.* The seeds of these trees or bushes contain an important proportion of polysaccharides. In the case of the three species, works regarding their medicinal use have been reported (Torres et al, 2011).

Table 2. Proximal composition of three seed endosperms: chirimoya (*Annona cherimola*), guanabana (*Annona muricata)* and papausa (*Annona diversifolia*) (g/100g D.W.)

	*Annona cherimola***	*Annona muricata*	*Annona diversifolia*
Humidity	4.43 ± 0.03	5.13 ± 0.01	4.89 ± 0.06
Protein	19.53 ± 0.04	18.74 ± 0.39	18.76 ± 0.13
Fiber	18.71 ± 2.19	16.29 ± 0.10	17.49 ± 0.59
Fat	35.75 ± 0.42	41.12 ± 0.08	39.82 ± 1.70
Ashes	2.65 ± 0.02	2.55 ± 0.03	2.42 ± 0.03
Nitrogen free extract*	23.36 ± 0.01	16.99 ±0.02	21.50 ± 0.03

Media ± Std. Deviation, 3 replicates. *Calculated by difference.
Adapted from Torres et al. (2011), except ** L. Corzo, Personal communication.

The procedure employed to get the polysaccharide consisted of solvent washing and dryness. Before that procedure, the seed is milled and passed through mesh 100.

After the cleaning process, the proximal composition of the three polysaccharides was that showed on table 2. As observed, between 16 and g/100 DW of the seed consist of fat-free polysaccharide.

On the other hand, the oil content for the three seeds fluctuated between 35 and g/100 DW. Some researchers have proposed the use of *A. cherimolla* oil, with the purpose of producing biodiesel.

The de-fated seeds have also been proposed to be used as cattle feed, because of its protein and fiber contents. In the case of the later, the contents can be between 18 and 19 g/100 DW (Fasakin et al., 2008).

Wu et al. (2009) reported protein contents for fenugreek, guar, tara and locust bean gums of 2.62, 3.46, 0.71 and 4.57%, respectively regarding the ashes, values between 0.72 and 1.5% were measured. All these figures were for galactomannans containing 9.77-12.37% of moisture.

Functional Properties

Plant polysaccharides have been employed in a wide range of applications. Very traditional gums, such as arabic, karaya and tragacanth gums, which in turn are exudates gums, are known from ancient times (Verbeken et al, 2003).

Their recent applications range from food (beverages bases, chewing gums, confectionary and frostings, dairy products analogs, fats and oils, gelatins, puddings and fillings, hard candy, etc.) to non-food applications (pharmaceutical industry, suspending agents, etching and plating solutions, foam stabilizers, dispersant in paint and insecticidal emulsions, etc).

Just to mention the market of one of those gums, Gum Arabic was produced in an annual average of 50,576 tons/year in the period of 1965-1969, but production felt down to an amount of 18,358 ton/year in the period of 1990-1994. (Verbeken et al, 2003).

Wu et al. (2009) have published an excellent review regarding four commercial galactomannans (*i.e.* fenugreek FG, guar GG, tara TG and locust bean LBG gums). They have included emulsion properties such as surface tension, emulsion capacity and emulsion stability, as well as rheological properties.

This work shows that galactomannans studied decrease surface tension of water in 4-14 mN/m, depending on the specific galactomannan. Besides, the four galactomannans were capable of emulsifying canola oil, forming stable emulsions. The order of emulsification capacity was GG>FG>TG>LBG which was the same order of intrinsic viscosity level.

Torres et al. (2011) have published recently the use of guar and locust bean gum, as well as a rhamnolipid produced by *Pseudomonas aeruginosa*, with the purpose of enhance oil recovery. They have reported the surface tensions of a group of synthetic polymers and the three natural polysaccharides in water solution (In presence and absence of brine). Also they reported two different assessments which simulate the polymer surfactants in suitable conditions.

The first is a foamability test in presence and absence of brine. The second one is a heat treatment consisting of 3 days at 75°C. In the three assessments, both the guar gum and the rhamnolipid had a very good development in comparison with the synthetic polymers. Figure 3 shows a profile of surface tension ST for guar and locust bean gums, as a function of polysaccharide concentration (% w/w).

It is clear that both gums lowered the ST value, but the behavior is not very predictable. Lower values were found for guar gum at 0.025 and 0.05 % (w/w).

Other properties for the selected galactomannans reported by Wu et al. (2009) are the intrinsic viscosity and molecular weight (form 14.20-15.80 dL/g and 2.08-3.23x106 g/gmol), for both properties, respectively.

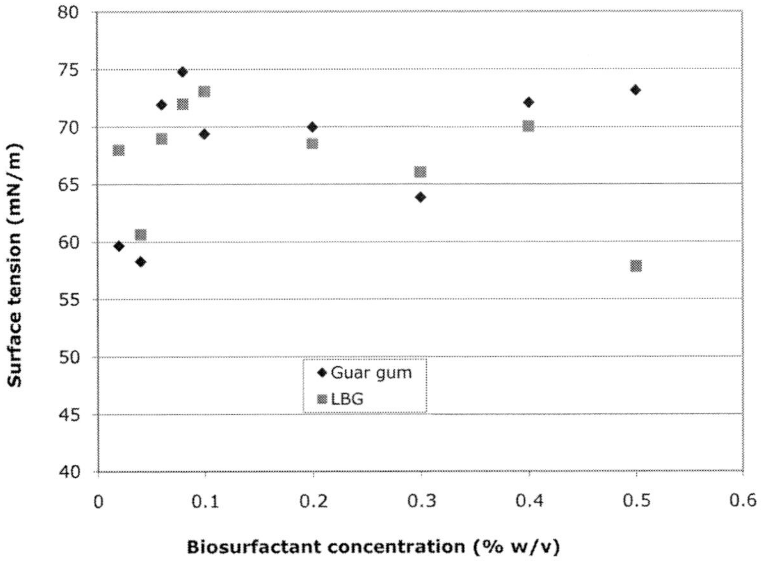

Figure 3. Surface tension for guar and locust bean gums as a function of their concentrations in water. Adapted from Torres et al., 2011.

POLYSACCHARIDES AS COAGULANT-FLOCCULANT AIDS

In the last decades, some authors have reported the use of natural products as biodegradable flocculants (Singh et al, 2000). Products range from guar, starch, amylopectin, sodium alginate, xanthan, carboxymethyl cellulose, and grafted polymers.

Bolto and Gregory (2007) have discussed the use of organic polyelectrolytes in water treatment, with different applications, including potable water treatment, polymers as coagulant aids, recycling of filter backwash waters and sludge thickening. The same authors proposed three mechanisms of action: 1) polymer adsorption, b) polymer bridging and c) charge neutralization.

Our research group has reported the use of natural polymers in the treatment of municipal and industrial wastewaters. Natural biopolymers employed comprehend galactomannans (such as guar, locust bean gum and mezquite seed gums), lecithin, xanthan gum, calcium alginate, *Opuntia ficus* mucilage and mixtures of synthetic polymers-natural products.

Torres et al. (1997) reported the pretreatment of high-load chemical-pharmaceutical industry wastewaters using either Al and Fe salts, or vegetal and microbial polysaccharides. Those wastewaters were characterized in terms of the total suspended solids (TSS= 40.4 mg/L), chemical oxygen demand (COD = 63,650 mg/L), alkalinity (Alk = 490 mg/L), and total dissolved solids (TDS = 14,555 mg/L). As observed, the wastewaters are highly contaminated, present a lot of organic material, color, solids and dissolved salts.

These wastewaters were treated by coagulation flocculation employing Al sulphate, ferric chloride, synthetic products (such as Bl-5086, Niad II-3, Bubond 65, Bufloc 528, etc) and natural products (such as sodium alginate, guar and locust bean gums, dextrane, and carboxymethyl-cellulose).

Best products were (in terms of the COD % removals), BL-5086 (40.6), guar gum (40.6), Niad II-3 (37.5), ferric chloride (33.4), and Niad II-4 (27.7).It is important to remark that those COD removals were obtained with different salt-polymers doses (from 400 to 1,200 mg/L).

So when the used dose is taken into account, the best COD removal/dose values wee for guar gum (6.76), BL-5086 (5.07), Niad II-3 (4.6) and Niad II-5 (4.67).

Recently, Carpinteyro-Urban et al. (2010), reported the treatment of a municipal wastewater containing organic matter (COD = 827 mg/L, BOD 444 mg/L), dissolved salts (conductivity = 1,870 μS), surfactants (MBAS = 4.14 mg/L), grease and oil (230 mg/L), a pH of 6.7 unites, Al and Fe in concentrations of 0.98 and 1.4 mg/L, respectively. Wastewaters were treated by coagulation-flocculation process, without any pH adjustment using guar, mezquite and locust bean gums, cosmedia guar, *Opuntia mucilage* and $FeCl_3$ as comparison product.

Figure 4 show the results of the coagulation-flocculation of those municipal wastewaters using 25-150 mg/L of LBG, guar, cosmedia and mezquite seed gums, *Opuntia mucilage*, in comparison with the use of $FeCl_3$, a Fe salt very used in coagulation-flocculation of wastewaters.

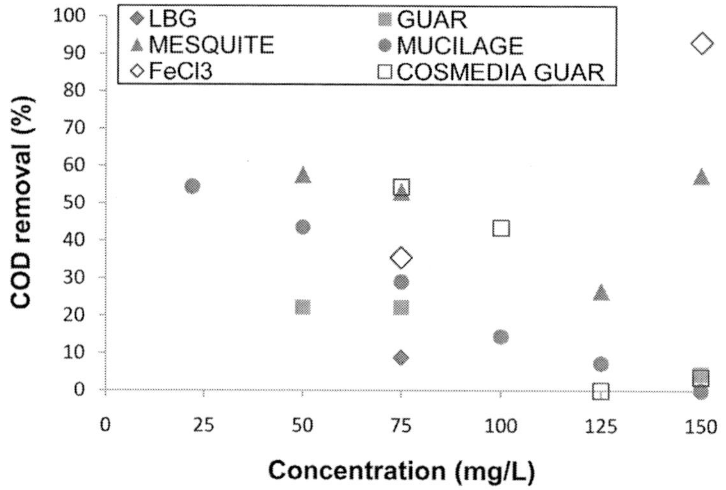

Figure 4. COD removal vs. Coagulant-flocculant concentration. Adapted from Carpinteyro-Urban et al, 2010.

As observed, most of the polymers behaved very well, but best results were observed when using mezquite seed gum at 50 mg/L (*c.a.* 60% of COD removal), very near from the Opuntia mucilage and cosmedia guar 25 mg/L test (*c.a.* 55% of COD removal). Cosmedia guar is a cationic guar gum derivative.

The chemical name for cosmedia guar is guar hidroxy propyl triammonium chloride HPTAC-guar and its molecule is shown at figure 5.

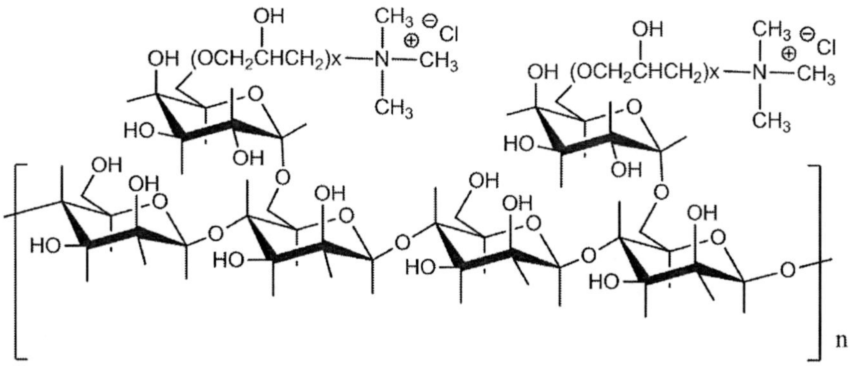

Figure 5. Hidroxy-propyl triammonium chloride guar (HPTAC-guar). The draw was prepared by M.A. Brito-Arias.

Advantages of using biopolymers include the fair COD, conductivity turbidity and color removals, but also the low production of sludges (if compared with $FeCl_3$ or $Ca(OH)_2$). These sludges would be more biodegradable (as the biopolymers are easily degradable) and also have the capacity of adsorption of metals and other compounds dissolved in the wastewaters.

Table 3 show another table from Carpinteyro et al. (2011), where three natural gums were employed in the treatment of a cosmetic industry wastewaters. It is clear that the three polysaccharides were very efficient in treating the low-medium and high loaded industrial wastewaters.

Table 3. Results for industrial wastewaters coagulation-flocculation using biopolymers

Polymer	Dose (mg/L)	COD load*	Final pH	Removal (%) Turbidity	Cond.	COD	Sludge produ-ction (mL/L)	COD removed (mg/L)	COD removed/ dose
Guar gum	150	Medium	7.04	11.47	9.72	9.09	275	933.33	6.2
	300	High	7.01	67.82	5.90	22.51	400	3000.00	10.0
	500	Low	6.65	25.86	20.13	0	50	0	0
LBG	150	Medium	7.08	0	5.90	0	275	0	0
	300	High	7.09	67.82	4.510	25.01	425	3333.40	11.1
	500	Medium	6.03	11.47	15.62	33.79	350	3466.73	6.9
Mucilage	150	High	6.68	49.56	0	23.76	450	3166.70	21.1
	300	Medium	6.32	16.39	0	0	400	0	0
	500	Low	6.12	18.96	0	38.59	150	2300.00	4.6

*COD load (mg COD/L): Low= 6,000, Medium= 10,200, High 13,300.pH initial value= 5. 6 units.
Adapted from Carpinteyro- Urban et al., 2011.

Form the mentioned table, it is remarkable that COD removals as high as 38.6% were observed when mucilage at 500 mg/L was employed with the lower COD load. The second best value corresponds to LBG at 500 mg/L for the medium COD load (33.79%); followed by mucilage at 500 mg/L with higher COD load (32.51%). For the assessments with $FeCl_3$, the maximum COD removal was as high as 47.7% (for a dose of 500 mg/L). Torres et al. (1997) reported COD removals up to 40.6% when using guar gum or BL5086 (a synthetic polymer).

Regarding the amount of produced sludge, the maximum value corresponds to mucilage at 500 mg/L for the higher COD load, with a value of 500 mL/L. As expected, higher values of produced sludge (400-500 mL/L) were observed for the more concentrated WW, followed for the group of assessments were COD load was medium (275-400 mL/L) and at the end for the low COD load WW, i.e. 50-150 mg/L. When using $FeCl_3$, sludge volumes of 350-425 mL/L were achieved for the medium COD load WW.

A major problem of textile industries is the generation of high volumes of wastewaters containing different components. In some of the textile production process, color is generated as it is very difficult to remove from waste waters. Sanghi et al. (2002) used *Cassia angustifolia* seed gum as a natural coagulant for discolouring dye solutions. In this work, artificial wastewaters containing three groups of dyes, i.e. acid Sendula red, direct Kahi green and reactive Remazol brilliant violet, were treated by coagulation with both the natural product and polyaluminium chloride (PAC). The natural product was very effective in removing the acid and direct dyes, but not to reactive dye solutions. It was also assessed the combination of PAC and seed gum with good results.

Mishra et al. (2003) employed fenugreek mucilage as flocculating agent for sewage treatment. Using the fenugreek mucilage to treat sludges, TSS and TDS of 95 and 22% were eliminated, respectively. Polymer doses were as low as 0.01-0.2 mg/L.

In other work, Zhang et al. (2006) assessed the use of cactus mucilage in the treatment of simulated wastewaters. They demonstrated that cactus mucilage by itself is capable of reducing the initial turbidity and up to 55% of COD removal. In combination with ferric chloride, cactus mucilage behaved very well, giving a 71% of COD removal and up to 90% of turbidity removal.

POLYSACCHARIDES AS SURFACTANTS IN SOIL WASHING

Some research groups have explored the use of natural polymers in the surfactant enhanced soil washing (Urum and Pekdemir, 2004; Conte et al, 2005; Mulligan et al, 1999; Rosenberg and Ron, 1999)

Some articles have reported the use of vegetal surfactants such as lecithin (Soeder et al, 1996), a natural surfactant obtained from the perycarp of the fruit *Sapindus mukurossi* (Roy et al, 1997), and the fruit of *Quilaya saponine* (Soeder et al, 1996)., aescine, lethicine, saponine and tannin (Urum and Packdemir, 2004).

Torres et al. (2007) have employed three surfactants, i.e. guar and locust bean gums and soy lecithin, in the washing of soils contaminated with high concentrations of diesel (98,000 mg/kg). They have found that the three polymers behaved very well to remove TPH-diesel amount with removal efficiencies up to 49 and 48% when using solutions of guar or locust bean gums, respectively.

The soil washing efficiencies were compared to that obtained when using TW80 or SDS (nonionic and anionic synthetic surfactants. The use of biopolymers was equal or more efficient than using the synthetic ones. Even, mixtures of natural and synthetic surfactants were assessed, but no better results were obtained. One fact that should be underlined at this point is that due to the viscous nature of the gum solutions, it is impossible to use these products at the same concentration used with synthetic surfactants. Even at concentrations between 0.001 and 0.005%, biopolymers were able to remove a big amount of TPH-diesel from the contaminated soils. Figure 7 shows the high capacity of guar and locust been gums in comparison with SDS to remove TPH from soil. Results are expressed as the grams of TPH removed by gram of surfactant, for different surfactant solutions. As observed, best results were obtained for guar gum, followed by LBG and SDS, at the end of the list. Guar and LBG were capable of removing up to 10 mg TPH/mg polymer, which is a quite high efficiency.

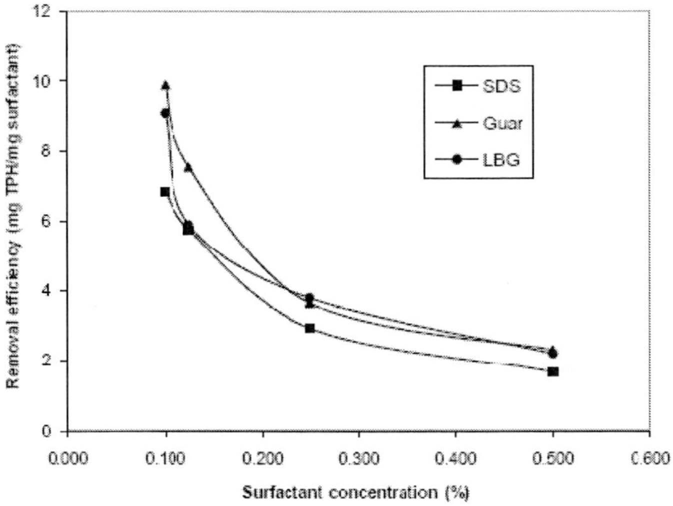

Figure 7. Removal efficiency (mg TPH/mg surfactant) as a function of the surfactant concentration for the three single surfactants). Adapted from Torres et al., 2007.

Polysaccharides are also very effective in removing metals form a contaminated soil. Torres et al. (2011) have reported the surfactant enhanced soil washing of and industrial soil using different synthetic surfactants and one biopolymer: mesquite seed gum. These soils are of industrial nature, and big quantities of Cu+Ni+Cd+Pb+Zn+As were found at different sites and depths of the contaminated site. For the sample treated in the mentioned work, a total concentration of about 45,500 mg/Kg was the initial contamination value. Table 5 shows the specific metal initial concentrations.

Also, in table five, the results obtained when washing the soil with tap water are presented. After that, the results of the soil washing with different surfactants at a concentration of 0.5% (w) are presented. The individual (per metal) and the global final metal concentrations (Cu+Ni+Cd+Pb+Zn+As) are presented.

Note that the surfactant that diminished in a higher extent some of the present metals, but mainly the global metal concentration is undoubtedly, mezquite seed gum. Besides, mezquite seed gum was used in a 1/5 concentration in respect to the rest of the surfactants. These findings demonstrate the suitability of using natural surfactants for surfactant enhanced soil washing, beside the other advantages yet mentioned. Biopolymers are produced by renewable sources; they are biodegradable and friendlier to environment.

Table 5. Final metal concentrations for the contaminated soils. All surfactants at a 0.5% w/w concentration, except mezquite gum, prepared at 0.1% w/w

Surfactant	Cu	Ni	Cd	Pb	Zn	As	Total metals Cu+Ni+Cd +Pb+Zn+As
Initial Soil	35,582.41	2,602.82	14.46	69.93	261.37	4,018.86	42,549.85
Water	7,605.85	537.2	8.78	62.9	41.97	3,622.41	11,879.11
Surfacpol 203	7,449.9	534.12	6.83	76.0	41.33	2,760.13	10,868.31
Emulgin W600	8,102.28	666.43	6.14	44.44	45.56	2,448.35	11,313.2
Canarcel 20	7,909.5	584.36	7.02	56.44	52.04	2,525.91	11,135.27
Canasol BJ35	9,453.92	676.3	7.16	55.97	69.36	2,912.81	13,180.52
SDS	8,250.03	638.95	9.07	64.33	52.93	3,173.93	12,189.24
Maranil Lab	7,392.94	542.34	8.38	59.51	78.56	2,779.11	10,860.80
Polafix CAPB	7,614.78	546.09	6.14	56.73	43.75	2,020.0	10,287.49
Mezquite seed gum	6,589.1	486.3	7.19	55.71	41.3	2,543.68	9,723.28
Average without initial soil	7,397.48	551.78	6.87	57.84	57.63	2,687.82	10,759.42

Adapted from Torres et al. (2011).

Torres et al. (2011) reported the surfactant enhanced soil washing of a methyl-parathion MP contaminated soil. Soils containing around 0.4 and 13 mg/L were treated with different solutions of anionic, nonionic, zwitterionic and natural surfactants. It is remarkable that a number of surfactants had a high capacity of MP removal. No matter which was the MP initial concentrations, the surfactant solutions removed between 63 and 98% of the pesticide. It is also true that removal efficiencies for the less and more MP concentrated soils were quite similar. It is clear that for the more concentrated soil, best results were obtained when using locust bean, guar and mesquite gums (removal efficiencies of about 98, 96 and 94%, respectively). In the case of the soil containing only 0.4 mg/kg, best results were found when using Surfacpol G, guar and mesquite gums, followed very near from Triton X-100 (MP removals of 91, 90, 90 and 88%). Removal due to locust bean gum was almost 88%. Tween 80 solutions only removed 63 and 75% of the present MP in the les and more concentrated soils, respectively. Emulgin 600 only was capable of removing around 54 and 82% of the MP in the same soils. Note that all experiments were carried out using 0.5% of the surfactant in every solution.

When washing the same soils only with the natural surfactants, the effect of the surfactant concentration was evaluated. Solutions with concentrations of 0.075, 0.05, 0.025 and 0.01% were employed. In previous studies, natural surfactants demonstrated to be useful at lower concentrations if compared with those employed when using synthetic ones. Figure 8 show those results. It is clear that MP removals were excellent for the three natural surfactants at the different concentrations. In the case of locust bean gum, removals between 99.5 and 99.7% were obtained. For guar gum, the removals were between 84.2 and 99.7%.

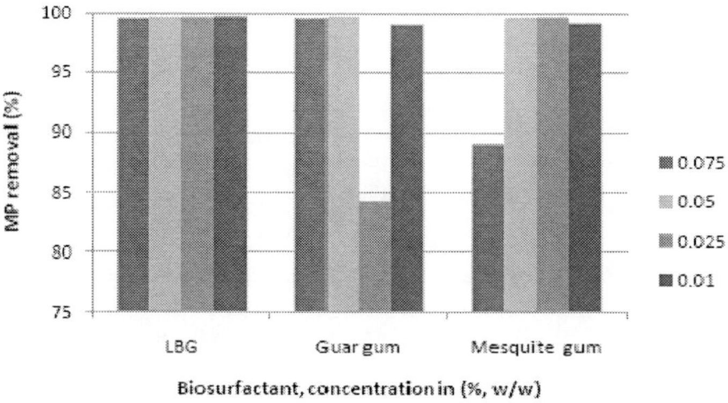

Figure 8. Methyl-parathion removal for three natural surfactants. Effect of the surfactant concentration % w/w). Adapted from Torres et al, 2011.

Definitely the worse point corresponds to a 0.025% biosurfactant concentration. Finally for mesquite gum, removals in the range of 89-99.6% were observed. In this case the worse point corresponds to a concentration of 0.075%. In general, natural surfactants were capable in removing the MP in high rates, even at low surfactant concentrations.

Polysaccharides in Surfactant Enhanced Biodegradation of Organic Compounds

Besides the use of surfactants (and in this case, vegetal surfactants) in removing the hydrocarbon compounds, metals or pesticides form a contaminated soil, it has also reported the possibility of using small amounts of surfactant in order to help autochthonous microorganisms to biodegrade organic compounds (such as petroleum hydrocarbons, pesticides, pharmaceutical products) from a contaminated soil.

Bonfanti (2009) have reported the use of three vegetal surfactants in the biodegradation of petroleum compounds present in the soil of an abandoned refinery in Mexico City. The soil came from an abandoned refinery located in Mexico City. The initial total petroleum hydrocarbon level was of 32,000 mg/kg, far from the Mexican legislation level of 6,000 mg/kg.

The three natural surfactants were guar, locust bean and mezquite seed gums in concentrations of 1, 10 and 100 mg/kg. These surfactants were compared to the use of Tween 80 and SDS, two nonionic and anionic synthetic products. Results showed that the three vegetal gums were capable of removing more TPH than the synthetic products. While the maximum TPH removal for the synthetic compounds was 31% (Tween 80@1 mg/kg), mezquite seed gum reached TPH removals of about 36% (at a concentration of 10 mg/kg). Note that the assessment without any surfactant reached a TPH removal of only 3.5%.

On the other hand, guar gum reached a 37% TPH removal with a concentration of 100 mg/kg, and LBG showed a maximum TPH removal of 25% at 1 mg/kg dose. Biodegradation rates were in the range of 16-119 mg/kg/day for the SDS and Tween 80, and of 90.5-137.4 mg/kg/day for the vegetal surfactants.

Table 6. Comparison of the use of synthetic and natural surfactants in the treatment of an oil-contaminated soil. Initial TPH concentration of about 32,000 mg/kg

Test	Surfactant	Concentration (mg/kg)	TPH removal (%)	Biomass (CFU/gr soil)	Biodegradation rate (mg/kg/day)
1	Blank	0	3,53	6,56E+06	13,42
2	SDS	1	4,25	7,91E+06	16,12
3	SDS	10	9,74	8,03E+06	37,00
4	SDS	100	28,65	2,42E+06	108,79
5	Tween80	1	31,34	Nd	119,02
6	Tween80	10	29,11	1,27E+07	110,56
7	Tween80	100	16,54	5,91E+06	62,80
8	Guar gum	1	23,84	6,73E+06	90,55
9	Guar gum	10	28,55	1,22E+06	108,41
10	Guar gum	100	37,16	7,28E+06	141,13
11	LBG	1	24,70	7,23E+06	93,80
12	LBG	10	16,96	1,09E+07	64,41
13	LBG	100	19,06	2,75E+07	72,38
14	Mezquite gum	1	31,17	1,06E+07	118,36
15	Mezquite gum	10	36,18	1,41E+07	137,42
16	Mezquite gum	100	35,63	1,29E+06	135,31

Adapted from Bonfanti et al. (2009). Nd: Not determined.

At the time, our research group is working with the surfactant enhanced biodegradation of pesticides (i.e. 2,4-Dichlorophenoxiacetic acid, DDT and methyl parathion) using synthetic and natural surfactants, where vegetal products show excellent levels of biodegradation.

CONCLUSIONS

Vegetal polysaccharides are emerging bio-products with a wide range of applications in the oil, food, pharmaceutics and textile industries. Due to the abundance of plants which contain cellulose, arabinoxylan, galactomannans, starches and other groups of polysaccharides worldwide, they are a never-ending source of new products.

In this chapter, the use of plant polysaccharides has been discussed in terms of two specific applications. First, gums can be employed as replacement of Fe and Al + synthetic polymers in the treatment of both municipal and industrial wastewaters. It has been shown that plant polysaccharides can successfully be employed to cover both roles of coagulant and flocculant. Removal efficiencies (COD, turbidity, salts content, etc) are very similar in some cases, superior to that showed by Fe and Al salts and/or synthetic polymers. Sludges are produced in a lower extent and they are presumably more biodegradable.

On the other hand, plant polysaccharides can be employed to wash contaminated soils, at low doses. It has been demonstrated that guar, locust bean, and mezquite seed gums, as well as *Opuntia mucilage*, lethicine, and other biopolymers are very suitable for the remediation of soils contaminated by petroleum hydrocarbons, metals or pesticides.

Besides, gums can be employed in the surfactant enhanced biodegradation of compounds present in contaminated soils, such as petroleum hydrocarbons or pesticides.

The research on these areas is a must in order to find new products and new applications for plant polysaccharides.

ACKNOWLEDGMENTS

The participation of Luis J Corzo-Rios in the extraction and characterization of some seed galactomannans is thanked. Part of this work has been supported by ICyT-DF Grant PICSO10-8 and SEP-CONACyT project grant 84084.Author thanks the San Juan Ixhuatepec (Estado de Mexico) wastewater treatment plant for the use of raw wastewaters. I also thank my students who have participated in the coagulation-flocculation project (J. Jaimes, S. Carpinteyro-Urban, A. Carrillo and A. Perez), or the contaminated soil washing project (E. Zamudio, P. Zavala, L. Bofanti, C. Caballero, F. Ramos, B, Lopez,C. Galindo) in the last 10 years. The review of the manuscript by M. A. Brito-Arias is acknowledged and thanked.

REFERENCES

Buckeridge M.E., Panegassi V.R., Rocha D.C., and Dietrich S.M.C. (1995) Seed galactomannan in the classification and evolution of leguminoseae. *Phytochemistry*. 388: 71-875.

Carpinteyro-Urban S., Yañez J. and Torres L.G. (2010) Coagulation-flocculation of wastewaters emplying guar, locust bean and mezquite gums, as well as *Opuntia ficus* mucilage. IAW Young Proffesionals Meeting. Juriquilla, Queretaro. Junio 2010.

Carpinteyro-Urban S., Yañez J., Vaca M. and Torres L.G. (2011) Use of biopolymers from vegetal origin in the coagulation-flocculation of a high-load cosmetic industry wastewater. Memories of the Water and Industry IWA Congress. Valladolid, España. 1-4 May, 2001.

Conte P, Agretto A., Spaccini R., and Piccolo A. (2005) Soil remediation: humic acids as natural surfactants in the washing of highly contaminated soils:. *Environmental Pollution135*: 515-522.

Fasakin A.O., Fehintola E.O., Obijole O.A. and Oseni O.A. (2008) Compositional analyses of seed of sour soup, *Annona muricata* L., as potencial animal feed supplement. *Scientific Research and Asay*. 3:521-523.

Kooiman P (1971) Structures of the galactomannans from seeds of *Annona muricata, Arenga saccharifera, Cococ nucifera, Convovolvus tricolor and Sophora japonica. Carbohydrate Research*. 20: 329-337.

Mishra A., Agarwal M. and Yadav A. (2003) Fenugreek mucilage as a flocculating agent for sewage treatment. *Colloid Polymer Science* 281: 164-167.

Mulligan C.N., Young N.R. and Gibbs B.F. (1999) On the use of biosurfactants for the removal of heavy metals form oil-contaminated soil. *Environmental Progress*. 18: 50-54.

Bonfanti L. (2009) Biodegradazione assistita con surfactanti naturali e sintetici di idrocarburi presenti nel suolo della ex Raffineria di Azcapotzalco, Messico. Politecnico di Torino. Corso di Laurea Spacialistica in Ingegneria per l´Ambiente ed il Territorio. Tesi di Laurea Magistrale. Trabajo experimental realizado en UPIBI.IPN, Mexico. Dirigida por Luis G Torres B. Mexico.

Mohamed A.A. and Rayas-Duarte P. (1995) Nonstarchy polycaccharide analysis of cotyledon and hull of Lapinus albus. *Ceral Chemistry* 72: 648-651.

Nishinari K., Takemasa M., Zhang H., Takahashi R. (2011) Storage plant polysaccharides: xyloglucans, galactomannans, glucomannans. In *Comprehensive glycoscience*. Kamerling J.P. (editor). Volume 4: . 613-622.

Otegui M.S. (2007) Endosperm cell walls: formation, composition and functions. *Plant Cell Monographies* 8: 159-177.

Rosenberg E. and Ron E.Z., (1999) High and low-molecular mass microbial surfactants. *Applied Microbiology and Biotechnology* 52:154-162.

Roy D., Kommalapati R.R., Mandava S.S., Valsaraj T. and Constant W.D. (1997) Soil washing potential of a natural surfactant. *Environmental Science and Techology*. 31:670-675.

Sanghi R., Bhatttacharya B and Singh V. (2002) *Cassia angustifolia* seed gum as an effective natural coagulant for decolourisation of dye solutions. *Green Chemistry* 4: 252-254.

Singh R.P., Tripathy T., Karmakar G.P., Rath S.K., Karmakar N.C., Pandey S.R., Kannan K. and Lan N.T. (2000) Novel biodegradable flocculants based on polysaccharides. *Current Science*. 78: 798-803.

Soeder C.J., Papaderos A., Kleespies M., Kneifel H., Haegel F-H., and Webb L. (1996) Influence of phytogenic surfactants quilaya saponin and soil lecithin) on bioelimination of phenanthrene and fluoranthene by three bacteria. *Applied Microbiology and Biotechnology*.44: 654-659.

Torres L.G., Jaimes J., Mijaylova P., Ramirez E. and Jimenez B. (1997) Coagulation-flocculation pretreatment of high-load chemical-pharmaceutical industy wastewater: Mixing aspects. *Water Science and Technology*. 36: 255-262.

Torres L.G., P. Zavala, M. Beltran, Mabel Vaca and R. Iturbe (2007) Combination of natural gums and synthetic surfactants for washing of a soil highly contaminated with crude. *Environmental Geosciences*. 14: 1-9.

Torres L, Moctezuma A., Avendaño J.R., Muñoz A. and Gracida J. (2010) Biosurfactants´s performance for enhanced oil recovery in comparison to that showedby synthetic products. *Journal of Petroleum Science and Engineering*.

Torres L.G., Ramos F., Avila M.A., and Ortiz I (2011) Removal of methylparathion by surfactant assisted soil washing and the subsequent wastewater biological treatment. *Journal of Environmental Science and Health*. Part A. Submitted.

Torres L.G., Lopez R.B., Beltran M. (2011) Removal of As, Cd, Cu, Ni, Pb, and Zn from a highly contaminated industrial soil using surfactant enhanced soil washing. *Journal of Physics and Chemistry of the Earth.* Published online.

Urum K.and Pekdemir T. (2004) Evaluation of biosurfactants for crude oil contaminated soil washing. *Chemosphere* 57:1139-1150.

Verbeken D., Dierckx S. and Dewettinck K. (2003) Exudates gums: occurrence, production, and applications. *Applied Microbiology and Biotechnology*. 63: 10-21.

Wu Y, Cui W., Eskin N.A.M., and Goff H.D. (2009) An investigation of four commercial galactomannans on their emulsion and rheological properties. *Food Research Interntational*. 42: 1141-1170.

Zhang J., Zhang F., Luo Y. and Yang H. (2006) A preliminary study on cactus as coagulant in water treatment. *Process Biochemistry*. 41: 730-733.

INDEX

B

C

D

E

F

G

H

I

J

K

L

N

O

P

Q

R

S

T

U

V

W

X

Y

Z